COHERENT OPTICAL ENGINEERING

PREFACE

This volume contains the invited lectures and seminars presented at the NATO Advanced Study Institute on Coherent Optical Engineering, sixth Course of the International School of Quantum Electronics, affiliated with the Centre for Scientific Culture E. Majorana, Erice, Sicily. The Institute was held at Villa Le Pianore (Lucca), Versilia, Italy, September 1-15, 1976.

The aim of the Institute was to provide a presentation of several optical measurement methods, including holographic and speckle interferometry, laser interferometry, laser telemetry, intensity correlation spectroscopy and Doppler velocimetry, white light and "moiré" techniques, and to describe the state of the art in the field of optical information processing. The Course was made self-contained by a series of lectures which gave the basic principles underlying Coherent Optical Engineering and reviewed the most important optical devices, such as laser sources, laser deflectors and modulators, detectors and recording media.

The members of the Organizing Committee were:

- F. T. Arecchi, D. Roess – Directors of the International School of Q. E.
- J. M. Burch, National Physical Lab., Teddington, U. K.
- V. Degiorgio, CNR and CISE, Milano, Italy
- A. W. Lohmann, University of Erlangen, Erlangen, Germany
- S. Lowenthal, Institut d'Optique, Orsay, France

In addition to those published here, the following lectures and seminars were also given at the Institute:

M. Bertolaccini : "Photoelectric Detectors"
J. M. Burch : "Holographic Interferometry-White Light and Moiré techniques - Laser Resonators"
R. Leighty : "Optical Processing in Photogrammetry"
W. Lukosz : "Fourier Optics"
V. Russo-Checcacci : "Optical Processing"
G. Toraldo di Francia : "Information Content of an Optical Image"

We wish to express our appreciation to the NATO Scientific Affairs Division whose financial support made this Institute possible. We also acknowledge the contribution of cultural institutions and industries as :

CISE
CNR (Italian National Research Council)
Deutsche Physikalische Gesellschaft
European Research Office
IBM Italia
PHILIPS Eindhoven
SIEMENS München

We finally thank Dr. V. Fossati-Bellani and the secretaries of the Divisione Elettronica Quantistica, CISE, Mrs. G. Ravini and Miss M. Oriani, for their precious assistance in the organization of the Institute and in the preparation of these proceedings, and Miss A. Camnasio of Servizio Documentazione, CISE, for her assistance during the School days.

F. T. Arecchi
V. Degiorgio

CONTENTS

Coherant Optical Engineering, F.T. Arecchi and V. Degiorgio (eds.)

AN INTRODUCTION TO HOLOGRAPHY

E.N. Leith

The University of Michigan
and
Environmental Research Institute of Michigan
Ann Arbor, Michigan U. S. A.

With the award of the 1971 Nobel Prize in physics to Dennis Gabor, holography reached a new pinnacle of prestige. Gabor's method, simple and elegant, solved in quite a general way the basic problem of recording the phase, as well as the amplitude, of a wave. In three principal papers between 1948 and 1951 [1], he developed the theory in considerable depth and offered convincing experimental results. Gabor's original purpose, to record electron waves and regenerate them at optical wavelengths, thereby compensating with optical techniques for the uncorrectable aberrations of electron lenses, is an historical point that today is of secondary importance. However, the holographic process has been revealed to have far more potential than one could, at that time, have imagined.

Interest in holography continued strong for several years afterward, and produced some notable pioneers, such as G.L. Rogers, M.E. Haine, J. Dyson, T. Mulvey, A.W. Lohmann, and others. Despite the initial impetus, however, interest in holography waned in the middle 1950's, although activity never completely ceased.

The principal reason for the loss of interest was the relatively poor imagery, due mainly to the well-known twin image, which occurs because the recording process is sensitive only to the intensity of the incident radiation. As a consequence, the reconstruction process not only recreates the original wave, it also creates a conjugate wave that, under collimated illumination, forms an image in mirror symmetry to the "true" image with respect to the plane of the hologram. Whichever image one elects to use, he must view if against the out-of-focus background of the other, and the result is a noisy image.

MODERN HOLOGRAPHY

In the early 1960's, several papers appeared that proved to be forerunners of the great explosion of activity that ushered in the next stage of holography. Juris Upatnieks and I announced at the October 1961 Optical Society of America meeting a number of new concepts, including the off-axis or spatial-carrier frequency method of holography, which removed the twin-image problem in a simple and practical way [2]. In this method, the reference and object waves are brought together at an angle, to form a rather fine fringe pattern. The resulting hologram, behaving like a diffraction grating, produces several non-overlapping diffracted orders. The zero-order wave produces the usual inseparable twin images which, in combination with other defects of in-line holography, result in poor imagery; but each first-order diffracted wave produces an image of high quality.

When, however, we extended the process to continuous-tone object transparencies instead of the black and white transparencies used in holography until then, another defect became prominent. The difficulty was the well known "artifact" problem of coherent light-each extraneous scatterer (for example, a dust particle) produces a wake of diffraction patterns that contaminate the resultant image. We surmounted this problem with diffused coherent illumination, which properly used, smooths the field produced by these scatterers.

The diffused-illumination hologram is different in appearance from ones made with undiffused illumination. The latter show the characteristic diffraction patterns of the original objects. These diffraction patterns bear a definite, although not always obvious, relation to the object. Also, broad, relatively low-spatial-frequency regions tend to retain their identity, since the defocusing process is most pronounced on high spatial-frequency components.

Holograms made with diffused illumination show no discernible diffraction patterns from the object. Rather, the diffused light, when scattered by the object, acquires the same uniform, grainlike structure that it had when emitted from the diffuser. This pattern is homogeneous and always has the same general appearance, regardless of the object, except for very simple object composed of a few discrete points.

Holograms made with diffused illumination have the property that each point, or each element, regardless of its size, scatters to all points of the hologram. Consequently, information about each portion of the object is recorded over the entire hologram, and each portion of the hologram, no matter how small, reproduces the entire object to within the limitations imposed by the size of the portion, which becomes the limiting aperture of the process.

The most significant property of the diffused-illumination hologram is that the images it produces can be seen without the need for an eyepiece. The virtual image can be seen by looking "through" the hologram as if it were a window, and the real image can be seen suspended in front of the hologram. If the diffuser had not been used in making the hologram, the reconstruction could not be thus observed. To explain why this is so, let us consider what happens when one observes a transparency illuminated from behind with a point source. Except for some scattering, the observer receives light only from the part of the transparency that lies on the line between the point source and the pupil of the eye; this part is usually negligible, However, if a diffuser is placed between the source and the transparency, light from all points of the transparency reaches the eye, and the whole transparency is seen.

This argument readily applies to the hologram case if one thinks of the hologram as reconstructing not only the transparency, but also the diffusing plate. Thus, the observer sees the reconstructed image as if it were illuminated by a diffuse source.

Finally, we went to the use of arbitrary, three-dimensional reflecting objects. The resulting are, optically speaking, a highly exact replication of the original object, fully three-dimensional and with full parallax.

Stability and Coherence Requirements

As the holograms became more sophisticated, the requirements on their construction became more severe. As is well known, holography today is generally carried out on a granite bench, and the techniques of interferometry are brought to bear; for example, optical elements are either massive or securely held so as not to vibrate. Yet, the early Gabor type holograms can be produced on an ordinary optical bench mounted on a very ordinary table. Where along the line have these great stability requirements arisen? In general, they have increased for each of the various steps we have enumerated.

Of course, some stability is required even for ordinary photography; if the camera moves too much, the image is blurred. It is easily shown that, for Gabor, or in-line holograms, the stability requirements are just those of conventional photography. As we go to the off-axis method, the stability requirements increase somewhat. The fringes are finer, and it takes less movement of, for example, the recording plate, to blur them out on the hologram. Going to diffuse illumination produces a still greater stability requirement, since in general, with diffused illumination, the hologram structure becomes much finer.

The really great jump, however, occurs when go to reflecting objects, and it is very easy to see why. If the object transparency moves during recording, the optical path of the transmitted light changes little, and the fringe pattern to be recorded also moves only slightly. But when light is reflected from the object, an object movement of a half wavelength of the illuminating light results in a phase change of the object beam of about 2π, and the fringe shift is a full cycle. The stability requirements have now become so severe that the holography experiment is driven to the granite bench. All of the previously noted forms of holography can be done perfectly well on an ordinary table, but reflecting elements in the system put an end to this simple state of affairs.

A similar evolution has occured with the coherence requirements. The holograms of Gabor were made with the Hg arc source, but of course such a source is not at all satisfactory for the holography of 3-D, reflecting objects.

Proceeding as before, we start with the coherence requirements for ordinary photography. Of course, there are none. From photography to Gabor's holography, there is a jump, since this form of holography has definite, although modest, coherence requirements. If the resolution elements on the object are of linear dimension d, and if each such element produces a diffraction pattern of linear dimension L on the hologram, then the coherence requirement is

$$\Delta\lambda = 4d/L$$

The quantity L/d we may call the expansion ratio. It is interesting that the coherence requirement is given by this single parameter. Calculations show the coherence requirements to be quite modest; for example, for an expansion ration of 100 (which represents an increase of spread function area by a factor 10^4 between object and hologram) and for $\lambda = 5000$ Angstroms, the source spectral width $\Delta\lambda$ can be as great as 200 Angstroms. The 5461 line of the Hg arc can be many times narrower than this.

When we proceed to off-axis holography, the fringes are finer, and there are more of them. Therefore, one might expect the coherence requirements to be greater. This is not necessarily true; the off-axis hologram can be thought of as containing the diffraction information on a spatial carrier, and it can be shown that the operation of placing a signal on a spatial carrier does not require monochromaticity. There should be ways of doing off-axis holography with the same coherence requirements as for in-line holography, and this is indeed the case. In particular, the use of a diffraction grating as a beam splitter leads to systems of off-axis holography with coherence requirements no more than that of the in-line systems.

For diffused illumination, the espansion ratios are greater, and therefore the coherence requirements are also greater. Finally, for 3-D reflecting objects, the coherence requirements take a tremendous jump; rays reflected from all parts of the object must interfere with the reference beam; thus, the object depth should be about twice the source coherence length. For an object of 10 cm depth, the coherence requirement is about 20 cm. This new requirement is so great that it makes the previous requirement small by comparison. In general, only the laser can effectively meet this requirement, and so it is only for the 3-D, reflecting object that a laser is required. All of the previously-noted types of holography can be done quite well with the Hg source, and indeed, our first successes in off-axis holography had been achieved with the Hg source.

Since hologram viewing involves essentially a retracing of the ray paths involved in making the hologram, we expect that these same arguments apply in viewing holograms. In general, this expectation is correct, but with several exceptions, the principal one being that holograms of 3-D, reflecting objects require no more coherence for viewing than holograms of transparencies; the reason is that the large optical path differences arising through reflection from 3-D objects do not occur in the viewing process, since this process does not involve any such reflections.

VOLUME HOLOGRAMS

In 1962, Yu.N. Denisyuk of the USSR introduced a new concept into holography, the "volume hologram", which combines holography with the Lippman color process [3]. Object and reference beams are introduced from opposite sides of the recording plate, and the resulting fringes are embedded within the emulsion as surfaces running nearly parallel to the emulsion surface, with half a wavelength spacing between them. Typically, the number of fringes in a cross section is about 50, although for very thick (a few mm) recording materials, there may be thousands. The twin image is eliminated by the thickness effect and, in addition, the holograms, because of their wavelength selectivity, can be viewed in white light derived from a point source. Denisyuk's work is a cornerstone of modern holography.

One could certainly combine our technique, involving separately-produced object and reference beams, with the Denisyuk technique. Suprisingly, this combination was first made only in the latter part of 1965, and almost simultaneously by three groups. Our group reported and demonstrated such holograms at the Spring 1966 meeting of the Optical Society of America, as did also G. Stroke and A. Labeyrie. However, N. Hartman of the Battelle Institute had been the first to do this and consequently was awarded the patent.

THE GREAT SURGE

By late 1964, holography had become probably the most active field of research in optics, engaging hundreds of groups throughout the world. Discovery and invention dominated the next three years; this period produced several techniques of color holography, hologram interferometry in its various forms, techniques for holographic imagery through scattering and aberrating media, and many other basic concepts. The vast potential that had been inherent in holography now emerged with astonishing force.

Holography was found capable of an astonishingly wide variety of tasks that were normally done in other ways. Holographic methods could be useful in optical metrology and offered some interesting possibilities for spectroscopy. Optical memories using holographic techniques seemed destined to make significant inroads into the huge computer-memory field. Optical reading and feature-recognition machines that used holography were visualized. Microscopy, at least in the visible wavelength range, seemed promising. With hologram interferometry, a wide variety of non-destructive testing techniques became available, ranging from early detection of fatigue failure, to the detection of "debands" in multilayered materials such as tires and honeycomb panels, to the determination of heat flow in transparent materials, to the study of bending moments and to the dynamic operation of audio speakers and other sound-transducing equipment. A most ingenious and unlikely application of hologram interferometry is the determination of the complex mode structure of a laser [4]. Merely to list the applications of hologram interferometry in reasonable completeness would fill a page. Even in conventional interferometry, it was found that one could apply to the hologram essentially all the techniques such as schlieren, dark field, and phase contrast-normally done with the actual object [5].

Holography can be used to detect and examine aerosol particles, such as atmospheric pollutants. Not only was holography unique for visual displays of many kinds, including portraiture, but it could also be used in instrumentation for conventional stereo imagery. Holographically produced optical elements showed promise for improving the performance of optical elements. The versatility of holography seemed limitless.

Some of these applications appear promising now, others not so. If only a few proved viable, holography will have a bright future.

REFERENCES

1) - D. Gabor, Nature 161, 771 (1948); Proc. Roy. Soc. (London) A197, 454 (1949); Proc. Phys. Soc. (London) B64, 449(1951).

2) - E. Leith, J. Upatnieks, J. Opt. Soc. Am. 52, 1123 (1962); 53, 1377 (1963); 54, 1295 (1964).

3) - Yu. N. Denisyuk, Sov. Phys. Dokl. 7, 543 (1962); Opt. Spectrosc. 15, 279 (1963).

4) - C. Aleksoff, Appl. Opt. 10, 1329 (1971).

5) - M. Horman, Appl. Opt. 4, 333 (1965).

Coherant Optical Engineering, F.T. Arecchi and V. Degiorgio (eds.)

AN INTRODUCTION TO QUANTUM OPTICS

F. T. Arecchi

Istituto Nazionale di Ottica, Firenze

and C. I. S. E., Milano, Italy

1 - Quantum Optics : a heuristic approach - Terminology and Numerology

1.1 Definition of Quantum Optics

The term "quantum electronics" was first used in 1959 at a conference dealing with the physical and engineering uses of the MASER (Microwave Amplifier by Stimulated Emission of Radiation) and MASER oscillators in high resolution spectroscopy and in the handling of electromagnetic signals at $\lambda \sim 1$ cm.

Half a year later, the first LASER was operated (L stays for light. The L has replaced the M because this time the generated radiation is at $\lambda \lesssim 1\ \mu m$).

Since 1960 the LASER has become a useful device in many areas of physics and technology.

In the spectral range of interest, it is more convenient to speak of Quantum Optics, rather than Quantum Electronics.

Quantum Optics can be approached from three points of view, namely:

i) physics of the stimulated emission processes;

ii) coherence and cooperative phenomena in radiation-matter interaction;

iii) application of the LASER related to its spectral purity.

We shall discuss the three aspects in sequence, defining the terms and giving the orders of magnitude.

1.2 Physics of the stimulated emission processes

If the e.m. cavity where we are considering the radiation-atom interaction is a rectangular cavity of sides X, Y, Z; volume V = X Y Z, then the solution of the wave equation, with periodic boundary conditions, yields the plane wave expansion for the field

$$E(x,y,z,t) = \sum E_k(t)\, e^{i(k_1 x + k_2 y + k_3 z)} \tag{1.1}$$

where
$$k_1 = n_1 \cdot 2\pi / X$$
$$k_2 = n_2 \cdot 2\pi / Y \quad (n_i = 1, 2, \ldots)$$
$$k_3 = n_3 \cdot 2\pi / Z$$

For each set of $k_{1,2,3}$ we have a different field configuration, or <u>mode</u>.

The dispersion relation imposes a constraint between frequency ω and amplitude $k = \sqrt{k_1^2 + k_2^2 + k_3^2}$ of the k vector

$$\omega = c\,k \tag{1.2}$$

In k space (Fig. 1.1) each mode occupies an elementary volume

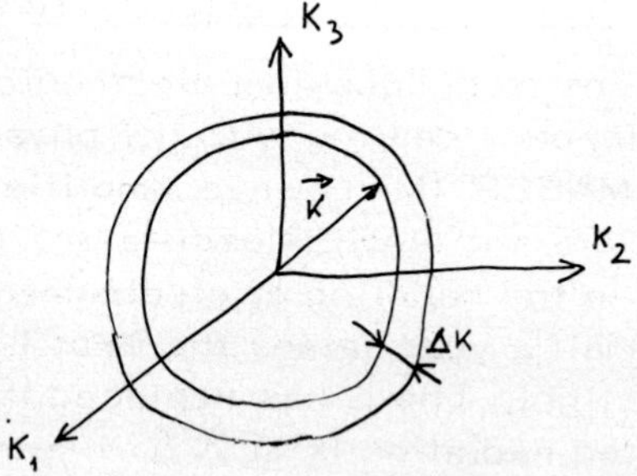

Fig. 1.1

$$\delta^3 k = \delta k_1 \; \delta k_2 \; \delta k_3 = \frac{(2\pi)^3}{V}. \tag{1.3}$$

In a spherical shell of radius k and thickness Δk there are

$$M = 2\,\frac{4\pi k^2 \Delta k}{\delta^3 k} = \frac{k^2 \Delta k}{\pi^2} V \tag{1.4}$$

modes. The extra-factor 2 accounts for the two possible polarizations for each k vector. Hence the mode density in frequency is

$$dM/_{d\omega} = \omega^2 \, V/\pi^2 \quad , \quad \text{or}$$

$$\frac{dM}{d\nu} = \frac{8\pi\nu^2}{c^3} V \tag{1.5}$$

If the cavity contains radiators (atoms on the walls or inside) in thermal equilibrium at a temperature T, then the electromagnetic energy density in the cavity is given by Planck's blackbody formula (1900)

$$\frac{dW}{d\nu} = \frac{8\pi\nu^2}{c^3} V \cdot h\nu \cdot \frac{1}{e^{h\nu/k_BT} - 1} \qquad (1.6)$$

$$= \frac{dM}{d\nu} \cdot h\nu \cdot \bar{n}_1(\nu)$$

Here

$$\bar{n}_1(\nu) = (e^{h\nu/k_BT} - 1)^{-1}$$

is the average photon number for each mode whose k vector lies on the spherical surface of radius $k = 2\pi\nu/c$.
The distinction between spontaneous and stimulated emission came in the 1917 Einstein's derivation of Eq. (1.6), as follows

Consider two relevant levels of an atom coupled by an optical transition. Each time the atom goes up (absorption) or down (emission), this is a one-photon exchange process.

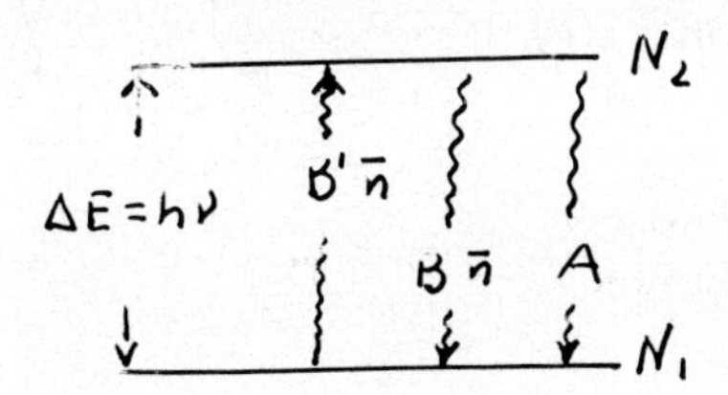

Fig. 1.2

The emission or decay process can be spontaneous (i.e. not triggered by photons) as well as stimulated (i.e., proportional to the photon number $\bar{n}$ at the frequency $\nu = \Delta E/h$); see Fig. 1.2.

If there is an ensemble of N atoms in thermal equilibrium, with N_2 in the upper state and $N_1 = N - N_2$ in the lower, then we have

$$N_1/N_2 = e^{\Delta E/k_BT} \qquad (1.7)$$

$$N_1 B' \bar{n} = N_2 B \bar{n} + N_2 A \qquad (1.8)$$

From this latter equation

$$\bar{n} = \frac{A/B}{\frac{B'}{B}\frac{N_1}{N_2} - 1}$$

By use of Eq. (1.7) and comparison with (1.6), we get

$$\frac{A}{B} = \frac{8\pi \nu^2}{c^3} V \tag{1.9}$$

$$B' = B . \tag{1.10}$$

This can be interpreted by representing the degrees of freedom of the e.m. field as boxes and the excited atom as linked to all of them as in Fig. 1.3

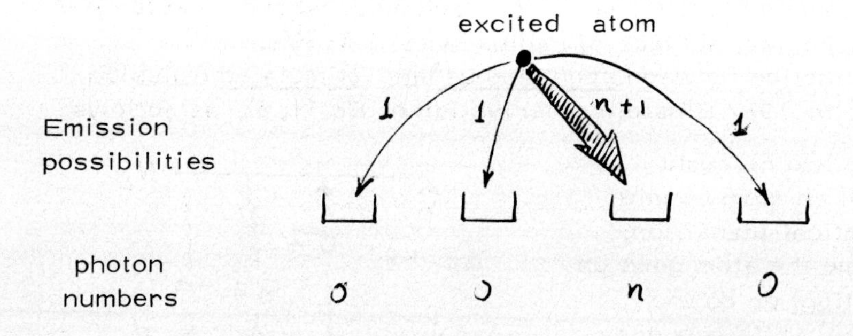

Fig. 1.3

With the probabilities there indicated, the stimulated emission probability into the mode is larger than the total spontaneous emission over Δ M modes (all those within the linewidth of the atomic emission) when

$$n > \Delta M. \tag{1.11}$$

In other words, it is not enough to have a mode with a large population ($n \gg 1$) but it is necessary to obey the specific condition (1.11) in order to observe stimulated effects. Hitherto we referred to the whole photon number in the cavity. It is more convenient to refer to the photon flux, given by $\phi_n = c\, n / V$ or in terms of the field amplitude and the frequency ω, by

$$\phi_n = \frac{c\, \varepsilon_0 E^2}{\hbar \omega} \quad \frac{\text{photons}}{\text{cm}^2 \text{ sec}} \tag{1.12}$$

This becomes smaller for high frequencies ω , unless we simultaneously increase the E field strength.

The stimulated transition probability per second per atom is given by

$$P_s = \sigma \Phi_n \tag{1.13}$$

where the cross section σ for the process can be evaluated on a purely classical basis. For a free electron, it has the Thomson value

$$\sigma_{th} = \frac{8}{3} \pi r_o^2 \sim 0.6 \times 10^{-24} \text{ cm}^2 \tag{1.14}$$

where

$$r_o = e^2 / (4 \pi \varepsilon_o m c^2)$$

is the classical electron radius.

For a bound electron, if the field frequency is resonant with the atomic transition, the cross section is

$$\sigma_{res} = \sigma_{th} \left(\frac{\omega_o}{\gamma} \right)^2 \tag{1.15}$$

At optical frequencies the decay rate γ can be as small as $10^{-5} \omega$, hence σ_{res} can be as large as $10^{10} \sigma_{th}$, (Fig. 1.4)

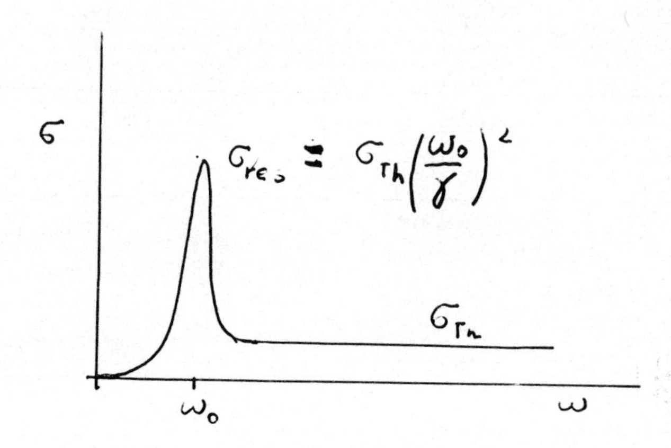

Fig. 1.4

Condition (1.11) with ΔM given by (1.4) is unfair. We try either to privilege some modes at constant σ, by a tuned cavity or to privilege a frequency by using the strong frequency dependence of σ res.

The former is done at low frequencies (microwaves)

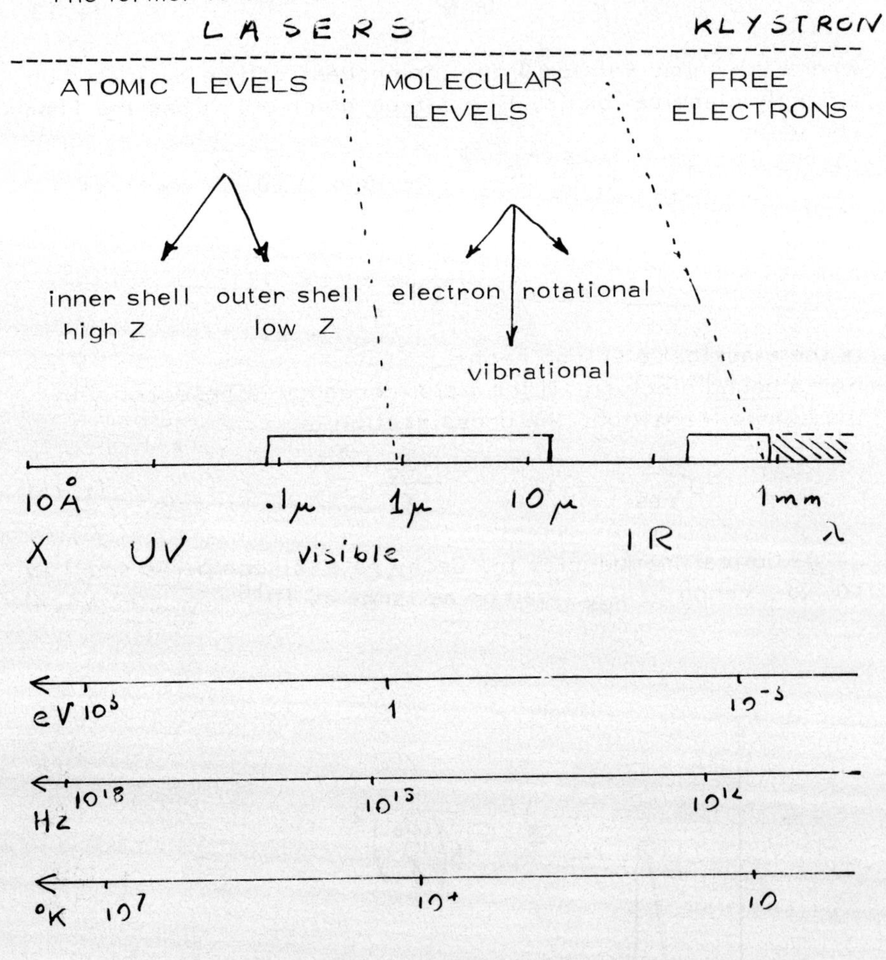

Fig. 1.5

the latter at high frequencies (from IR up).

By Eqs. (1.12) and (1.13), the condition (1.11) can be rephrased in terms of σ as $\sigma > \Delta M/\phi \;\propto\; \omega^3/E^2$
This shows that for high frequencies and constant σ, stimulated emission can be obtained only by very high fields. Hence above 10^{12} Hz it is no longer convenient to extract energy from free electrons (with $\sigma = \sigma_{Th}$), even though oustanding examples are offered of e.m. fields in the visible and U.V. generated by free electrons (Cherenkov radiation, synchrotron radiation). In Fig. 1.5 we report the regions where most lasers are available. It is useful to compare the λ scale with the eV, Hz and °K scales.

To put Eq. (1.15) in a different way, we recall that if γ is due only to radiation damping (no collision broadening or other things), then from classical electro-dynamics

$$\omega/\gamma \sim \lambda/r_0$$

hence

$$\sigma_{res} \sim \lambda^2/4\pi \tag{1.15}$$

Notice that the cross section is not the square of the electron radius nor the square of the Bohr's orbit but the squared wavelength which in the visible and U.V. is much bigger.

One arrives at the same numerical value by a semiclassical approach. Take the interaction Hamiltonian

$$H' = -\vec{d}\cdot\vec{E} \tag{1.16}$$

with a classical $\vec{E}$ field and a dipole operator whose matrix element between the two atomic states is $\langle 1|d|2\rangle = \mu$ ($\mu \sim 10^{-27}$ C x cm for an allowed transition). Then by Fermi's golden rule the transition rate is

$$W = \frac{2\pi}{\hbar^2}\left(\frac{dn}{d\omega}\right)_{max}\mu^2 E^2 \tag{1.17}$$

putting $(dn/d\omega)_{max} = 2/(\pi\Delta\omega)$ (Lorentzian curve) and dividing by the photon flux $\phi_n = c\,\varepsilon_0 E^2/(\hbar\omega)$, one has

$$\sigma = \frac{W}{\phi_n} = \frac{4\omega}{\hbar c\,\varepsilon_0\,\Delta\omega}\,\mu^2\ (cm^2) \tag{1.18}$$

This value is numerically as $\lambda^2/4\pi$

Dividing w by the photon number in the cavity

$$n = \varepsilon_0 E^2 V / \hbar \omega_0$$

one has the coupling constant per atom and per photon

$$g = \frac{1}{\hbar \varepsilon_0 V} \frac{\omega}{\Delta\omega} \mu^2 \qquad (1.19)$$

that we shall later use in the fully quantized interaction Hamiltonian (Fig. 1.6)

$$H = \hbar g (a^+ S^- + a S^+) \qquad (1.20)$$

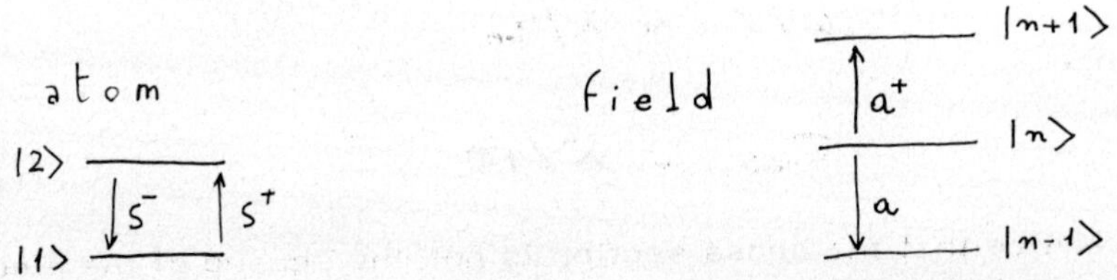

atom

$|2\rangle$ S^- S^+ $|1\rangle$

field

$|n+1\rangle$ a^+ $|n\rangle$ a $|n-1\rangle$

$S^{\pm}$ = Pauli operators $\qquad$ a^+, a = Bose operators

Fig. 1.6

Let us now go back to condition (1.11). In order to keep a large photon number within the cavity, we must have small losses. This can be obtained by confining the radiating atoms within two mirrors which act as a photon store.

This way, the "ingredients" necessary for a laser are

i) an <u>active medium</u> i.e. a collection of atoms having a transition at the selected frequency
ii) a <u>pump</u> or excitation mechanism to take the atoms away from thermal equilibrium, up to excited state
iii) a cavity made of two facing mirrors with high reflectivity (this is called in optics a Fabry-Perot interferometer).

Such a cavity is resonant for those frequencies corresponding to the standing wave condition (Fig. 1.7)

$$m \lambda / 2 = L$$

which amounts to a minimum frequency separation

$$\Delta \nu_{m,m+1} = \frac{c}{2L} \tag{1.21}$$

For these resonances, the escape time of photons is

$$T_d = \frac{L}{c} \frac{1}{1-R} \tag{1.22}$$

that is, a transit time L/c multiplied by a loss factor 1/(1-R), R being the mirror reflectivity

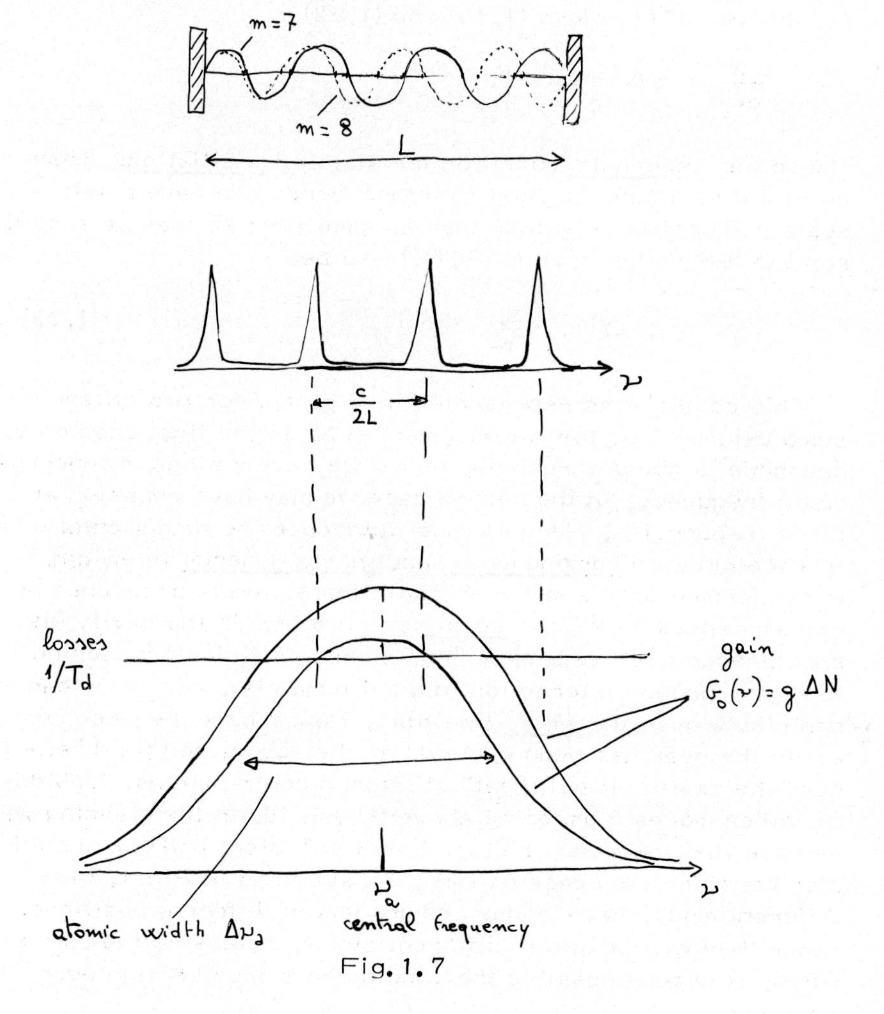

Fig. 1.7

Let us call $N = N_2 - N_1$, the difference of population between upper and lower state in the active medium, as <u>population inversion</u>.

Then, neglecting the spontaneous contributions (i.e. assuming Eq. (1.11) fulfilled) the rate equation for photons in the cavity will be

$$\frac{dn}{dt} = \left.\frac{dn}{dt}\right|_{gain} - \left.\frac{dn}{dt}\right|_{loss}$$

Or, by use of relations (1.19) and (1.22),

$$\frac{dn}{dt} = g \Delta N n - \frac{n}{T_d}$$

Hence the <u>threshold</u> condition for starting <u>oscillations</u> from an initial spontaneous <u>seed</u> (in other words : to have a self - sustained oscillator rather than an amplifier; so that we should say LOSER rather than LASER!) will be

$$g \Delta N \geq 1/T_d \qquad (1.23)$$

This condition is represented in Fig. 1.7 for two different pump values, i.e. for two different ΔN. In the first case only one mode is above threshold, hence we have a single monochromatic frequency. In the second case we may have emission at three frequencies. Here we must introduce the fundamental difference between <u>homogeneous</u> and <u>inhomogeneous</u> linewidth. In the former case a monochromatic transition is broadened by circumstances which are <u>equal</u> for all atoms in the cavity (as spontaneous lifetime broadening, atomic collision broadening in a gas, phonon interaction in a solid matrix). All atoms can contribute over the <u>whole</u> linewidth. Hence, once the mode nearest to the peak has been excited, as the associated field "sweeps" the cavity, it will "eat" all atomic contributions, forbidding the other modes from going above threshold. In the standing wave case this frequency picture is not sufficient and one should also consider the space pattern. As sketched in Fig. 1.8 two different modes have nodes and maxima in different positions, hence they will "exploit" different atoms, releasing the competition. It is then possible the simultaneous laser action over many modes.

The inhomogeneous line broadening corresponds to different frequency locations of different atoms. This can be due, e.g., to Doppler shift in a gas where thermal agitation gives a distribution of velocities to the molecules and hence a distribution of Doppler shifts

$$\Delta\omega = \omega \, V_k/c$$

where V_k is velocity component along the k vector.

Another inhomogeneity occurs in a crystal where active ions are exposed to a crystal field which changes from site to site, contributing with a spread in the stark shifts.

For an inhomogeneous line, different modes can go above threshold even without a standing wave pattern.

In general if $c/2L$ is much smaller than the atomic linewidth $\Delta\nu_a$ there are many independent laser lines, without phase relations.

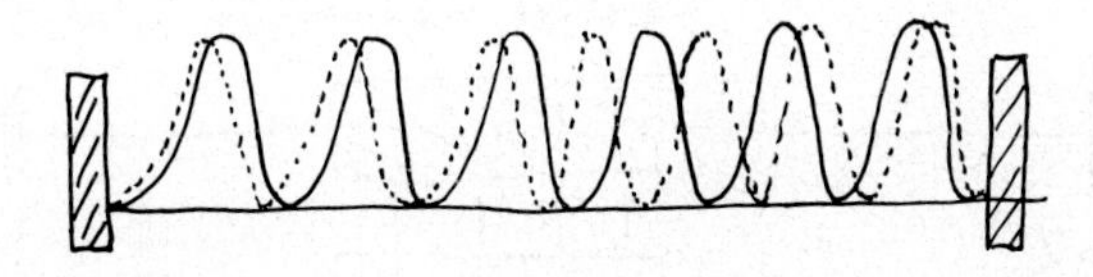

Fig. 1.8

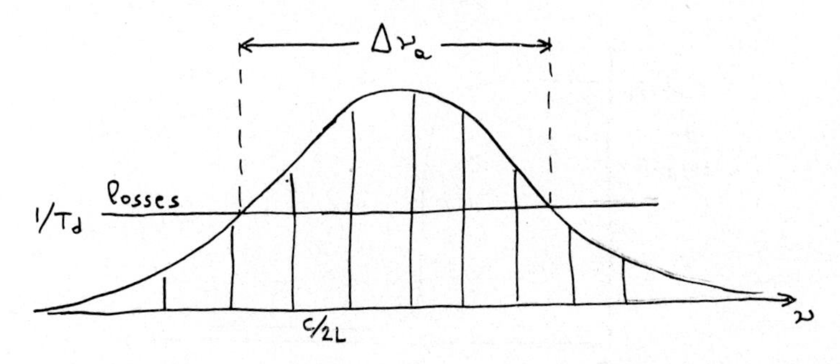

Fig. 1.9 a)

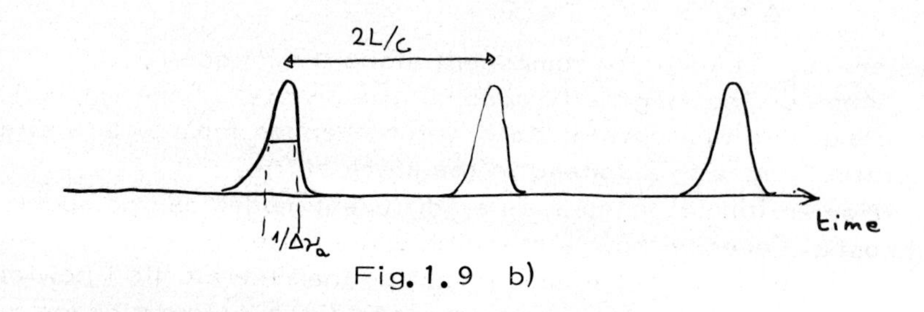

Fig. 1.9 b)

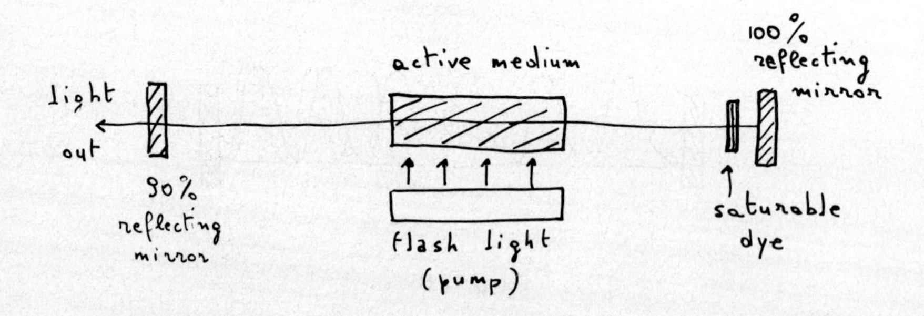

Fig. 1.9 c)

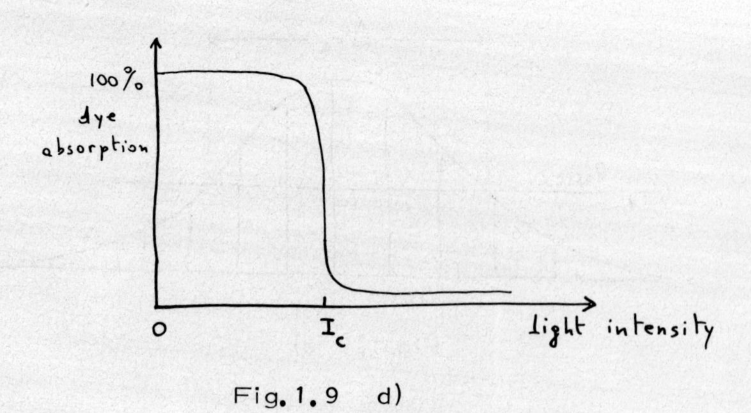

Fig. 1.9 d)

τ_s dye level scheme

Fig. 1.9 e)

In Fig. 1.9 we have shown the schme of mode locking operation. In that figure, the several parts have the following meaning.

a) frequency picture of a many-mode Laser
b) if the different laser fields have fixed phase relations, they act as the different Fourier components of a train of pulses, each lasting $1/\Delta\nu_a$ and separated by 2 L/c. b) is the Fourier transform of the amplitude spectrum a), provided the phases are all equal
c) practical scheme of a mode-locked laser.
Besides the three main ingredients (active medium, incoherent light to excite the atoms at the upper level, mirrors) there is also a saturable dye which becomes transparent at a critical light intensity I_c (see d)). All the standing waves of the different modes will self-adjust their phases to have a maximum when the dye is transparent. Transparency is then lost with a decay time $\tau_s \ll 2L/c$ and then recovered after a transient 2L/c. This corresponds to having a narrow pulse bouncing forth and back between the two mirrors. Notice that, from b) the pulse duration is $1/\Delta\nu_a$

In the scheme of Fig. 1.9 we have thus explained how to lock in phase the laser lines, in order to make short pulses (as short as the uncertainty relation permits, i.e. $1/\Delta\nu_a$).

By using a Doppler broadened atomic line in a gas (like in a He-Ne, or in an A^+ laser), then

$$\Delta\nu \sim \frac{\nu}{c}\sqrt{\frac{K_B T}{M}} \sim 10^9\ Hz$$

hence

$$t_{pulse} \sim 1\ ns.$$

Using ions of a transition element imbebbed in a crystal or glass matrix, as Cr^{3+} in Al_2O_3 (ruby), or Nd^{3+} in glass, one may have large $\Delta\nu_a$ because the 3 d or 4 f electron is strongly affected by the crystal field. A large $\Delta\nu$ a can also be achieved in the case of complex dye molecules in a liquid solution because of the overlapping among many vibrational and rotational levels.
It is nowadays easy to achieve

$$\Delta\nu_a \sim 10^{13} \quad \text{Hz}$$

and hence

$$t_{pulse} \sim 0.1 \text{ p sec}$$

Notice that the range of picosecond times can be attained only by techniques as in Fig. 1.7, and not by electronic shutters.

1.3 Stimulated emission and nonlinear optics (NLO)

In Fig. 1.3 we have represented the difference between spontaneous and stimulated emission. With reference to the Hamiltonian (1.20), and looking only at the field creation operator, it is well known that it acts on an m-photon state as follows

$$a^+ |n\rangle = \sqrt{n+1}\, |n+1\rangle \tag{1.21}$$

Therefore the transition rate for an emission process will be proportional to

$$\text{spontaneous} \quad : \quad |\langle 1 | a^+ | 0 \rangle|^2 = 1$$

$$\text{stimulated} \quad : \quad |\langle n+1 | a^+ | n \rangle|^2 = n+1$$

Also in higher order processes as those studied in NLO we can have a spontaneous and a stimulated version. Take a parametric process implying three Bose fields as in Fig. 1.10

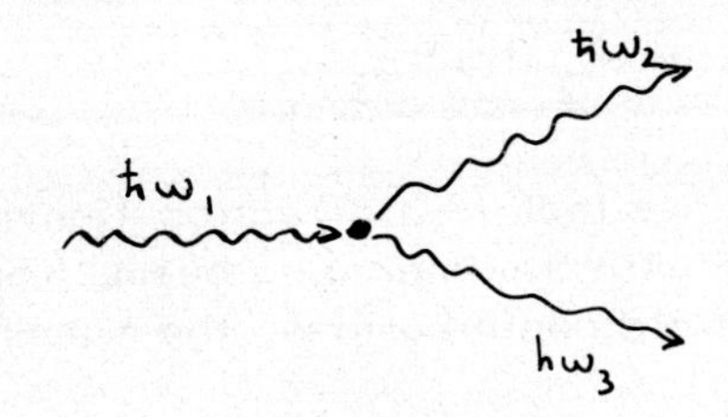

Fig. 1.10

The model Hamiltonian for the process in Fig. 1.10 is

$$H = \hbar g\,(a_1\, a_2^+\, a_3^+ + a_1^+\, a_2\, a_3)$$

and the transition rate will have matrix elements as

$$|\langle 1 | a_2^+ | 0\rangle|^2 = 1$$ spontaneous emission in the field 2

or

$$|\langle n_2+1 | a_2^+ | n_2\rangle|^2 = n_2+1$$ stimulated emission in the field 2

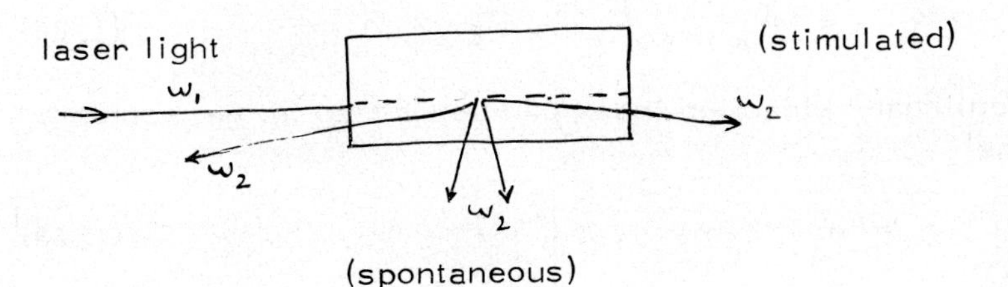

Fig. 1.11

In the first case, we look at 90°, and we collect point-like processes, having to satisfy the conservation of energy :

$$\omega_1 = \omega_2 + \omega_3$$

In the second case, we look in a direction (forward or backward) almost collinear with the impinging beam. Here, in order to add coherently the field contributions, the momentum matching condition

$$k_1 = k_2 + k_3$$

has also to be satisfied.

In a similar way we may desribe usual light propagation in a transparent medium as an elastic two-photon process. Since the scattered contributions sum in phase, it is more convenient to speak of a linear polarization and a refraction index

$$P_i = \varepsilon_o \chi^{(2)}_{ij} E_j \; , \; n_{ij} = \sqrt{\chi^{(2)}_{ij}} \qquad (1.22)$$

rather than stimulated emission in the scattered channel. Similarly, there are 4 photon processes leading to "<u>self-actions</u>" in the propagation of a large e.m. field, that is, self-focusing, self-defocusing, self modulation in phase (self-broadening) and amplitude (self-steepening). These non linearities on the same light beam are described by a nonlinear polarization index as

$$P_i = \chi^{(4)}_{ijkl} E_j E_k E_l \qquad (1.23)$$

The nonlinear refraction index can be written in the isotropic case as

$$n = n_o + n_2 |E|^2 \qquad (1.23)'$$

In a liquid of anisotropic molecules, self actions stem from orientation of the molecules due to interaction with the induced dipole moments (high frequency Kerr effect). In a liquid of isotropic molecules, or in solids and gases, self actions are due to distortion of the electron cloud.

In Table 1.1 we show some examples of NLO processes.

TABLE 1.1

Nature of the quanta		Name of the process
2	3	
light	molecular vibrations	Raman
light	optical phonons in solids	Raman
light	acoustical phonons in solids	Brillouin
light	sound waves in liquids	Brillouin
light	light	parametric conversion (sum or difference of frequency, second harmonic generation, etc.)

1.4 Coherence and cooperative phenomena

As shown in Fig. 1.3, stimulated emission explains the privileged filling of a given mode, that is, a narrowing in the frequency spectrum and in the spectrum of possible directions (monochromaticity and directionality). This amounts to increasing the spectral purity, and use can be made of it in physics and technology (linear spectroscopy, holography, plasma production and compression by powerful laser pulses). But all this has very little to do with coherence.

Stimulated emission is a first order effect, i.e. it provides a linear source for Maxwell eqs.

Each mode has still a harmonic oscillator dynamics, that is, it is like a particle in a parabolic potential well, with an equilibrium statistical distribution given by Maxwell-Boltzmann. To have a sizeable amount of energy $|E_0|^2$ one has to increase the "temperature" i.e. the excitation, thus broadening the distribution and increasing the entropy as well (Fig. 1.12).

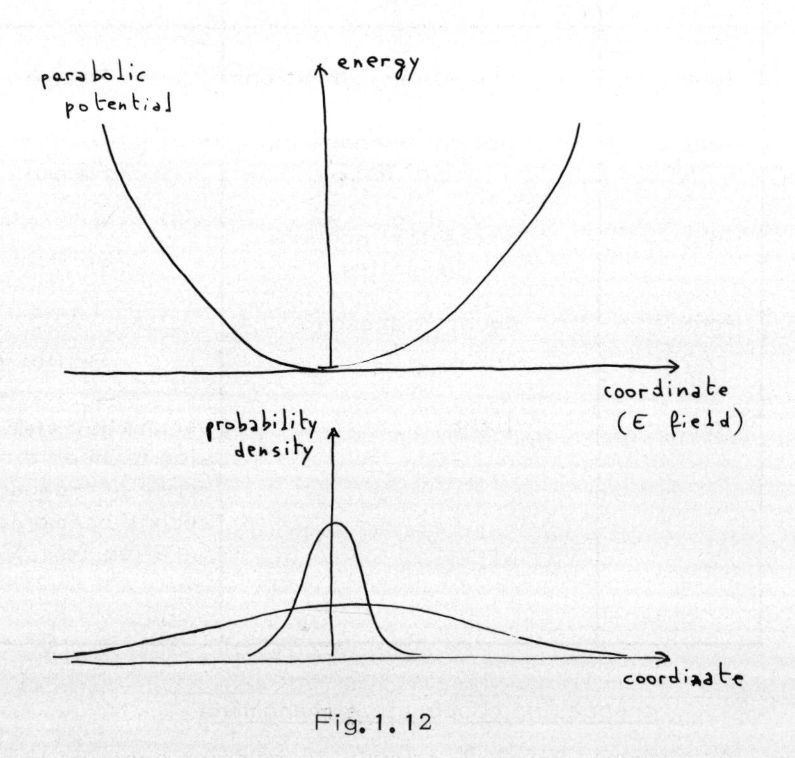

Fig. 1.12

However as the field E increases, one must consider high order processes, besides the one photon emission, as e.g. the three-photon process of Fig. 1.13 which gives a quadratic field correction in the gain

$$G = G_0 - \beta|E|^2 \tag{1.24}$$

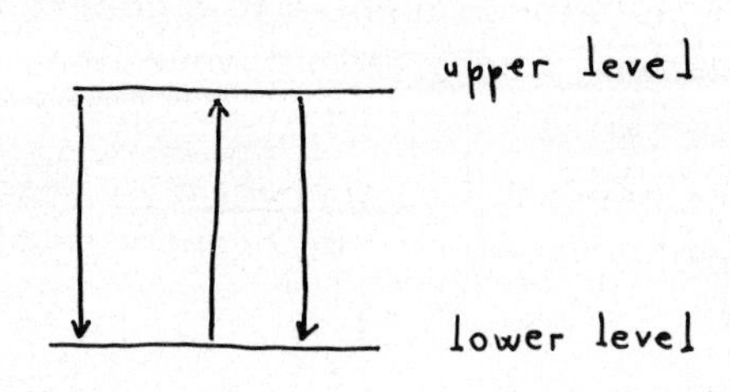

Fig. 1.13

This is equivalent to a cubic polarization

$$P = G_o E - \frac{1}{3} \beta |E|^2 E$$

and to a quartic free energy

$$W = - P. E = - \frac{G_o}{2} |E|^2 + \frac{\beta}{12} |E|^4$$

As E increases, the quartic potential well becomes steeper and steeper (Fig. 1.14), so that a useful E_o can be reached with a little amount of spread, or statistical fluctuations, around it.

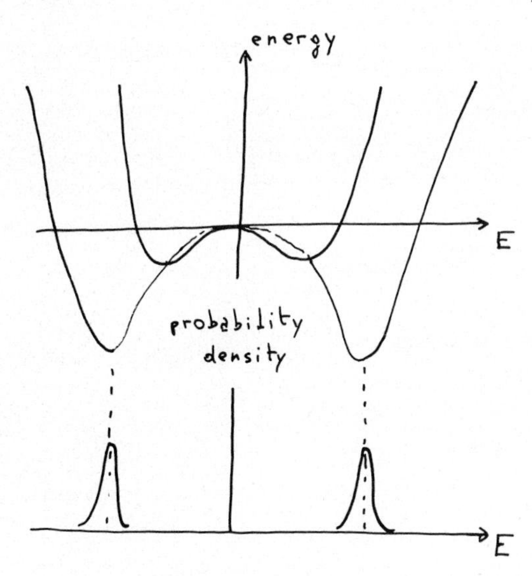

Fig. 1.14

We call coherent this highly excited field state without noise. The field can be described with very good approximation by a c-number with constant amplitude and phase. Such a field can bring the induced atomic dipoles to a coherent motion in which the phase relations among atomic wave functions are kept for long times.

This is the basis for a coherent nonlinear spectroscopy which sheds information on fine properties of atoms and molecules.

Coherant Optical Engineering, F.T. Arecchi and V. Degiorgio (eds.)

SOME FUNDAMENTAL PROPERTIES OF SPECKLE*

J.W. Goodman
Department of Electrical Engineering
Stanford University
Stanford, California 94305
U.S.A.

INTRODUCTION

Objects illuminated by light from a highly coherent CW laser are readily observed to acquire a peculiar granular appearance. Figure 1 shows a typical pattern observed in the image of a uniformly white reflecting object. This extremely complex pattern bears no obvious relationship to the macroscopic properties of the object illuminated. Rather it appears chaotic and unordered, and is best described quantitatively by the methods of probability and statistics.

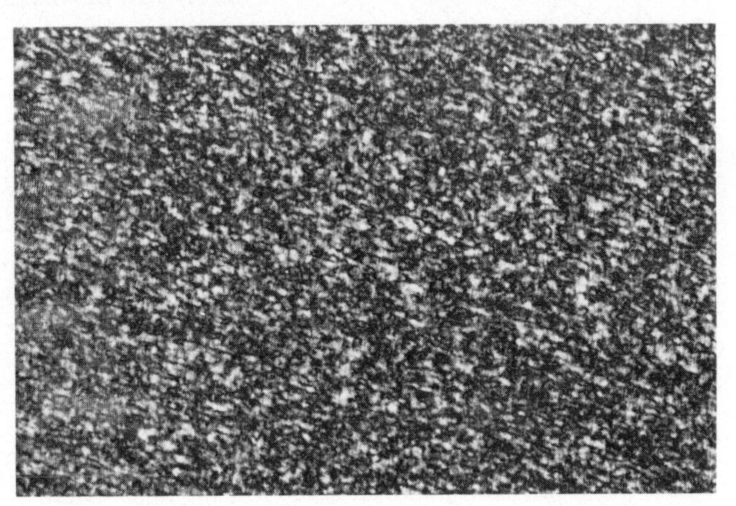

Fig. 1: A typical speckle pattern

*Work sponsored by the Office of Naval Research

The origin of this granularity was quickly recognized by early workers in the laser field[1,2]. The vast majority of surfaces, man-made or natural, are extremely rough on the scale of an optical wavelength. Under illumination by coherent light, the wave reflected from such a surface consists of contributions from many independent scattering areas. Propagation of this reflected light to a distant observation point results in the addition of these various scattered components with relative delays which may vary from several to many wavelengths, depending on the microscopic surface and the geometry (see Fig. 2).

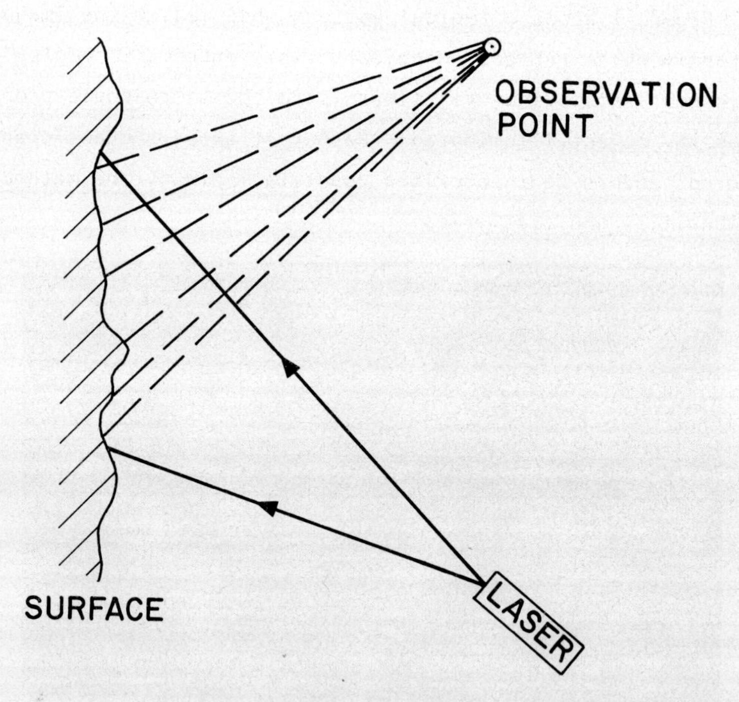

Fig. 2: Speckle formation in the free-space geometry

Interference of these dephased but coherent wavelets results in the granular pattern we know as speckle. Note that if the observation point is moved, the path lengths traveled by the scattered components change, and a new and independent value of intensity may result from the interference process. Thus the speckle pattern consists of a multitude of bright spots where the interference has been highly constructive, dark spots where the interference has been highly destructive, and irradiance levels in between these extremes. Accordingly, we observe a continuum of values of irradiance which has the appearance of a chaotic jumble of "speckles".

While the origin of speckle is perhaps easiest to discuss in the free-space reflection geometry of Fig. 2, with some additional work its appearance in the imaging geometry of Fig. 3

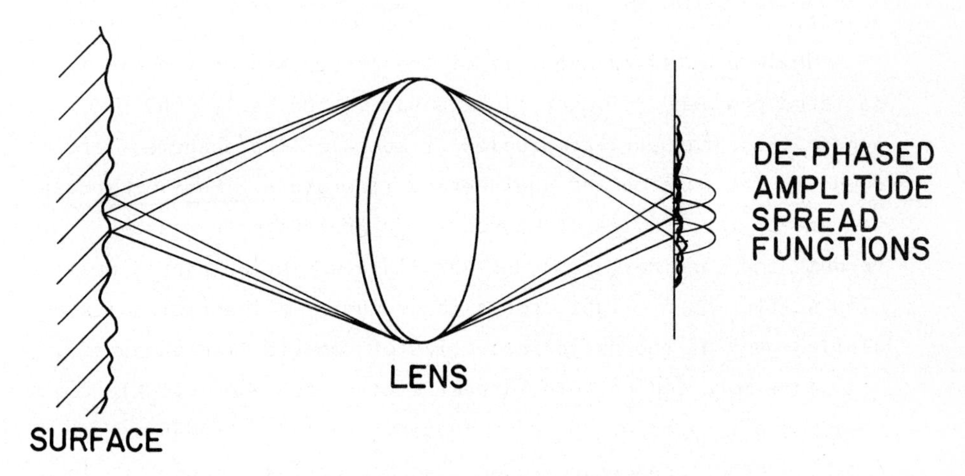

Fig. 3: Speckle formation in the imaging geometry

can also be explained. The image formed at a given point in the observation plane consists of a superposition of a multitude of complex amplitude spread functions, each arising from a different scattering point on the surface of the object. As a consequence of the roughness of this surface, the various amplitude spread functions add with different phases, resulting again in a complex pattern of interference, or a speckle pattern superimposed on the image of interest.

The appearance of speckle is not limited to imagery formed with reflected light. If a photographic transparency is illuminated through a diffuser, the wavefront passing through the transparency has a highly corrugated and extremely complex structure. In the image of such a transparency we again find large fluctuations of irradiance caused by the overlapping of a multitude of de-phased amplitude spread functions. While most of the discussions in this lecture are presented in terms of the reflecting geometries of Figs. 2 and 3, the conclusions apply equally well to the transmission geometry, provided the wavefront transmitted by the transparency satisfies the same basic assumptions applied to the wavefront reflected from a rough object.

While a detailed analysis of the properties of speckle patterns produced by laser light began in the early 1960's, nonetheless, far earlier studies of speckle-like phenomena are found in the physics and engineering literature. Mention should be made of the studies of "coronas" or Fraunhofer rings by Verdet[3] and Lord Rayleigh[4]. Later, in a series of papers dealing with scattering of light from a large number of particles, Laue[5] derived many of the basic properties of speckle-like phenomena.

In a more modern vein, direct analogs of laser speckle are found in all types of coherent imagery, including radar astronomy[6], synthetic-aperture radar[7], and acoustical imagery[8]. In addition, statistical phenomena entirely analogous to speckle are found in radio-wave propagation[9], the temporal statistics of incoherent light[10], the theory of narrowband electrical noise[11], and even in the general theory of spectral analysis of random processes[12]. As a consequence of the ubiquitous nature of the random interference phenomenon, the term "speckle" has taken on a far broader meaning than might have been envisioned when it was introduced in the early 1960's.

The purpose of this lecture is to provide an introduction to the more important properties of speckle patterns, in hopes that this background will enable the reader to more fully understand the effects of speckle in coherent optical systems. The reader wishing to acquire a more detailed understanding of the statistical properties of speckle patterns may consult reference 13. Ultimately, a completely rigorous understanding of laser speckle requires a detailed discussion of the properties of electromagnetic waves that have been reflected from rough surfaces[14]. However, a good intuitive feeling for the properties of speckle can be obtained without an extremely detailed examination of the physics of the scattering process.

SPECKLE AS A RANDOM-WALK PHENOMENON

Due to our lack of knowledge of the detailed microscopic structure of the surface from which the light is reflected, it is necessary to discuss the properties of speckle patterns in statistical terms. The statistics of concern are defined over an ensemble of objects, all with the same macroscopic properties, but differing in microscopic detail. Thus if we place a detector at position (x,y,z) in the observation plane of Fig. 2 or Fig. 3, the measured irradiance is not exactly predictable, but we can describe its statistical properties over an ensemble of rough surfaces.

To aid in the development of a statistical model for speckle, we initially make some simplifying assumptions. Namely, we suppose that the field incident at (x,y,z) is perfectly polarized and perfectly monochromatic. Under such conditions we can represent this field by a complex-valued analytic signal of the form

$$u(x,y,z;t) = A(x,y,z)\exp(i2\pi\nu t) \quad , \tag{1}$$

where ν is the optical frequency and $A(x,y,z)$ is a complex phasor amplitude,

$$A(x,y,z) = |A(x,y,z)|\exp[i\theta(x,y,z)] \quad . \tag{2}$$

The directly observable quantity is the irradiance at (x,y,z), which is given by

$$I(x,y,z) = \lim_{T\to\infty} \frac{1}{T} \int_{-T/2}^{T/2} |u(x,y,z;t)|^2 dt = |A(x,y,z)|^2 \, . \tag{3}$$

Perhaps the most important statistical property of a speckle pattern is the probability distribution of the irradiance I. How likely are we to observe a bright peak or a dark null in the irradiance at a given point? This question can be answered by noting the similarity of the problem at hand to the classical problem of the random walk, which has been studied for nearly 100 years[4,15,16].

The complex amplitude of the field at (x,y,z) may be regarded as resulting from the sum of contributions from many elementary scattering areas on the rough surface. Thus the phasor amplitude of the field can be represented by

$$A(x,y,z) = \sum_{k=1}^{N} |a_k|\exp(i\phi_k) \tag{4}$$

where $|a_k|$ and ϕ_k represent the amplitude and phase of the contribution from the kth scattering area and N is the total number of such contributions. Figure 4 illustrates this phasor addition.

Now we make two important assumptions about the contributions from the elementary scatterers.

(1) The amplitude a_k and the phase ϕ_k of the kth elementary phasor are statistically independent of each other and of the amplitudes and phases of all other elementary phasors (i.e., the elementary scattering areas are unrelated and the strength of a given scattered component bears no relation to its phase);

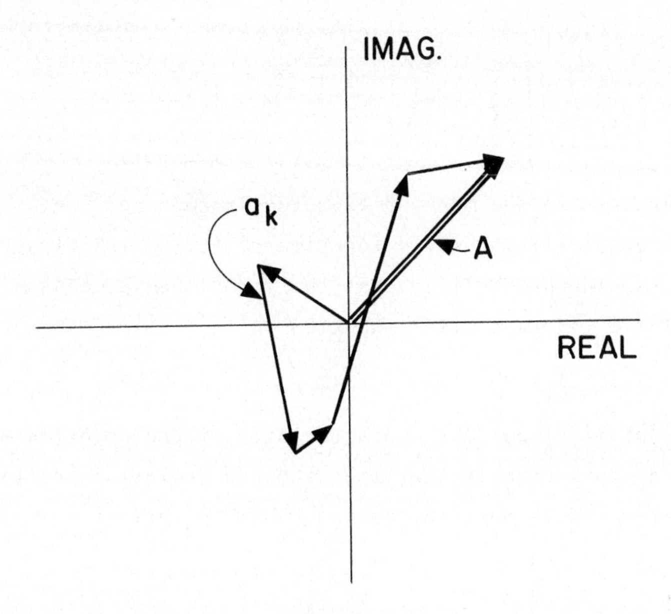

Fig. 4: Random walk in the complex plane

(2) The phases ϕ_k of the elementary contributions are equally likely to lie anywhere in the primary interval $(-\pi,\pi)$ (i.e., the surface is rough compared with a wavelength, with the result that phase excursions of many times 2π radians produce a uniform probability distribution on the primary interval.)

With these two assumptions, the similarity of our problem to the classical random walk in a plane becomes complete. Without presenting the detailed mathematical proofs, which can be found elsewhere, (e.g., Ref. 13), we simply state the conclusions of the random walk analysis. Provided the number N of elementary contributions is large, we find:

(1) The real and imaginary parts of the complex field at (x,y,z) are independent, zero mean, identically distributed gaussian random variables; and

(2) The irradiance obeys negative exponential statistics,

i.e., its probability density function is of the form

$$p(I) = \begin{cases} \dfrac{1}{\bar{I}} \exp\left(- \dfrac{I}{\bar{I}}\right) , & I \geq 0 \\ 0 & \text{otherwise} \end{cases} \quad (5)$$

where $\bar{I}$ is the mean or expected irradiance.

The probability density function of Eq.(5) is illustrated in Fig. 5. The probability that the irradiance exceeds a given threshold I_t is a similar function

$$P(I > I_t) = \exp(-I_t/\bar{I}) , \quad I_t \geq 0 . \quad (6)$$

A fundamentally important characteristic of the negative exponential distribution is that its standard deviation precisely equals its mean. Thus, the contrast of a polarized speckle pattern, as defined by

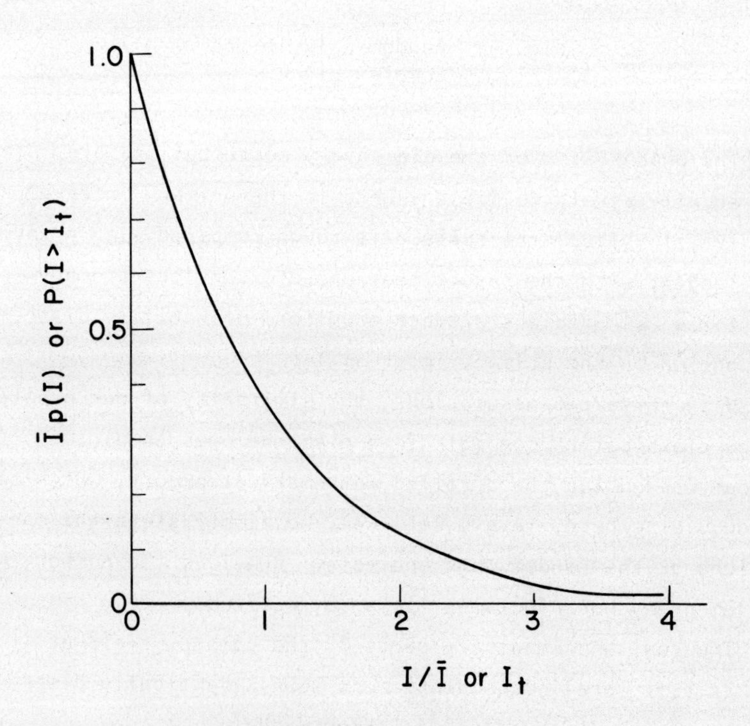

Fig. 5: Probability density function of a polarized speckle pattern

$$C = \sigma_I/\bar{I} , \quad (7)$$

is always unity. Herein lies the reason for the subjective impression that the variations of irradiance in a typical speckle pattern are indeed a significant fraction of the mean.

In the free space geometry of Fig. 2, the assumption that the number N of elementary contributions is large is generally well satisfied. However, in the imaging geometry of Fig. 3, this assumption can be violated if the imaging system is nearly capable of resolving the elementary scattering areas. Nonetheless, in the vast majority of practical applications the number N is indeed very large. The predicted negative exponential statistics of irradiance have been well-verified experimentally[17,18,19,20].

It is perhaps also worth emphasizing that, once the surface has a roughness which is sufficient to produce phase excursions comparable with 2π radians, negative exponential statistics result, and further increases of roughness produce no perceptible changes of irradiance statistics. Thus to a good approximation, all surfaces that are rough on the scale of a wavelength produce the same form of irradiance statistics, regardless of just how much rougher than this limit they may be.

SUPPRESSION OF SPECKLE

Although a number of beneficial uses can be made of speckle, it usually is more of a hindrance than a help. The presence of speckle in an image reduces the ability of a human observer to resolve fine detail. The presence of speckle in the signal detected by an optical radar system can reduce the probability of target detection and/or cause the radar system to lose track. Thus in most cases of practical interest, suppression of speckle is a goal towards which we aspire. How is it possible to reduce the fluctuations present in a detected speckle pattern?

The answer to this question follows from the fundamental result of probability theory that the sum of M identically distributed, real-valued, uncorrelated random variables has a mean value which is M times the mean of any one component, and a standard deviation which is $\sqrt{M}$ times the standard deviation of one component. Thus if we add M uncorrelated speckle patterns on an irradiance basis, the contrast of the resultant speckle pattern is reduced in accord with the law

$$C = \frac{\sigma_I}{\bar{I}} = \frac{1}{\sqrt{M}} \tag{8}$$

Uncorrelated speckle patterns can be obtained from a given object by means of time, space, frequency, or polarization diversity. For example, reflection from a surface such as non-glossy paper generally involves multiple scattering, with a considerable amount of depolarization resulting. For such a surface, the speckle patterns observed through a polarization analyzer will change detailed structure dramatically as the analyzer is rotated through 90°. The irradiance in a speckle pattern is, of course, the sum of the irradiances contributed by two orthogonal linear polarization components. Hence when complete depolarization of the reflected wave occurs, the contrast of the total speckle pattern is $1/\sqrt{2}$, rather than unity.

Pure spatial diversity occurs, for example, when a reflecting surface is illuminated by several different lasers from different angles. If the angles of illumination are sufficiently separated, the path length delays experienced by each of the reflected beams will be different enough to generate uncorrelated speckle patterns[21]. Since the lasers are mutually incoherent on the time scale of this experiment, the M speckle patterns add on an irradiance basis, with a consequent reduction of contrast by $1/\sqrt{M}$. This particular example is suggestive of the more general fact that illumination of an object by a sufficiently extended incoherent source suppresses speckle.

A second way of changing optical paths (in wavelengths)

traveled by a reflected wave is to change the optical frequency of the illumination[22]. Thus, if the surface is illuminated by light with M separate frequency components of equal strength, and if the separation of these frequency components is sufficiently great, M uncorrelated speckle patterns will result, with addition on an irradiance basis. In a reflection geometry, with angles of incidence and reflection near normal to the surface, the separation $\Delta\nu$ required to produce uncorrelated speckle is approximately[23]

$$\Delta\nu \cong \frac{c}{2\sigma_z} \quad , \tag{9}$$

where c is the velocity of light and σ_z is the standard deviation of the surface height fluctuations. This particular example is suggestive of the more general fact that illumination of an object with sufficiently extended spectral bandwidth suppresses speckle.

Time diversity is most readily applied in the transmission geometry. If a transparency object is illuminated through a diffuser, then motion of that diffuser results in a continuous changing of the speckle pattern in the image. A time exposure in the image plane then results in the addition, on an intensity basis, of a number of uncorrelated speckle patterns, thus suppressing the contrast of the detected speckle pattern.

We close this discussion of speckle suppression by describing an important negative result. Suppose that, with a single ideally monochromatic laser, we simultaneously illuminate two separated regions on a rough reflecting object in the free-space geometry of Fig. 2. These regions are physically separated by a distance large compared to the correlation distance of the surface itself, so the microscopic surface structures in the two areas can be considered to be independent. Either of these two areas illuminated individually will produce a speckle pattern with unity contrast. If the two areas are illuminated simultaneously by the same monochromatic source, will the superposition of the two independent speckle patterns reduce the contrast of the resultant pattern?

The answer is an emphatic no, for the two speckle patterns add on a complex amplitude basis, rather than the intensity basis described earlier. Each of the complex fields contributed by one of the surface areas is a random walk in the complex plane. The addition of two random walks simply results in a third random walk with more steps. As long as the number of steps in each of the component random walks is large, the first-order statistics of the combined random walk are indistinguishable in form from those of the two components. Thus the statistics of the irradiance fluctuations remain negative exponential, and we conclude that addition of speckle patterns on a complex amplitude basis does not reduce the speckle contrast.

THE DISTRIBUTION OF SCALE SIZES IN A SPECKLE PATTERN

A second property of a speckle pattern which greatly influences its effects on optical system performance is the coarseness of the granularity in the pattern. In actuality, a speckle pattern consists of peaks and nulls of many different scale sizes, so any measure of coarseness must of necessity indicate an average distribution. A very suitable measure is the so-called Wiener spectrum of a speckle pattern, which is a measure of the average strengths of all possible spatial frequency components of the pattern. The usual way to find the Wiener spectrum is to first calculate the autocorrelation function of the pattern, and then to Fourier transform that pattern in accord with the Wiener-Kinchine theorem[24]. Here we prefer to take a more physical approach to the problem.

Considering first the free-space propagation geometry of Fig. 2, we suppose that the speckle pattern of concern is detected in a plane with coordinates (x,y). Any specific detected speckle pattern consists of a continuum of Fourier components. Focusing attention on a specific vector spatial frequency, $\vec{\nu} = (\nu_X, \nu_Y)$, if we square the amplitude of this Fourier component and average over an ensemble of microscopically different objects, the resulting number is the value of the Wiener spectrum at frequence $\vec{\nu}$.

An ideal Fourier component of the intensity may be thought of as a sinusoidal fringe-component of the speckle pattern. Now we ask how a sinusoidal fringe of vector frequency $\vec{\nu} = (\nu_X, \nu_Y)$ can be generated in the speckle pattern. The answer is that such a fringe must of necessity arise by simple interference of light reflected from two points on the object separated by a vector spacing

$$\vec{s} = (s_X, s_Y) = (\lambda z \nu_X, \lambda z \nu_Y) \quad , \tag{10}$$

where z is the distance from the reflecting surface (assumed approximately planar) to the observation plane.

As illustrated in Fig. 6, there are many pairs of points on the object separated by a given vector spacing $\vec{s}$; all such pairs contribute a fringe component at the same frequency $\vec{\nu}$.

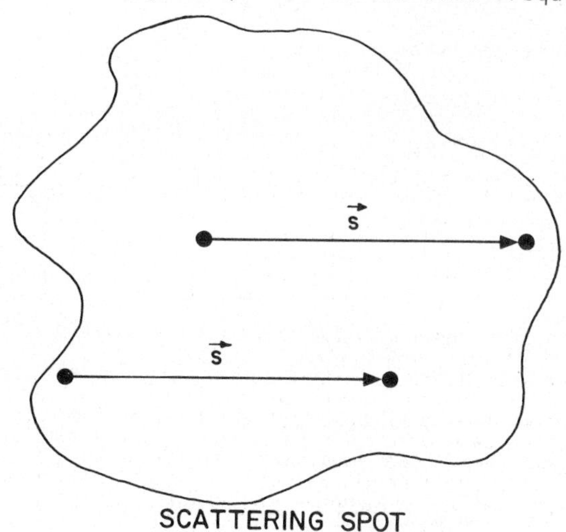

Fig. 6: A single vector spacing $\vec{s}$ is embraced by the scattering spot in many different ways.

Assuming that the correlation area of the surface itself is small compared with the size of the illuminated spot on the object, to an excellent approximation these many fringes at frequency $\vec{\nu}$ add with random spatial phase. Over the ensemble of reflecting objects, the mean-square strength of the fringe at $\vec{\nu}$ is simply the sum of the mean square strengths of the component fringes. Thus for a uniformly bright object on a dark background, the value of the Wiener spectrum at $\vec{\nu}$ is proportional to the number of ways the vector spacing $\vec{s}$ is embraced by the object. An equivalent statement, which also holds for objects of non-uniform brightness, is that the shape of the Wiener spectrum is identically the same as the shape of the autocorrelation function of the object radiance distribution. A slight qualification of this statement must be made for the zero-spatial-frequency component of spectrum. Since irradiance is a non-negative quantity, the speckle pattern always rides on constant mean irradiance level, with the result that the Wiener spectrum contains a δ-function component at the origin. A more detailed analysis shows that the Wiener spectrum $\mathcal{W}(\vec{\nu})$ is given mathematically by

$$\mathcal{W}(\vec{\nu}) = (\bar{I})^2 \left\{ \delta(\vec{\nu}) + \frac{\displaystyle\iint_{-\infty}^{\infty} R(\vec{\xi})R(\vec{\xi}-\lambda z\vec{\nu})d\vec{\xi}}{\left[\displaystyle\iint_{-\infty}^{\infty} R(\vec{\xi})d\vec{\xi}\right]^2} \right\} \qquad (11)$$

where $\bar{I}$ is the mean irradiance in the speckle pattern, $R(\vec{\xi})$ represents the radiance of the object at coordinate $\vec{\xi}$, and $\delta(\vec{\nu})$ is a two dimensional Dirac delta function at the origin of the $\vec{\nu}$ plane.

For the case of a uniformly bright square object of width L, the form of the Wiener spectrum is

$$\mathcal{W}(\vec{\nu}) = (\bar{I})^2 \left\{ \delta(\vec{\nu}) + \left(\frac{\lambda z}{L}\right)^2 \Lambda\left(\frac{\lambda z}{L}\nu_X\right) \Lambda\left(\frac{\lambda z}{L}\nu_Y\right) \right\} \qquad (12)$$

where $\Lambda(x) = 1 - |x|$ for $|x| \leq 1$, zero otherwise. This spectrum is shown in Fig. 7. Note that the speckle pattern contains no frequency components higher than $L/\lambda z$ in the ν_X- and ν_Y-directions, a consequence of the fact that no spacings larger than L can be embraced by the object in these two directions.

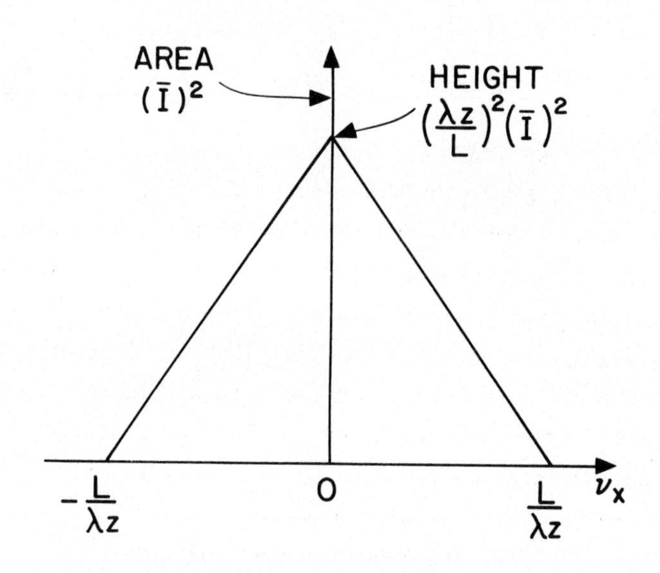

Fig. 7: Cross section of the Wiener spectrum of a speckle pattern arising from a rectangular scattering spot

For the imaging geometry of Fig. 3, the argument leading to the Wiener spectrum must be changed slightly. In this case we note that a fringe of spatial frequency $\vec{\nu}$ is generated in the image speckle pattern by interference of light from two points on the lens aperture separated by vector spacing $\vec{s} = (\lambda z \nu_X, \lambda z \nu_Y)$, where now z is the distance from the lens to the image plane. Thus the relative mean-square strengths of fringe patterns of different spatial frequencies are found by determining the degree to which corresponding vector spacings are embraced by the lens aperture. The result is a continuous component of the Wiener spectrum which is proportional to the diffraction-limited optical transfer function of the imaging system. For a circular lens of diameter D , the Wiener spectrum of the image speckle pattern is[13]

$$\mathcal{W}(\vec{\nu}) = (\bar{I})^2 \left\{ \delta(\vec{\nu}) + 2\left(\frac{\lambda z}{D}\right)^2 \cdot \frac{2}{\pi}\left[\cos^{-1}\left(\frac{\lambda z}{D}\nu\right) - \left(\frac{\lambda z}{D}\nu\right)\sqrt{1 - \left(\frac{\lambda z}{D}\nu\right)^2}\right]\right\} \tag{13}$$

for $\nu \leq D/\lambda z$, zero otherwise where $\nu = |\vec{\nu}|$.

The general conclusions to be drawn from these arguments are that, in any speckle pattern, large scale-size fluctuations are the most populous, and no scale sizes are present beyond a certain small-size cutoff. The distribution of scale sizes in between these limits depends on the autocorrelation function of the object radiance distribution in the free-space geometry, or on the autocorrelation function of the pupil function of the imaging system in the imaging geometry.

SPECKLE FROM "SMOOTH" SURFACES

In recent years, considerable interest has arisen in the properties of speckle patterns generated by "smooth" surfaces, i.e., surfaces with an r.m.s. roughness that is less than a wavelength. This interest was largely sparked by the work of Fujii and Asakura[25], who first demonstrated that the contrast of a speckle pattern in the image of a diffuser depends on the r.m.s. wavefront deviation introduced. For extremely smooth surfaces, the contrast of the speckle pattern is near zero while as the surface roughness increases, the contrast asymptotically approaches the value unity characteristic of a fully developed speckle pattern.

Pederson[26] explained this dependence of speckle contrast

on surface roughness by assuming that the light reflected from a rough surface consists of the sum of a specular component and a diffuse component, with the strengths of the specular and

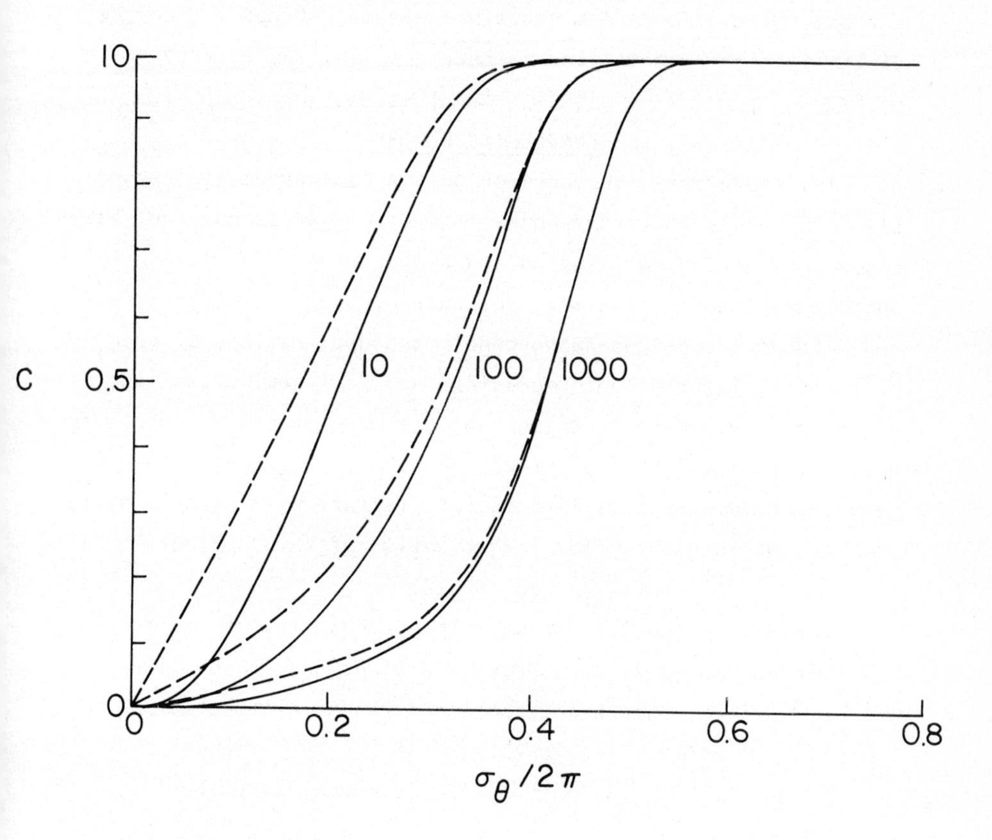

Fig. 8: Speckle contrast as a function of r.m.s. phase deviation, gaussian surface correlation function assumed

diffuse components depending on the surface roughness. Goodman[27] further pointed out that the statistics of the diffusely transmitted field, as it appears in the image plane, are gaussian but non-circular. Figure 8 shows the dependence of speckle contrast on the r.m.s. wavefront deviations. The dotted curves are the predicted dependences when the diffuse fields are assumed to obey circular complex gaussian statistics.

The solid curves correspond to the more correct prediction that takes account of the non-circular nature of the diffuse fields in the image. The parameter N represents the number of surface correlation areas contained within one resolution area of the imaging system. For further details, see references 13 or 27.

CONCLUDING REMARKS

This paper has touched upon only a few fundamental properties of speckle patterns. Many interesting areas have not been discussed, including: second and higher order statistics of speckle patterns; statistics of sums of speckle patterns, with and without a coherent background; and applications of speckle patterns to information processing, non-destructive testing, and astronomy. Many of these subjects are covered in detail in a recent book on the subject of speckle[13]. Hopefully, the background provided by this introductory lecture will be helpful to the student in his further investigation of this subject.

REFERENCES

1. J.D.Rigden and E.I.Gordon, "The granularity of scattered optical maser light", Proc. I.R.E. 50, 2367-2368 (1962).
2. B.M.Oliver, "Sparkling spots and random diffraction", Proc. I.E.E.E. 51, 220-221 (1963).
3. E. Verdet, Ann. Scientif. l'Ecole Normal Superieur, 2, 291 (1865).
4. J.W. Strutt, (Lord Rayleigh), "On the resultant of a large number of vibrations of the same pitch and of arbitrary phase", Phil.Mag. 10, 73-78 (1880).
5. M. von Laue, Sitzungsber. Akad. Wiss. (Berlin) 44, 1144 (1914); Mitt. Physik Ges. (Zurich) 18, 90 (1916); Verhandl. Deut. Phys. Ges. 19, 19 (1917).
6. P.E.Green, Jr., "Radar measurements of target scattering properties", in Radar Astronomy, J.V.Evans and T.Hagfors, eds., McGraw-Hill Book Co., New York, 1968, pp.1-77.
7. E.N.Leith, "Quasi-holographic techniques in the microwave

region", Proc. I.E.E.E. 59, 1305-1318 (1971).

8. P.S.Green (ed.) Acoustical Holography, Vol. 5, Plenum Press, New York, N.Y. 1974.

9. J.A.Ratcliffe, "Some aspects of diffraction theory and their application to the ionosphere", in Reports on Progress in Physics, Vol. 19, A.C. Strickland, Ed., The Physical Society, London, 1956, pp.188-267.

10. L.Mandel, "Fluctuations of photon beams: the distribution of photoelectrons", Proc. Phys. Soc. (London) 74, 233-243 (1959).

11. D.Middleton, Introduction to Statistical Communication Theory, McGraw-Hill Book Co., New York, N.Y., 1960.

12. W.B.Davenport and W.L.Root, Random Signals and Noise, McGraw-Hill Book Co., New York, N.Y., 1958.

13. J.W.Goodman, "Statistical properties of laser speckle patterns", in Laser speckle and related phenomena (Topics in Applied Phys., Vol. 9), J.C. Dainty, Ed., Springer-Verlag, Heidelberg, 1975, pp.9-75.

14. P.Beckmann and A.Spizzichino, The Scattering of Electromagnetic Waves from Rough Surfaces, Pergamon/Macmillan, London, New York, 1963.

15. K.Pearson, A Mathematical Theory of Random Migration, Draper's Company Research Memoirs, Biometric Series III, London, 1906.

16. J.W.Strutt (Lord Rayleigh), "On the problem of random vibrations, and of random flights in one, two, or three dimensions", Phil. Mag., 37, 321-347 (1919).

17. M.A.Condie, An Experimental Investigation of the Statistics of Diffusely reflected Coherent Light, Thesis (Dept. of Electrical Engineering) Stanford University, Stanford, California (1966).

18. J.C.Dainty, "Some statistical properties of random speckle patterns in coherent and partially coherent illumination",

Opt. Acta, 17, 761-772 (1970).

19. T.S. McKechnie, "Measurement of some second order statistical properties of speckle", Optik, 39, 258-267 (1974); Statistics of Coherent Light Speckle Produced by Stationary and Moving Apertures, Ph.D. Thesis, Dept. of Physics, Imperial College, London (1974).
20. F.T. Arecchi, "Measurement of the statistical distribution of Gaussian and laser sources", Physical Review Letters, 15, pp. 912-916, December 1965.
21. D. Leger, E. Mathieu and J.C. Perrin, "Optical surface roughness determination using speckle correlation technique", Appl.Opt., 14, 872-877 (1975).
22. N. George and A. Jain, "Space and wavelength dependence of speckle intensity", Appl.Phys., 4, 201-212 (1974).
23. G. Parry, "Some effects of surface roughness on the appearance of speckle in polychromatic light", Opt.Comm., 12, 75-78 (1974).
24. L.I. Goldfischer, "Autocorrelation function and power spectral density of laser-produced speckle patterns", J.Opt.Soc.Am., 55, 247-253 (1965).
25. H. Fuji and T. Asakura, "Effect of surface roughness on the statistical distribution of image speckle intensity", Opt.Comm., 11, 35 (1974).
26. H.M. Pederson, "The roughness dependence of partially developed, monochromatic speckle patterns", Opt.Comm. 12, 156 (1974).
27. J.W. Goodman, "Dependence of image speckle contrast on surface roughness", Opt.Comm., 14, 324 (1975).

Coherant Optical Engineering, F.T. Arecchi and V. Degiorgio (eds.)

IMAGE FORMATION WITH COHERENT LIGHT - A TUTORIAL REVIEW

Brian J. Thompson
College of Engineering and Applied Science
University of Rochester
Rochester, New York

INTRODUCTION

Image formation as it has been historically performed and evaluated is an incoherent process. This occurred because the object to be imaged was either self-luminous such as a star or a thermal source, or the object was illuminated by light from a source of significant angular size. Naturally there are special examples that run counter to this general statement. In a sense the Michelson stellar interferometer can be thought of as a rather specialized imaging device, although normally it is evaluated as an interferometer. In this instrument, the degree of coherence that does exist in the field of any finite source is actually measured to determine the extent of the source. Testing methods such as the Foucault test and the various Schlieren systems as well as the phase contrast microscope cannot be operated if they are strictly incoherent. The Abbe theory of vision in a microscope presupposes that diffraction is an important process in the nature of the image formation; diffraction is fundamentally a coherent process.

In recent years this situation has changed significantly since there are now a variety of optical systems that require coherent light for their operation. The holographic imaging process is a two step coherent imaging method; optical processing techniques require a coherent imaging system. The laser is often used for illumination purposes, merely to produce sufficient energy over the object or because a narrow band of frequencies is required.

The purpose of this tutorial paper is to review the important differences between incoherent, partially coherent, and coherent image formation and see how this affects the performance of optical systems.

THEORETICAL REVIEW

Partially Coherent Fields

Let us describe an optical field by a function $V(x,t)$, where x and t are the coordinates in position and time. This function is, in fact, a single Cartesian coordinate of the electric field vector. The measurable quantity associated with this optical field is the intensity $I(x)$ given by

$$I(x) = \langle V(x,t)V^*(x,t)\rangle , \quad (1)$$

where the sharp brackets signify a time average and the star denotes a complex conjugate.

It is important to understand the result of adding two optical fields. The optical field $V_R(x,t)$ that is produced by the addition of two optical fields is simply given by

$$V_R(x,t) = V(x_1,t) + V(x_2,t). \quad (2)$$

The suffixes 1 and 2 are used to distinguish the two fields. The resultant intensity is then

$$I_R(x) = \langle V(x_1,t)V^*(x_1,t)\rangle + \langle V(x_2,t)V^*(x_2,t)\rangle + \langle V(x_1,t)V^*(x_2,t)\rangle + \langle V^*(x_1,t)V(x_2,t)\rangle . \quad (3)$$

The first two terms in equation (3) are simply the intensities $I(x_1)$ and $I(x_2)$ respectively according to the definition of equation (1).

The second pair of terms in equation (3) are the complex conjugates of each other. Taking the first of these terms we note that it is a cross-correlation function. When this cross-correlation function is zero in the time average, then the two fields are incoherent and equation (3) reduces to the well known form that the intensity is simply the sum of the individual intensities.

This time average cross-correlation function can, of course, be more than zero. It is clearly a measure of the relationship between the fields; this relationship, of course, is the coherence that exists between the fields. We define the mutual intensity function $\Gamma(x_1,x_2)$ as this time-averaged cross-correlation function; thus

$$\Gamma(x_1,x_2) = \langle V(x_1,t)V^*(x_2,t)\rangle \quad . \quad (4)$$

The fourth term in equation (3) is then $\Gamma^*(x_1,x_2)$. It is convenient to normalize equation (4) in the following way and define the complex degree of coherence $\gamma(x_1,x_2)$ as

$$\gamma(x_1,x_2) = \frac{\langle V(x_1,t)V^*(x_2,t)\rangle}{\sqrt{I(x_1)I(x_2)}} \quad . \quad (5)$$

Then

$$0 \leqslant | \gamma(x_1,x_2) | \leqslant 1. \quad (6)$$

When $|\gamma(x_1,x_2)| = 1$, then the two fields are coherent with respect to each other. For values of $|\gamma(x_1,x_2)|$ between 0 and 1, then the two fields are partially coherent. Notice that in this analysis the fields have been assumed to have a relatively narrow spectral width, i.e., they are quasi-monochromatic. The process could be repeated with a time delay coordinate τ added; then the time-averaged cross-correlation term is the mutual coherence function $\Gamma(x_1,x_2,\tau)$ where

$$\Gamma(x_1,x_2,\tau) = \langle V(x_1,t)V^*(x_2,t+\tau)\rangle \quad . \qquad (7)$$

Returning to the coherent situation we can write the spatial and temporal variations of the field as separate functions.

$$V(x_1,t) = \psi(x_1)e^{-2\pi i\nu t} \quad , \qquad (8)$$

where $\psi(x_1)$ is the complex amplitude of the field and ν is the optical frequency. The resultant intensity of a field is then written as the product $\psi(x_1)\psi^*(x_1)$. When two fields are added, the cross-correlation term is now the cross-correlation of the complex amplitudes.

Naturally this is a very brief review of the background. The reader will find some of the references helpful in understanding the theory and meaning of partial coherence.[1-4]

Image Formation

The process of image formation can be evaluated by propagating the optical field according to the appropriate wave equation. We will merely quote the results of such a process here. Let an image be formed of an object field by a single lens whose amplitude impulse response (the complex amplitude in the image of a point object) is

$$K\left(\frac{\xi}{z_2} + \frac{x}{z_1}\right)$$

and x is the coordinate in object space, ξ is the coordinate in image space, and z_1 and z_2 are the object and image space respectively. In the partially coherent case the mutual intensity function is propagated. Thus the mutual intensity in the image, $\Gamma_{im}(\xi_1,\xi_2)$, is given by

$$\Gamma_{im}(\xi_1,\xi_2) = \iint \Gamma_{ob}(x_1,x_2)e^{\frac{ik(x_1^2-x_2^2)}{2z_1}} K\left(\frac{\xi_1}{z_2} + \frac{x_1}{z_1}\right) K^*\left(\frac{\xi_2}{z_2} + \frac{x_2}{z_1}\right) d\xi_1, d\xi_2. \qquad (9)$$

where $\Gamma_{ob}(x_1,x_2)$ is the mutual intensity function of the optical field after reflection or transmission of an incident partially coherent field.

The resultant intensity is found by setting $\xi_1=\xi_2$ in equation (9).

The two limits of equation (9) give the incoherent and coherent results respectively; thus

$$I_{im}(\xi) = \int I_{ob}(x)\ S(\frac{\xi}{z_2} + \frac{x}{z_1})\ dx, \tag{10}$$

where $S(x) = K(x)K^*(x)$.

$$\psi_{im}(\xi) = \int \psi_{ob}(x) e^{\frac{ikx^2}{2z_1}} K(\frac{\xi}{z_2} + \frac{x}{z_1})\ dx, \tag{11a}$$

$$I_{im}(\xi) = \left| \int \psi_{ob}(x) e^{\frac{ikx^2}{2z_1}} K(\frac{\xi}{z_2} + \frac{x}{z_1})\ dx \right|^2. \tag{11b}$$

The incoherent result is the familiar one that the imaging system is linear in intensity and the image intensity is formed by a convolution of the object intensity with the intensity impulse response S(x) (see equation (10)). The coherent result is quite different since it is linear in the complex amplitude (equation (11a)) not in the intensity. The partially coherent case is linear in the mutual intensity function. The incoherent limit does then represent a rather special (and somewhat fortunate) situation since it happens to be linear in the quantity that is detected - the intensity.

It will also be noted that equations (9) and (10) are not spatially stationary because of the quadratic phase term inside the integral. However, this is avoided if a two lens system is used to balance out that quadratic phase term with one of equal and opposite sphericity.

Finally it should be stressed that the amplitude impulse response is the important function that describes the imaging system in general. The intensity impulse response is only useful in the special, but important circumstance, of incoherent illumination. The transfer function of an optical system is defined as the Fourier transform of the intensity impulse response and hence it only has value in the incoherent limit.

IMAGE FORMATION WITH PARTIALLY COHERENT LIGHT

A discussion of the difference between the classical incoherent imaging process and that with coherent and partially coherent light is most effectively done by means of examples. First of all it should be stressed that the effects are mainly important with high quality optical systems. A poor optical system will give a poor image no matter what the coherence properties of the illumination! It is then appropriate to discuss diffraction limited systems. Figure 1 shows the amplitude impulse response function (a), the intensity impulse response function, and a photography of the intensity impulse response for a diffraction limited system with a circular aperture. The amplitude impulse response is, of course, the Fourier transform of the aperture function and hence in this example has the form

$$2J_1\left(\frac{kr\xi}{z_2}\right)/\left(\frac{kr\xi}{z_2}\right),$$

where $k = 2\pi/\lambda$, r is the radius of the circular aperture of the lens, and ξ is the spatial coordinate in the image plane and in this special case can be considered a radial coordinate. The functional form of the amplitude impulse response is going to control the nature of the image formed. The fact that it has negative regions produces interesting effects when the convolution is performed; further effects are produced by the nonlinearity of all but the incoherent process.

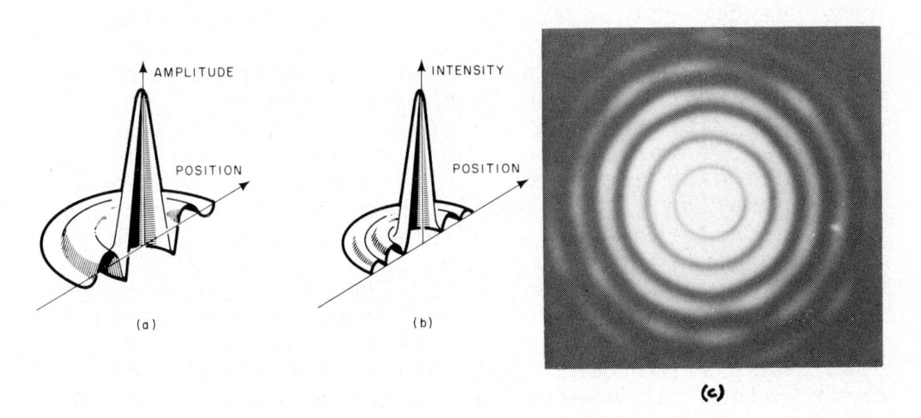

Figure 1

Response functions of a diffraction limited imaging system.
(a) the amplitude impulse response
(b) the intensity impulse response
(c) a photograph of the intensity impulse response

Image of Two Points

The image of a two point object is quite instructive even though it is simple. Figure 2a shows a comparison between the image intensity distribution of two points at the Rayleigh separation; as expected, when the two points are illuminated incoherently the two points are resolved in the image. On the other hand, if the two points are illuminated coherently and in phase, then there is no resolution; if the illumination is coherent but out of phase, then there is excellent resolution. Finally, if the two points are illuminated coherently and in phase quadrature, then the result is identical to the incoherent result. (Some further details of the two point image problem can be found in the literature.[4,5,6,7])

Naturally these curves show results for specific cases. In fact, there is a continuous family of curves as the degree of coherence between the two points is changed. Figure 2b shows the family of curves for two points as the degree of coherence is changed in steps of 0.1 from 0 to +1.0. The curve for $\gamma=0$ is the incoherent result. The separation here is slightly greater than that for the Rayleigh criterion.

The most dramatic result in the two point image problem is not the change in resolution but the effect on the measurement of the separation of the two points. This is well illustrated in Figure 2c; here the two points are well resolved in both the coherent and the incoherent cases. The vertical dashed lines show the true position of the two points; when the two points are incoherent, the peaks in the intensity distribution are in the correct location; when the two points are coherent and in phase the two peaks are displaced from their correct location and the separation is too small; when the two points are coherent and exactly out of phase the peaks in the resultant intensity distribution are again displaced but in the opposite sense; finally, if the two points are coherent and in phase quadrature, the resultant intensity is the same as the incoherent case. The positional error described here is not a constant error, since it depends on the actual separation of the two points. This effect is produced by the addition of the two amplitude impulse response functions which have negative regions, thus moving the positions of the two maxima in the resultant intensity profile slightly.

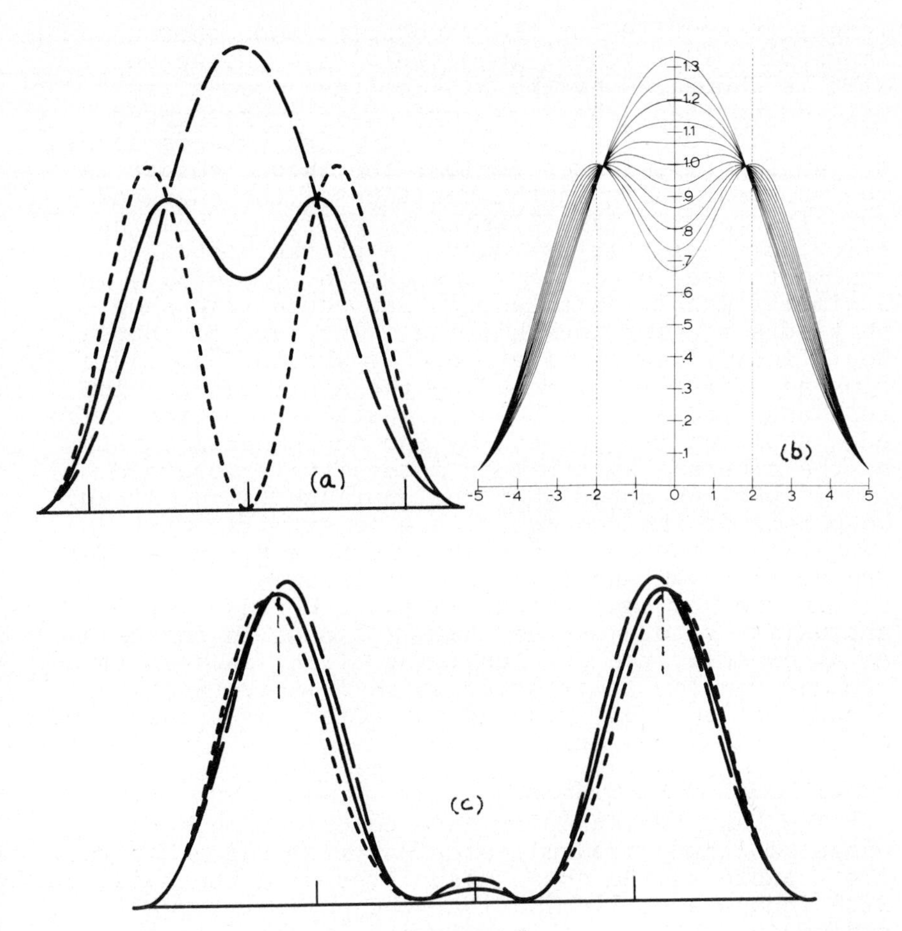

Figure 2

The image of two points with partially coherent light

(a) a comparison between the image intensity when the two points are incoherent (Rayleigh limit) (———), coherent and in phase (— —), coherent and out of phase by π (- - - -)

(b) the image intensity distribution as the degree of coherence is varied from 0 to 1.0 in steps of 0.1

(c) the image of two well resolved points when two points are incoherent, coherent and in phase, and coherent and out of phase by π.

Image of a Slit

The image of an edge, a bar, and a slit object can all be discussed together, since they also show similar effects. Figure 3a gives the resultant intensity profile for the incoherent and coherent image of a fairly well resolved slit. The actual slit is indicated by the broken line. Two effects will be noticed; the first effect is a pronounced ringing at each edge of the slit which extends across the entire illuminated region; the second effect is the apparent shift in the position of the edges of the slit towards the illuminated region; hence the slit looks a little narrower. A photograph of the coherent image of a slit is shown in Figure 3b. In a similar manner an edge or a bar would also show this edge ringing and shifting. The bar, of course, would look slightly wider than it should. The edge ringing effect is produced by the amplitude impulse response function not being a positive function. The edge shifting is produced by the basic nonlinearity of the coherent system when intensity is recorded.

It will be noted that the acutance (slope of edge response) of the coherent image is significantly better than the corresponding incoherent image. This can be an advantage in some circumstances.

In the low contrast limit the coherent system approximates to a system that is linear in intensity. As an example, the edge shifting effect is seen to be removed for low contrast edges as illustrated in Figure 3c; however, ringing now exists at either side of the true edge location.

Images of Phase Objects

The last two examples have concerned objects whose amplitude transmittance was real and positive, i.e., there was no phase associated with the object. With incoherent illumination, the phase information associated with the object is not retained in any event. With coherent illumination any phase that is associated with the object is retained, which complicates the image. For example, a pure phase object would not produce any variation of the intensity in the image plane if illuminated incoherently. With coherent illumination of uniform phase the resulting image intensity has the form

$$I_{im}(\xi) = \left| \int e^{i\phi(x)} K\left(\frac{\xi}{z_2} + \frac{x}{z_1}\right) dx \right|^2 \tag{12}$$

where $\phi(x)$ is the actual variation of phase with position. There will be a variation in the image intensity depending upon the actual value of $\phi(x)$. No detailed discussion of this type of imaging will take place here except to discuss an example.

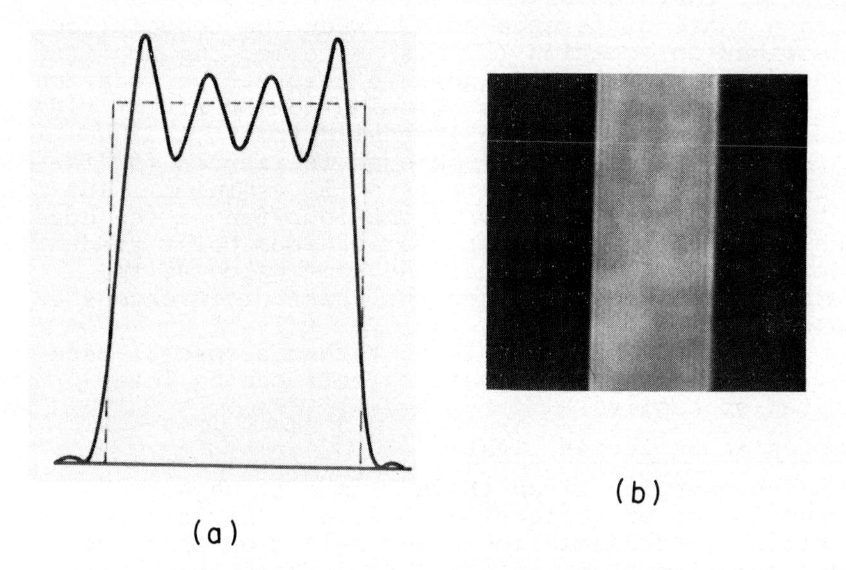

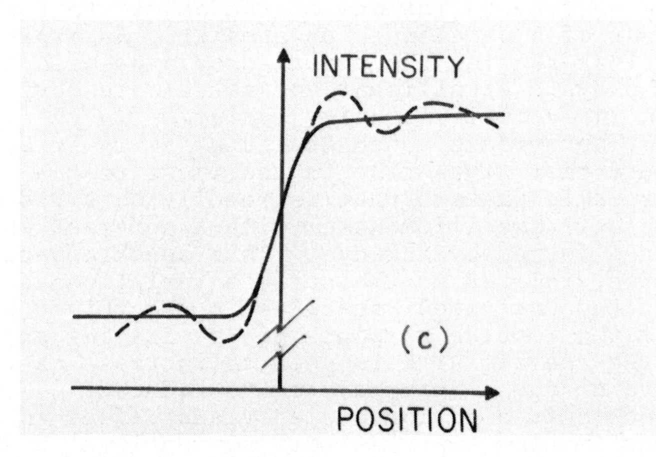

Figure 3

An illustration of the edge ringing phonomenon

(a) the image of a coherently illuminated slit - the original slit is shown for comparison
(b) photograph of the image of a coherently illuminated slit; this photograph is not intended to match the result shown in (a)
(c) the image of a low contrast edge with coherent (--) and incoherent (——) illumination.

Consider that the object consists of a phase edge with a phase difference of π. When the convolution described in equation (12) is performed the amplitude and hence the intensity goes to zero at the location of the edge with uniform illumination on either side and hence the phase edge is easily seen. As the phase difference across the edge is reduced (or increased) from π then there is still a minimum (but not a zero) of intensity at the location of the edge. The value of intensity at that minimum point gradually increases until it reaches the same value as the surrounding intensity when the phase difference goes to zero or 2π.

Naturally this example is rather a special case and not all images of phase objects can be interpreted so easily.

Images in Reflected Light

The examples given above relate to objects in transmission or reflection; however, they are most apparent in transmission. In reflection there is a much more important effect that is associated with the phase differences introduced into a reflected beam by surface variations on the object that are on the order of a wavelength or greater. As explained in the last section the phase associated with the object plays a significant and not always obvious role in the actual intensity variation in the final image. The surface roughness is a fairly irregular structure that gives rise in the image to the well-known speckle pattern that is readily observed when most objects are illuminated with a coherent beam and the image formed by the eye. This speckle pattern is, of course, related to the surface variations and the size of the individual speckles is determined by the size of the impulse response of the imaging system; hence the speckle size is seen to increase as the aperture of the imaging system is reduced.

Speckle is a serious deleterious effect in all coherent imaging systems, both conventional and holographic. The only successful approaches to the problem are to reduce the temporal coherence of the illumination or use redundancy techniques, either spatial redundancy (multiple images) or temporal redundancy (time average with a time varying function introduced into the beam).

DESIGN OF COHERENT IMAGING SYSTEMS

We will leave aside the problem of phase imaging and the special problems of speckle and discuss the possibility of designing a coherent system so that the edge ringing problem can be controlled.

It will be recalled that the ringing seen in the images of edges and bars and slits as well as the mensuration problem in the image of two points was caused by the particular form of the diffraction limited amplitude impulse response. That is, a lens was used that had been designed to give good performance in an incoherent imaging mode. It does not seem unreasonable then to ask what should be the appropriate form of the amplitude impulse response; clearly it should be a real and positive function. A number of such amplitude impulse response functions satisfying this condition can be achieved by appropriate modification of the amplitude transmittance of the diffraction limited pupil function (apodisation). It will not be possible to review this in detail here, but a recent review will be of value to the reader.[9] One example will be sufficient to illustrate the idea.

A Gaussian amplitude impulse response will satisfy the condition set out above. Since the lens aperture function and the amplitude impulse response are a Fourier transform pair, then the lens aperture function required might be a Gaussian function. In fact, the required Gaussian aperture function will have to be truncated by the actual physical aperture. Figure 4 illustrated this process by means of a series of one-dimensional calculations. Figure 4a shows the actual Gaussian aperture function (amplitude transmittance of the aperture); the dashed line shows the unmodified aperture. The amplitude impulse response is shown in Figure 4b, and it, too, is a Gaussian function. The intensity distribution in the image of two points is shown in Figure 4c; the vertical dashed lines indicate the positions of the two points. It will be noted that the peaks are in the correct location and hence the error that would have been present with the unmodified aperture has been removed. Finally, in Figure 4d, the intensity distribution in the image of a slit is shown; no ringing is observed. The unmodified aperture function would have produced the image shown in Figure 3a.

There is only one other interesting situation where the ringing can be controlled and the increased acutance taken advantage of. If the object to be imaged is a binary object and it is to be recorded on film, then the density versus log exposure curve of the film can be used to advantage. The exposure time and film combination is chosen so that the maximum density on the film is achieved at an intensity level

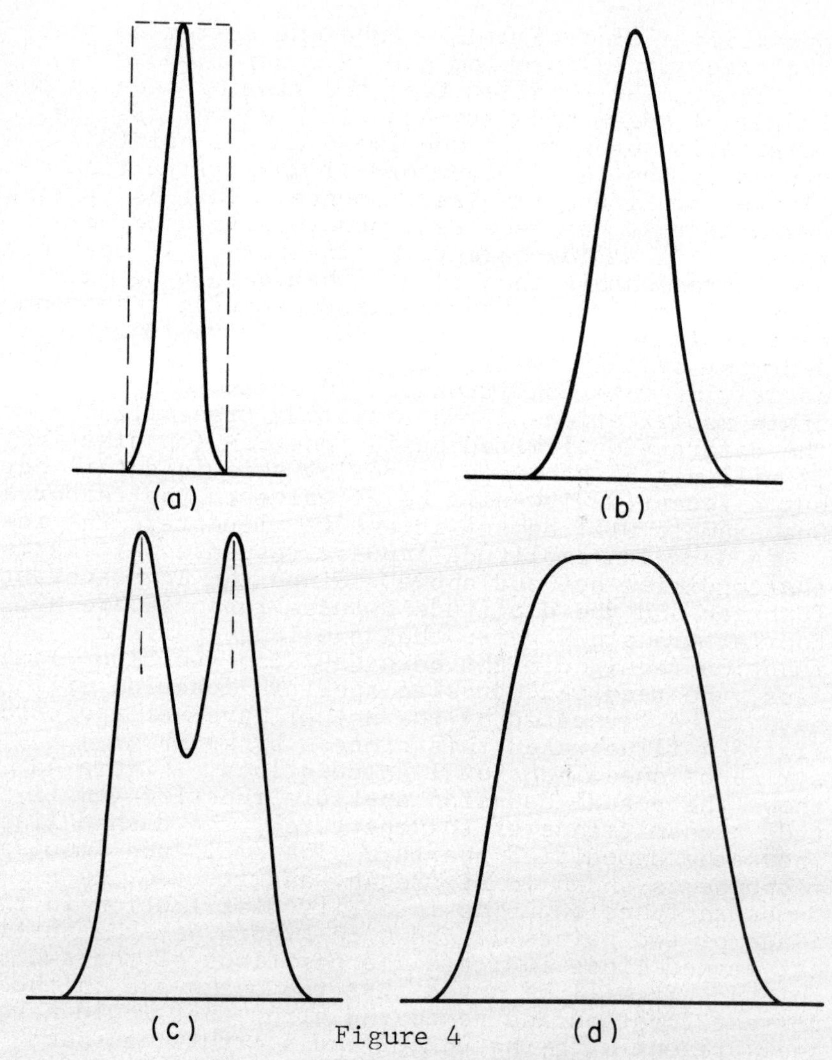

Figure 4

Control of ringing by apodisation

(a) the Gaussian amplitude transmittance of the aperture; the dashed line is the unapodised physical aperture
(b) the amplitude impulse response function produced by the aperture function shown in (a)
(c) image of two points with system with Gaussian apodisation; the vertical dashed lines indicate the correct position of the two points
(d) image of a slit with system with Gaussian apodisation; this should be compared with Figure 3a.

just below the level of the ringing; then the intensity variations which are the ringing all produce the maximum density. This has been successfully used in systems used for microcircuit imaging.

CONCLUSIONS

Naturally, in the space allowed for this article it has only been possible to give some background to, and a brief overview of coherent imaging. The subject is still developing and hopefully new techniques will be forthcoming to further understand image formation with partially coherent light and perhaps take advantage of some of the special properties that exist.

REFERENCES

1. Born, M. and Wolf, E., 1965, Principles of Optics, 3rd Ed. (Pergamon Press), Chapter X.
2. Hopkins, H. H., 1968, Advanced Optical Techniques, Ed. A.C.S. Van Heel (John Wiley and Sons, New York), Chapter 6, 189.
3. Beran, M. and Parrent, G. B., 1964, Theory of Partial Coherence (Prentice Hall). (Also reprinted by Soc. Photo-Optical Instrumentation Engineers, California.)
4. Parrent, G. B. and Thompson, B. J., 1969, Physical Optics Notebook (Soc. Photo-Optical Instrumentation Engineers, California).
5. Thompson, B. J., 1962, Progress in Optics, Vol. VII, Ed. E. Wolf, North-Holland.
6. Grimes, D. N. and Thompson, B. J., 1967, J. Opt. Soc. Amer. 57, 1330.
7. Goodman, J. W., 1968, Introduction to Fourier Optics (McGraw-Hill), 129.
8. Considine, P. S., 1966, J. Opt. Soc. Amer. 56, 1001.
9. Thompson, B. J. and Krisl, M. E., 1976, Proc. Int. Conference on Image Analysis and Evaluation, Ed. R. Shaw, SPSE publication, in press.

Coherant Optical Engineering, F.T. Arecchi and V. Degiorgio (eds.)

RECORDING MATERIALS FOR HOLOGRAPHY AND OPTICAL DATA PROCESSING

Klaus Biedermann
Department of Physics II and
Institute of Optical Research
Royal Institute of Technology
S-100 44 Stockholm 70
Sweden

WAVEFRONT RECORDING AND RECONSTRUCTION

A wavefront, at any point in space and time, is characterized by amplitude and phase of the electromagnetic field. It is not possible to record a wavefront directly, since detectors are sensitive to intensity, the square of the amplitude. Gabor, however, has shown that amplitude and phase can be recorded and reconstructed by means of an interferometric principle:the wavefield to be recorded is brought to interfere with a coherent reference field of known amplitude and phase; in this way, phase relations are rendered as intensity distributions. What is needed is a recording medium that, in this first step, is sensitive to the radiation used and capable of recording the microscopic interference structures into which information on the complex wevefield is encoded. Upon this exposure, the medium has to undergo a change of its optical properties, to the end that, in the second step, incident light be spatially modulated in such a way that a wavefront of certain amplitude and phase distribution is generated. Hence, a hologram may be defined as the physical realization of a boundary condition that forces one field to assume the values of a second field.

In order to obtain an idea of the demands holograms and complex spatial filters make on the recording material, we may give a condensed derivation of the basic equations of wavefront recording and reconstruction. Assume a wavefield component impinging under an angle θ to the normal to the recording plane (ξ, η). Along the ξ-direction it can be described by

$$\vec{U}(\xi, t) = A(\xi) \cdot \exp\left(i \cdot \frac{2\pi}{\lambda} \cdot \xi \cdot \sin\theta + i\omega t\right) \quad (1)$$

At any point, the interference of a signal field U_s and a coherent reference field U_r of the same polarisation yields a time-average irradiance

$$I = |U|^2 = (U_s + U_r) \cdot (U_s + U_r)^+$$
$$= U_s U_s^+ + U_r U_r^+ + U_s U_r^+ + U_s^+ U_r \qquad (2)$$

If we assume a recording medium which, within a certain interval, generates amplitude transmittance τ_A as a linear function of exposure, $E = I \cdot t$ (t is exposure time),

$$\tau_A = \tau_o + \beta \cdot E \qquad (3)$$

we obtain a hologram of amplitude transmittance

$$\tau_A = \tau_o + \beta \cdot t \cdot (U_s U_s^+ + U_r U_r^+ + U_s U_r^+ + U_s^+ U_r) \quad (4)$$

Upon reconstruction, the hologram is irradiated by a replica of the reference field

$$U = U_r \cdot \tau_A = U_r \cdot \tau_o + \beta \cdot t \cdot (U_s U_s^+ U_r + U_r I_r + U_s \cdot I_r + U_s^+ U_r^2) \qquad (5)$$

The term $U_s \cdot I_r \cdot \beta \cdot t$ is proportional to the original signal field U_s, which, in this way, has been reconstructed.

DEMANDS ON THE RECORDING MATERIAL

As to the demands on the material, we have to take a quantitative look at the irradiance distribution in Eq. (2) by substituting U_s and U_r from Eq. (1) into Eq. (2).

$$I = A_s^2 + A_r^2 + 2 \cdot A_s \cdot A_r \cdot \cos \left(2\pi \xi \cdot \frac{\sin\theta_s - \sin\theta_r}{}\right)$$
$$= I_s + I_r + 2 \cdot \sqrt{I_s \cdot I_r} \cdot \cos (2\pi \nu \xi) \qquad (6)$$

The reference wave has introduced a spatial carrier frequency ν, whose value is determined essentially by the angle between the propagation directions of the two interfering fields,

$$\nu = \frac{\sin\theta_s - \sin\theta_r}{\lambda} \quad [\text{cycles/mm}] \qquad (7)$$

E.g., $\nu = 1580. \ (\sin\theta_s - \sin\theta_r)$ [cycles/mm] for $\lambda = 632, 8$ nm of the He-Ne-laser; with $\vartheta_s \approx 30^\circ$ and $\vartheta_r = -30^\circ$, the spatial frequencies to be recorded are centered around $\nu \approx 1580\,[^c/mm]$. In a Lippmann hologram, $\vartheta_s \approx 90$ and $\vartheta_r = -90°$, the fringes in the depth of a photographic emulsion of index of refraction $n \approx 1.63$ represent a spatial frequency around $5150\,[^c/mm]$. Hence, a decisive requirement on a material for recording holograms and complex spatial filters is extremely high resolution.

Among the other demands are high diffraction efficiency, low scattered flux, and linearity.

Diffraction efficiency is defined as the part of the light flux in the reconstructing field diffracted into the desired field

$$\eta = \frac{\Phi_{+1}}{\Phi_{inc}} \tag{8}$$

Diffraction efficiency is a function of the modulation of the grating.

If the relation between exposure and amplitude transmittance is not a linear one, higher order terms will appear in Eq. (4). For extended objects, the term $U_s U_s^+$ is the autocorrelation function of the object field, also called the intermodulation term, which gives rise to a low-frequency, speckled, irradiance distribution in the recording plane. Nonlinear transfer generates, among other terms, terms containing products of $U_s U_r^+$ and $U_s U_s^+$, which superimpose spurious fields onto the desired reconstructed field. Reconstructions from phase holograms, i.e. recordings that lead to a complex or imaginary amplitude transmittance

$$\tau_A(\xi,\eta) = |\tau_A(\xi,\eta)| \cdot \exp(i\varphi(\xi,\eta))$$

are inherently nonlinear. In practice, with all recording media, linear transfer is approximated at the expense of diffraction efficiency by using low modulation, i.e. by choosing the intensity in the reference field stronger by a factor K=4 to K=25.

When the reconstructing field strikes the hologram, in addition to the fields generated at the hologram structure, also light scattered randomly at the grains and other inhomogeneities of the hologram and its support will be superimposed onto the reconstructed field. Even if this scattered flux is relatively weak, its effect can be very disturbing, since the scattered light is coherent to the reconstructed field and, hence, gives rise to interference and speckles. In the presence of an average scattered irradiance $\langle I_N \rangle$ from a multitude of randomly phased amplitudes, the signal-to-noise ratio is given by the deterministic image irradiance I_i to the standard deviation of the total irradiance 11

$$\frac{I_i}{\sigma} = \frac{I_i}{\langle I_N \rangle} \Bigg/ \sqrt{1 + \frac{2I_i}{\langle I_N \rangle}} \qquad (9)$$

It is convenient to express the scattered flux spectrum by the angular distribution of light scattered into a solid angle of 1 x 1 $[c/mm]^2$ two-dimensional spatial bandwidth [12].

TYPES OF HOLOGRAMS

According to the modulating properties and the thickness of the recording medium, a distinction is made between, in principle, eight different types of holograms. Fig. 1 shows these in a schematic way.

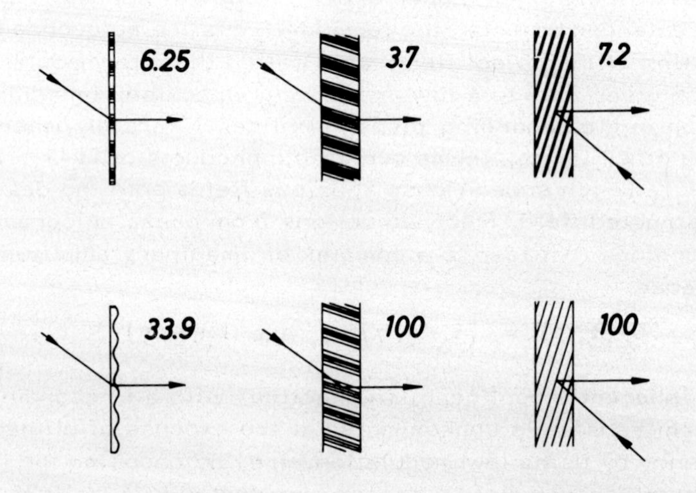

Fig. 1 - Types of holograms with theoretical upper limits for diffraction efficiency in per cent. [8]

The upper line illustrates holograms that absorb light and hence change the amplitude of an incident light field. In the lower line, the phase of the field is changed.

In the left column, "thin" transmission holograms are indicated, together with numbers for the maximum diffraction efficiency η theoretically achievable with sinusoidal gratings of modulation one. The same figures hold for thin reflection holograms (which are not shown explicitely).

For "thick" holograms, the volume nature of these recording media has to be taken into account. The Bragg condition implicates that wavelength and direction of incidence of the reconstructing beam must be identical to those of the reference beam in the recording step. Angular and wavelength sensitivity of thick holograms are properties of interest for data storage, multiple and colour hologram recording and white-light reconstruction. A quantitative parameter for the distinction between "thin" and "thick" holograms is the parameter Q

$$Q = 2\pi\nu^2 \cdot T \cdot \frac{\lambda}{n}, \qquad (10)$$

where T is the thickness of the medium and n its mean index of refraction. Holograms with $Q > 10$ have to be treated as "thick", while for $Q < 3$, holograms can be approximated as plane diffraction gratings [13, 5].

CHARACTERISTIC PARAMETERS OF SILVER HALIDE PHOTOGRAPHIC EMULSIONS

We shall take silver halide emulsions as an example when discussing the characteristics of hologram recording materials. Silver halide emulsions have a long tradition. They can be manufactured with the high resolution necessary for holography, sensitized to the red light of He-Ne and ruby lasers (λ = 633 nm and 694 nm); their sensitivity can be greater than that of other light-sensitive materials by several orders of magnitude, achieved, however, by means of the wet developing process.

Fig. 2 gives an idea of the spectral sensitivity, and hence the suitability for different laser wavelengths, for several commercially available emulsions.

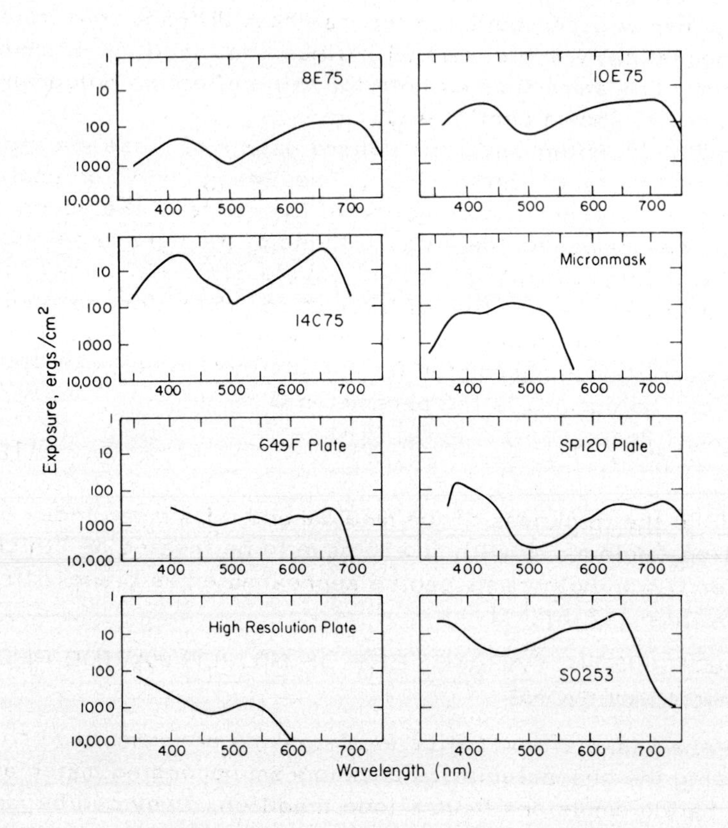

Fig. 2 - Spectral sensitivity curves for holographic emulsions made by Agfa-Gevaert (upper four) and Eastman Kodak (lower four). The ordinate indicates the exposure required for amplitude transmittance $\tau_A \approx 0.4$ (or $D \approx 0.8$) after development. (1 erg/cm^2 = 0.1 μJ/cm^2). (From H. M. Smith [3]).

The transfer of a signal from exposure to density or transmittance can be treated in a two-step model [10, 14]. Light spread within the material upon exposure lowers the modulation of the signal finally recorded, as sketched in Fig. 3.

This change of modulation is suitably described by means of the modulation transfer function (MTF), $M(\nu)$. In the second step, effective exposure is transferred to density D or amplitude transmittance τ_A ($|\tau_A| = \tau_i = 10^{-D/2}$).

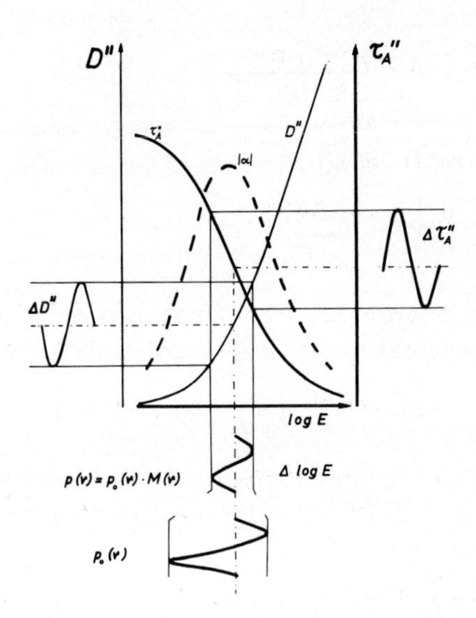

Fig. 3 – The transfer from exposure to density D or amplitude transmittance τ_A. Also given is $|\alpha|$, the derivative of the τ_A – log E curve [14].

From Eq. (6) we obtain the effective exposure

$$E(\xi) = \overline{E} \cdot (1 + p_o(\nu) \cdot M(\nu) \cdot \cos(2\pi\nu\xi)) \tag{11}$$

where we have introduced the modulation

$$p_o = \frac{2 \cdot \sqrt{I_s \cdot I_r}}{I_s + I_r} \tag{12}$$

The mean exposure E is transferred to mean amplitude transmittance $\overline{\tau}_A$, while, for small exposure modulation $\Delta E = E \cdot p_o(\nu) \cdot M(\nu) = \overline{E} \cdot p(\nu)$, the transmittance excursion $\Delta\tau$ (twice the amplitude) will follow from the first term in the Taylor series expansion of the transmittance-exposure relationship

$$\Delta\tau = 2 \cdot p(\nu) \cdot \bar{E} \cdot \left(\frac{d\tau}{dE}\right)_{\bar{E}}$$

$$= 2 \cdot \log e \cdot p(\nu) \cdot \left(\frac{d\tau}{d \log E}\right)_{\bar{E}}$$

$$= 2 \cdot 0.434 \cdot p(\nu) \cdot \alpha(\bar{E})$$

With $\eta = \dfrac{\phi_{diff.}}{\phi_{inc.}} = \dfrac{(\Delta\tau)^2}{16}$,

we obtain for the dependence of diffraction efficiency on the relevant emulsion parameters of thin amplitude holograms

$$\eta = \left[\frac{1}{2} \cdot 0.434 \cdot \alpha(\bar{E}) \cdot p_o(\nu) \cdot M(\nu)\right]^2 \qquad (13)$$

It is interesting to note that diffraction efficiency is proportional to the square of $\alpha(\bar{E})$, the derivative of the τ_A-log E function at the working point,

$$\alpha(E) = \left(\frac{d\tau_A}{d \log E}\right)_{\bar{E}} = \frac{1}{2} \ln 10 \cdot \tau_A(\bar{E}) \cdot \gamma(\bar{E}) \qquad (14)$$

where $\gamma(\bar{E})$ is the derivative of the traditional D-log E-curve [14]. Fig. 4 shows τ_A-log E and α^2-log E curves for a number of commercially available materials. The maximum value of α^2 is of influence on the diffraction efficiency achievable, its position along the logE-axis indicates the sensitivity of the material to radiation of λ = 633 nm.

For all emulsions, maximum η occurs at $\tau_A \approx 0.4$ (a criterion for developing plates for maximum η [15]); with respect to sensitivity, Fig. 4 is in agreement with Fig. 2. With hypersensitization treatments, a modest increase of film speed (by a factor of about 2) can be achieved [16].

Since, according to Eq. (13), the MTF at the carrier frequency enters into η with its square, it is a very important parameter. Methods for measuring MTF at high spatial frequencies are described in [4, 14, 17].

In Fig. 5, typical MTF values for five emulsions are compiled.

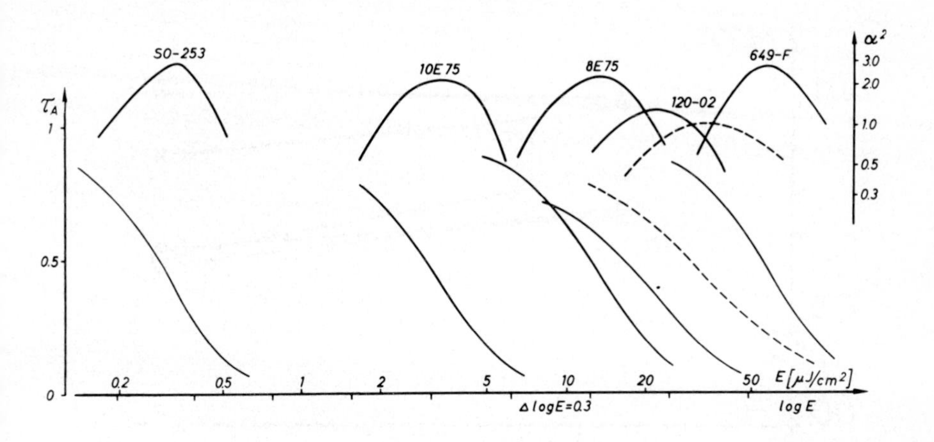

Fig. 4 – τ_A-log E and α^2-log E curves of five emulsions for $\lambda = 633$ nm.
Development : Kodak SO-253 and Agfa-Gevaert Scientia ("Holotest") 8E75 in Kodak D-19 for 5 min. at 20°C; Scientia 10E75 and Kodak 649-F in Afga 80 for 5 min. (these two developers are almost equivalent); Kodak 120-02 in D-19 for 6 min. (dashed curve) and 8 min. (full curve). From [4].

Development effects (adjacency effect), occurring especially with low-γ development, can cause an apparent increase of the MTF beyond 1 and distort the effect of spatial fulters [18].

As an example for light scattering in developed photographic emulsions, Fig. 6 shows scattered flux spectra of three plates for holography. We note that scattered flux is higher for plates of higher sensitivity. Furtheron, samples exposed to laser radiation show increased scattered flux compared to samples exposed to incoherent light of the same wavelength. This additional stray light stems from diffraction at a speckle structure exposed into the emulsion by coherent radiation scattered during exposure ("diffusion mottle"). This effect contributes also to the generation of scattered flux concentrated to rings as observed with thick recording media (e.g. [8, 19]).

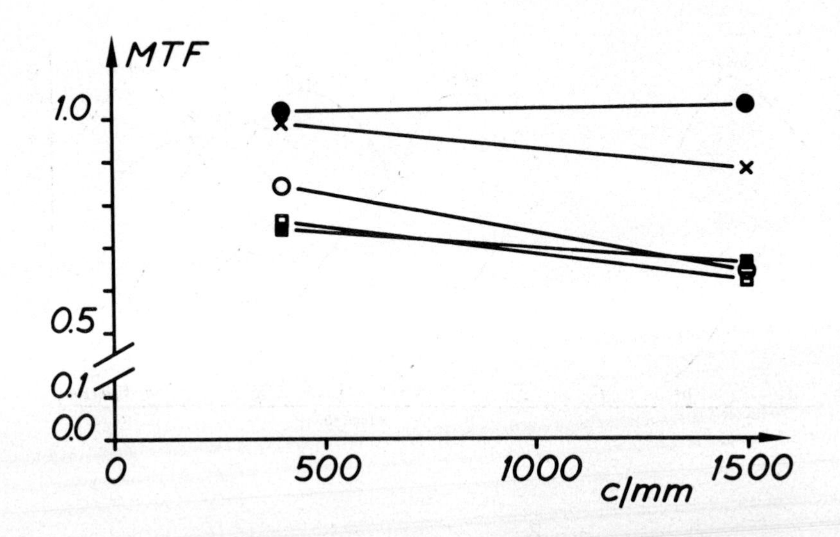

Fig. 5 - MTF values for $\lambda = 633$ nm at $\nu = 400$ c/mm and 1500 c/mm :

Kodak Plate 120-02; Agfa-Gevaert Scientia 8E75
Kodak S0-253; Scientia 10E75; Kodak 649-F. From [4].

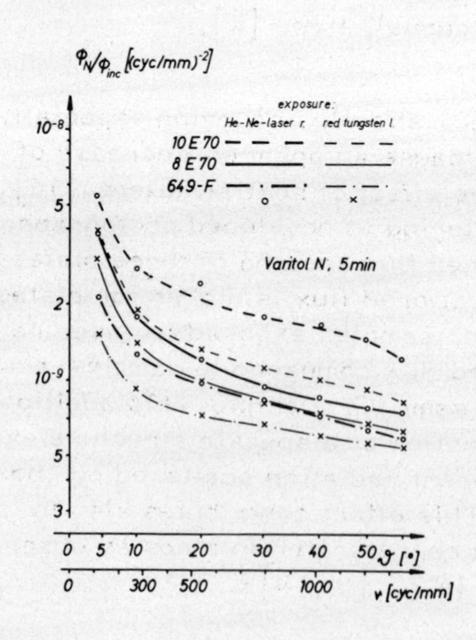

Fig. 6 - Scattered flux spectra at $\lambda = 633$ nm of three plates for holography exposed by He-Ne-laser radiation (o, heavy lines) and red tungsten light (x, light lines). Mean transmittance is $\tau_i = 0.2$ (From [12]).

PHASE HOLOGRAMS

As indicated in Fig. 1, while thin and thick amplitude gratings theoretically may diffract up to 6.25 per cent of the incident light and 3.7 per cent, respectively, the corresponding figures for phase gratings are 33.9 and 100 per cent:

$\eta = J_1^2 (\Delta\varphi)$ for thin sinusoidal phase gratings

$\eta = \sin^2 (\Delta\varphi/2)$ for thick sinusoidal phase gratings

$\eta \approx \frac{1}{4} \cdot (\Delta\varphi)^2$ for $\Delta\varphi \lesssim 0.6 \approx 2\pi \cdot 0.1 \cdot \lambda$ in both cases,

where $\Delta\varphi$ is the amplitude of the sinusoidal phase modulation.

Phase holograms in photographic layers can be obtained by oxidizing the photographic silver to dielectric compounds like AgC1, AgBr, AgI, $AgHgC1_2$, $Ag_4Fe(CN)_6$. The last one, as an example, is very easy to achieve by bathing ("bleaching")the developed and fixed plate for several minutes in a solution of 15–30 g potassium ferricyanide in 1 liter of water. More, and more sophisticated receipes are found in, e.g., Refs. [1 - 4, 7]. Instead of α^2 (E), it is now γ^2 (E) to which diffraction efficiency is related, and the working point for maximum η is typically at densities between 2.5 and 4 before bleaching [16].

NON-SILVER RECORDING MATERIALS

The parameters of a number of recording materials are compiled in tabular form in Table I, which is reprinted from Ref. [8]. The reference numbers to original work given in the last column are those of Ref. [8], where also some more comments on the different materials are found. The monograph, Ref. [4], is expected to become the standard reference for hologram recording materials.

Hologram recording materials

Material	Spectral sensitivity λ[nm]	Exposure for max η [mJ/cm²]	Spatial frequencies [cycles/mm]	Type of grating	Max. diffract. efficiency of sine grating η[%]	Development	Erasable	Remarks, typical applications	Typical references
Silver-halide emulsions	Sensitizer 400–700 (< 1300)	(10^{-6})... 10^{-4}... 10^{-1}	(100)... 500...10 000	Plane/vol. abs.	5	x (wet)	—	Commercially available, high sensitivity, panchromatic, latent image. Colour hol., reflection hol.	[29–31]
Silver-halide emulsions bleached				Plane/vol. phase	20...50	(bleach)			[44, 53–60]
Dichromated gelatin	350–520 (sens. 633)	2...30	1000 3000	Plane phase Vol. phase	30 90	x (wet)	—	High S/N, read-only memories, diffusers	[59, 61–65]
Photoresist	441 458 488	10 100...300 8000	< 3000	Plane phase blazed refl.	30 70...90	x (wet)	—	Precision optics, spectrosc. gratings, replication master, data storage	[59, 66, 67 7, 8, 69–74, 68]
	488	5	500	Plane phase	15	heat, 1 min			
Photoplastic (thermopl.) materials	Sensitizer pan., 633	(10^{-3})... 10^{-1}	Bandpass ~ 400–1000	Plane phase	6...15	x (heat, sec)	x (page)	Recyclable. Read-write memories, hologr. interferometry	[59, 75–82]
	1150	~ 500	~ 500	Plane phase	~ 1				[80]
Photopolymers	UV 458 Sensitizer	1 2000 10...40	Bandpass ~ 200–1500	Vol. phase	10...85	— (fix. : light)	—	Hol. interferometry. Read-only memories, thick optical elements	[59, 83–86]
Photodegradable polymers (polymethyl methacrylate, cellulose acetate butyrate)	(UV), (sensit.)	1000... 20 000	< 2000	Vol. phase (thickness μm...mm)	50...95	— (fix. : UV, min.–days)	—	Read-only memories, thick optical elements	[48, 59, 87–90]
Photochromic film, glass, crystals	Activ. 300–450 Bleach. 550–700	200...5000	1000... 10 000	Vol. abs.	1.2	—	x	Read-and-write memories, hol. interferometry	[91–102]
Electro-optical crystals (metal doped $LiNbO_3$, SBN) (undoped $LiNbO_3$)	(488)	(1...50...) 1000 (10^6)	> 1000	Vol. phase	(1...)60	— (fix)	x (heat, total)	Read-and-write memories	[46, 47, 103–108]
Magneto-optic materials	(Heat) (694)	10...100 (puls, < μsec)	> 1000	Polaris.	< 0.01	—	x (page)	Read-and-write memories	[109, 110]

Table I. (Reprinted from Ref. [8]. The numbers of the references given in the last column are those of [8]).

REFERENCES

Monographs on holography and optical data processing, which treat recording materials:

1) R. J. Collier, C. B. Burckhardt, L. H. Lin, "Optical Holography", Academic Press, 1971
2) W. T. Cathey, "Optical Information Processing and Holography", John Wiley & Sons, 1974
3) H. M. Smith, "Principles of Holography, 2nd Edition", John Wiley & Sons, 1975

Monograph and review articles on recording materials :

4) "Holographic Recording Materials", H. M. Smith Ed. Topics in Applied Physics, Springer-Verlag (forthcoming)
5) E. G. Ramberg, "Holographic Information Storage", RCA Review 33 (1972) 5 - 53
6) J. Bordogna, S. A. Keneman, J. J. Amodei, "Recyclable Holographic Storage Media", RCA Review 33 (1972) 227 - 247
7) W. S. Colburn, R. G. Zech, L. M. Ralston, "Holographic Optical Elements", Techn. Report AFAL-TR-72-409, Harris Electro-Optics Center of Radiation, Ann Arbor, Mich., 1973
8) K. Biedermann, "Information storage materials for holography and optical data processing", Opt. Acta 22 (1975) 103 - 124 (with 122 references incl. references of Table I of this text).

Monographs on silver-halide photographic materials :

9) C. E. K. Mees, T. H. James, Eds., "The Theory of the Photographic Process", 3rd ed., Macmillan, 1966
10) H. Frieser, "Photographic Information Recording - Photographische Informationsaufzeichung", Focal Press - R. Oldenbourg Verlag, 1975

Further references cited in this paper :

11) J. W. Goodman, "Film-Grain Noise in Wavefront-Reconstruction Imaging", J. Opt. Soc. Am. 57 (1967) 493 - 502
12) K. Biedermann, "The Scattered Flux Spectrum of Photographic Materials for Holography", Optik 31 (1970) 367 - 389
13) H. Kogelnik, "Coupled Wave Theory for Thick Holographic Gratings", Bell Syst. Techn. J. 48 (1969) 2909 - 2947

14) K. Biedermann, "A Function Characterizing Photographic Film that Directly Relates to Brightness of Holographic Images", Optik 28 (1968) 160 - 176
15) K. Biedermann, K. A. Stetson, "Adjusting Development Time to Influence the Characteristic of Holograms", Phot. Sci. and Eng. 13 (1969) 361 - 370
16) K. Biedermann, "Attempts to Increase the Holographic Exposure Index of Photographic Materials", Appl. Opt. 10 (1971) 584 - 595
17) S. Johansson, K. Biedermann, "Multiple-Sine-Slit Microdensitometer and MTF Evaluation for High Resolution Emulsions", Appl. Opt. 13 (1974) 2280 - 2287, 2288 -2291
18) K. Biedermann, S. Johansson, "Development effects and the MTF of high-resolution photographic materials for holography", J. Opt. Soc. Am. 64 (1974) 862 - 870
19) M. R. B. Forshaw, "Explanation of the Diffraction Fine-Structure in Overexposed Thick Holograms", Opt. Comm. 15 (1975)

Some more references :

20) A. Graube, "Infrared holograms recorded in high-resolution photographic plates with the Herschel reversal", Appl. Phys. Lett. 27 (1975) 136 - 137
21) H. M. Smith, M. H. Sewell and J. R. King, "Real-time holographic interferometry: a system", Appl. Opt. 15 (1976) 729 - 733
22) T. Kubota, T. Ose, M. Sasaki and K. Honda, "Hologram formation with red light in methylene blue sensitized dichromated gelatin", Appl. Opt. 15 (1976) 556 - 558
23) W. S. Colburn, E. N. Tompkins, "Improved Thermoplastic-Photoconductor Devices for Holographic Recording", Appl. Opt. 13 (1974) 2934 - 2941
24) R. Moraw, "Thermoplastschichten für Echtzeitbildwandler", Feinwerktechnik & Messtechnik 84 (1976) 50 - 52
25) B. L. Booth, "Photopolymer Material for Holography", Appl. Opt. 14(1975) 593 - 601
26) W. J. Tomlinson, "Volume holograms in photochromic materials", Appl. Opt. 14 (1975) 2456 - 2467
27) J. W. Burgess, R. J. Hurditch, C. J. Kirkby and G. E. Scrivener, "Holographic storage and photoconductivity in PLZT ceramic materials", Appl. Opt. 15 (1976) 1550 - 1557

Coherant Optical Engineering, F.T. Arecchi and V. Degiorgio (eds.)

LIGHT DEFLECTION

by

Robert V. Pole and Kurt E. Petersen
IBM Research Laboratory
San Jose, California 95193

INTRODUCTION

In most practical applications of lasers, there is a need for modulation and deflection of its output. While neither modulation nor deflection is easily achieved, the deflection presents the laser engineer with a far more formidable challenge. The root of all the difficulties lies in the properties of the photon--it has no mass and carries no charge. Hence, there is no way of influencing its direction of propagation by means of an external field, as is the case with the electron, for example. In fact, the only way a photon can be influenced is by slowing down its velocity of propagation. It is, therefore, no wonder that all conceivable schemes of both modulation and deflection reduce to finding ways how changes in the propagation velocity could be converted to changes in phase, amplitude, or direction.

The technical literature of the past twenty years is replete with rather ingenious light deflection concepts. Unfortunately, only a few of these concepts have matured into useful laboratory devices and even fewer into commercially exploited products.

In this monograph, we shall concentrate on those methods of light deflection which have resulted in useful devices or, at least, bear a promise of becoming such in the near future.

BASIC PARAMETERS OF A DEFLECTOR

The fundamental quantity which characterizes every deflector is the <u>number of resolvable spots</u>, N. It is sometimes referred to also as its <u>spatial resolution</u>, or simply the <u>resolution</u>. The number of resolvable spots is a ratio of the achievable deflection angle and the angular spread of the deflected beam. If an attempt is made to increase the deflection angle by means of a suitable optical system (say, an inverted telescope), the angular spread will increase in the same proportion and vice versa. Thus, the resolution N of a given deflector is, in principle, a conserved quantity; and it cannot be changed by an ideal optical system. In practice, it can always be degraded by the aberrations of the system--but never increased.

Thus:

$$N = \phi/\alpha \tag{1}$$

where ϕ is the deflection angle and α is the angular spread.

While the deflection angle ϕ depends generally on the geometry and/or the physical properties of the deflector, the spread α depends on the coherence of the source, quality of the intervening elements, as well as on the geometry of the device. Thus, since the deflection angle cannot be increased beyond certain limits in any given device, it is always desirable to minimize the spread by choosing the most coherent source available and by employing the optical elements with minimum aberrations. It is, therefore, not surprising that optical deflectors are useful only when employing lasers as their light source.

The other important parameter relates to the speed with which the deflector can operate. If the deflector is designed to operate in the random access mode, this parameter t is referred to as the access time of the deflector.

The ratio:

$$R = N/t \tag{2}$$

then denotes the deflection capacity or the <u>deflection rate</u> of the device. More commonly (often out of necessity), the deflectors are operated in the raster scanning mode. In this case, t denotes the line (duration) time and R from (2) again the deflection or scanning rate measured in spots (or bits) per unit time.

A graphical summary[1] of the performance characteristics of three of the most widely used classes of light deflectors is shown in Fig. 1. Specific galvanometer, acousto-optic (A-O), and electro-optic (E-O) devices will be discussed in the following sections.

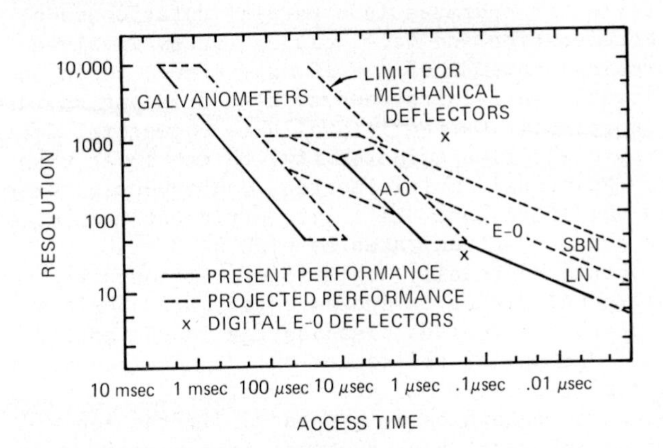

Figure 1. Deflector Performance[1]

MECHANICAL DEFLECTORS

Mechanical deflectors are, for obvious reasons, by far the most widely used and the most highly developed light deflection devices. Not only are they based on simple principles, but they are also quite efficient and effective. Their main drawback is in the inertia of their moving parts which limits them primarily to constant speed, raster-type scanning applications, although smaller devices, such as the torsional mirror, are capable of limited random access performance.

Deflection by means of mechanical motion can be achieved by utilizing any of the three basic optical phenomena: reflection, diffraction, or interference. Since the exploitation of the interference phenomenon requires very minute physical motions, the deflectors based on the phenomenon are not truly mechanical and will, therefore, be described later.

The three most important high performance mechanical deflection devices are: (1) the torsional or galvanometric deflector, (2) the rotating polygonal mirror, and (3) the rotating holographic diffraction deflector. Since the torsional deflector operates in a partial rotation mode, the large acceleration and deceleration forces involved set a theoretical maximum of R = 10 Mspots/sec, based on maximum allowable material stresses, heat dissipation, and optical distortion.[1] Currently available commercial devices have scanning capacities a factor five to ten lower than this limit. Polygonal, multi-faceted, continuously rotating mirrors, on the other hand, have less stringent mechanical limitations and deflection rates as high as R = 30 Mspots/sec should be readily achievable. Although the rotating polygonal mirror is indeed a very high performance light deflector, one serious disadvantage is the cost of fabrication. Mechanical tolerances on the machined facets (flatness, parallelism, etc.) are very tight.[2] Multi-faceted, holographically generated diffraction gratings offer important advantages in this respect. Recorded in a photosensitive material coated on the inside surface of a glass ring, each individual grating deflects the beam over an angle in a manner similar to the rotating mirror.[3] Since the grating is produced holographically, however, most of the system optics and even system aberrations can be incorporated in the hologram itself, greatly simplifying the external optics.

ACOUSTIC DEFLECTORS

The first acousto-optic deflector was reported in 1965;[4] it employed Bragg diffraction from ultrasonic waves in water at a frequency from 40-50 MHz. The device had 73 resolvable spots and a diffraction efficiency of 16% with 200 mW of electrical input. In 1966, the use of phased transducers to track the Bragg angle and increase the resolution was demonstrated by Korpel et al.[5] Their television system had 200 resolvable spots and 17 MHz bandwidth and virtually 100% efficiency with 1.1 W of electrical input; the acousto-optic medium was water.

Acousto-optic deflectors diffract light through the elasto-optic effect. Electrically generated acoustic waves produce periodic variations in the index of refraction of the material, which then behaves as a grating. Device performance characteristics depend strongly on material parameters such as the photoelastic constant, acoustic velocity, density, acoustic attenuation coefficient,[6] and the maximum available crystal size. Equally important is the device geometry. Multiple transducers in a phased-array configuration can significantly increase bandwidth and resolution,[5] while surface wave transducers together with integrated optic techniques allow better control over device characteristics and a more efficient acousto-optic interaction geometry.[7,8]

Rapid advancements in techniques and materials have taken place over the past ten years. Resolutions of over 1,000 spots have been predicted[9] for $LiNbO_3$, GaP, $Sr_{0.75}Ba_{0.25}Nb_2O_6$, Tl_3AsS_4, and TeO_2. Gorog et al.,[10] were able to obtain a diffraction efficiency of 50% at a resolution of 500 spots and television deflection rates with only 50 mW of electrical drive power in TeO_2. More recently, deflection of optical guided waves by acoustic surface waves has been demonstrated, with a promise of higher efficiencies and overall superior performance.[11]

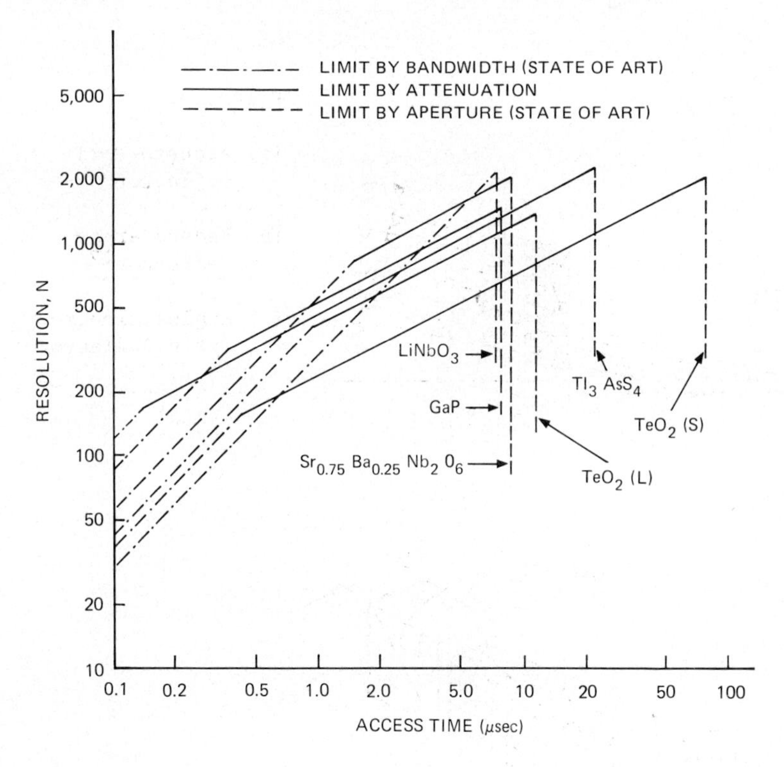

Figure 2. Acousto-Optic Deflector Materials[9]

INTERFEROMETRIC DEFLECTORS

Although, strictly speaking, the phenomenon of interference is present in any light deflection process considered so far, the two devices described in this section are termed interferometric because the interference is the

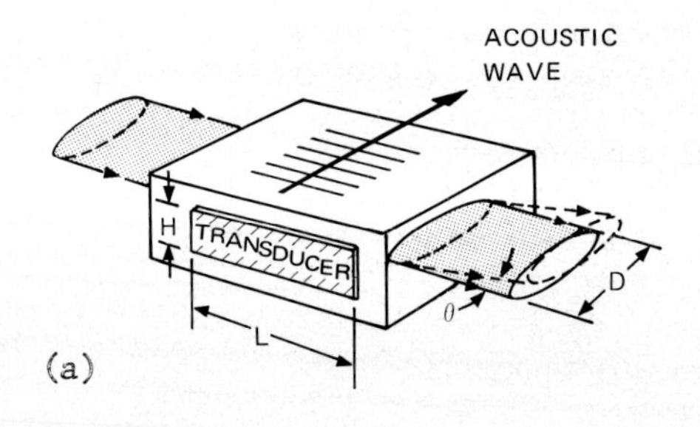

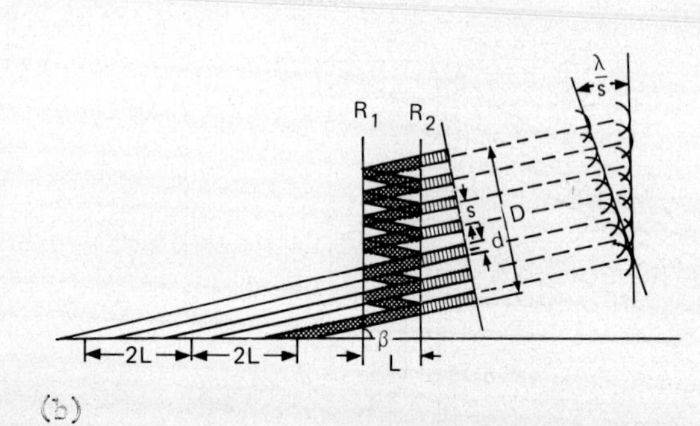

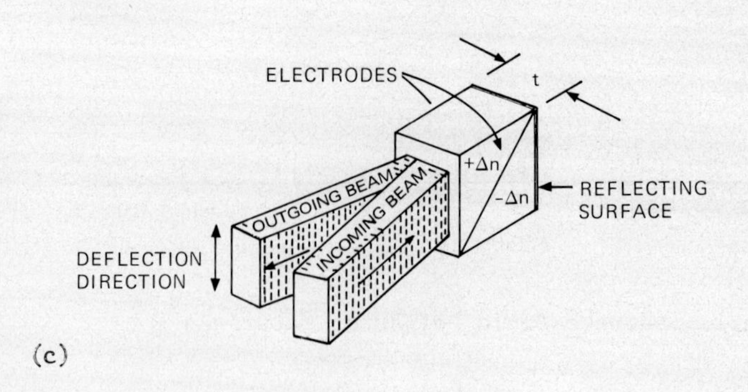

Figure 3.

(a) Acousto-Optic Deflector[6]

(b) Phased Array Deflector[12]

(c) Analog Electro-Optic Deflector[17]

primary process used for deflection. The first of the two devices, proposed by Korpel[12] in 1965 and referred to as the Phased Array Scanner, was apparently stimulated by the existence of a similar concept in the microwave field.

The second device, which we shall call the Fabry-Perot scanner, was first described by Buck and Holland in 1966.[13]

In the phased array scanner, a narrow, collimated beam is multiply reflected between two parallel mirrors and partially coupled out at each bounce through one of the mirrors. The combined wavefront emerging from the device will be a coherent superposition of individual leaky waves. If the phases of the individual beams are varied, the wavefront will tilt away from the original direction. In this way, mechanical variations in the mirror spacing will result in angular variations in the direction of the emerging beam. An approximate analysis indicates that under optimum conditions, the resolution of this deflector will not exceed the number of internal beam reflections, 70 to 80 spots.

The Fabry-Perot scanner also consists of two mirrors, one of which is slightly tilted with respect to the other. When the mirrors are uniformly illuminated, the tilt is adjusted so that two interference fringes at most are visible in the field. Now, by changing the optical pathlength between the interferometer by fractions of a wavelength (either mechanically or electro-optically), the fringes move across the field. The moving fringe can easily be converted to a moving spot by a cylindrical lens. As in the case of the phased array scanner, resolution is limited by beam walk-off; the light tends to walk-off the structure due to the mirror tilt (or due to the multiple reflections in the phased array approach). Spatial resolutions of only about 50 seem to be the practical limit.

ELECTRO-OPTIC DEFLECTORS

Although all electro-optic deflectors are based on an electric field-induced change in the index of refraction of a material, two primary modes of operation can be distinguished. Digital E-O deflectors employ an electro-optic material to change the polarization state of the incident beam together with a birefringent material oriented in such a way that a given angular or lateral displacement is produced between the two polarizations.[14]

Arrays of such binary stages have been demonstrated which will deflect the beam to 2^n (where n is the number of stages) possible locations, with n as large as 20 (i.e., 1024 × 1024 positions). Analog E-O deflectors vary the field continuously across a prism made from an electro-optic material, oriented to obtain angular variations in the light beam which correspond to the field-induced changes in the index of refraction.

Three common birefringent polarization displacement configurations used in digital deflectors are the split angle deflector, the Wollaston prism deflector, and the total internal reflection (TIR) deflector.[15] The split angle and Wollaston prism deflectors take advantage of polarization-dependent refraction angle in birefringent materials (e.g., typically 9.2° for $NaNO_3$ in the split angle mode). In the TIR technique, a birefringent plate is oriented such that one polarization passes through the plate, while the other is reflected off the plate. The Wollaston prism configuration has the advantage that pathlength compensation is not necessary.

In terms of resolution and access time, digital light deflectors exhibit exceedingly good characteristics due to the fact that the index change is essentially instantaneous with the application of the voltage. The system constructed by Meyer et al.,[16] for example, resolved over 10^6 points in a 1024 × 1024 array with less than 1 μsec access time. The maximum optical aperture in this device was less than 1 cm^2. Unfortunately, however, the complexity of the system is very discouraging. Twenty states of high quality, high precision, and delicately aligned optical components are driven by twenty high voltage, high speed pulsers (voltages between 2 and 8 kV). The optical pathlength from laser to screen is almost 6 meters. High cost and large physical size (primarily of the power supplies) seem to limit such systems to very special applications.

Analog E-O prism deflectors have been extensively discussed by Lee and Zook,[17] and reviewed by Zook.[1] Whatever the configuration of these types of devices, which can be quite complicated when many prisms and multiple pass methods are used, or quite sophisticated when integrated optics techniques are employed,[11] the basic idea is the same. The top portion of the beam experiences a different phase retardation through the system than the bottom portion (which is observed as a change in propagation direction), and this phase retardation is adjusted continuously by applying a voltage to a prism-like E-O material. Important performance criteria are the maximum voltage which can be applied to the crystal (again in the kilowatt range), the system aperture and other geometrical factors, and, of course, the electro-optic coefficient. Using optimum geometries and making considerations for maximum available crystal sizes, Zook[1] has estimated that the maximum spatial resolution for three common E-O materials (KD*P, $LiNbO_3$, SBN) is near 1,200 (for crystal apertures of 7, 2, 1 cm, respectively).

REFERENCES

1. J. D. Zook, "Light Beam Deflector Performance: A Comparative Analysis," Appl. Opt. 13, 875 (1974).
2. J. M. Fleischer et al., "Laster Optical System of the IBM 3800 Printer," Digest of Papers, Conference on Laser and Electro-Optical Systems, May 1976, San Diego, California (OSA, Washington, D.C.).
3. R. V. Pole and H. P. Wollenmann, "Holographic Laser Beam Deflector," Appl. Opt. 14, 976 (1975).
4. A. Korpel et al., "An Ultrasonic Light Deflection System," IEEE J. Quan. Elec. QE-1, 60 (1965).
5. A. Korpel et al., "A Television Display Using Acoustic Deflection and Modulation of Coherent Light," Proc. IEEE 54, 1429 (1966).
6. D. A. Pinnow, "Guidelines for the Selections of Acousto-optic Materials," J. Quan. Elec. QE-6, 223 (1970).
7. L. Kuhn et al., "Deflection of an Optical Guided Wave by a Surface Acoustic Wave," Appl. Phys. Lett. 17, 265 (1970).
8. R. V. Schmidt, "Acousto-optic Interactions Between Guided Optical Waves and Acoustic Surface Waves," IEEE Trans. on Sonics and Ultrasonics, SU-23, 22 (1976).
9. I. C. Chang, "Acousto-optic Devices and Applications," IEEE Trans. on Sonics and Ultrasonics, SU-23, 2 (1976).
10. I. Gorog et al., "A Television Rate Laser Scanner, II; Recent Developments," RCA Review 33, 667 (1972).
11. C. S. Tsai et al., "Acousto-optic Bragg Diffraction Using Multiple Surface Acoustic Waves and Electro-optic Deflection/Modulation Using Tilted Electrodes," Paper TuA2-1, Tech. Digest, Integrated Optics Meeting, Salt Lake City, January 1976 (OSA, Washington, D.C.).
12. A. Korpel, "Phased Array-Type Scanning of a Laser Beam," Proc. IEEE 54, 1419 (1966).
13. W. E. Buck and T. E. Holland, "Optical Beam Deflector," Appl. Phys. Lett. 8, 195 (1966).
14. V. J. Fowler and J. Schlafer, "A Survey of Laser Beam Deflection Techniques," Proc. IEEE 54, 1437 (1966).

15. W. Kulcke et al., "Digital Light Deflectors," Proc. IEEE 54, 1419 (1966).
16. H. Meyer et al., "Design and Performance of a 20-Stage Digital Light Beam Deflector," Appl. Opt. 11, 1732 (1972).
17. T. C. Lee and J. D. Zook, "Light Beam Deflection with Electro-optic Prisms," IEEE J. Quan. Elec. QE-4, 442 (1968).

Coherant Optical Engineering, F.T. Arecchi and V. Degiorgio (eds.)

OPTICAL MODULATORS

E. H. Turner

Bell Telephone Laboratories
Holmdel, New Jersey 07733

In order to modulate on optical wave one must be able to vary the amplitude, phase or frequency of this wave in a controllable fashion. This may be done by acting directly on the source or by using a separate modulating element to operate on an existing wave. Direct modulation accomplished by varying the electrical input to a laser has the great apparent advantage of simplicity and the fact that the information to be put on the wave is generally in electrical form. One case in which this method of modulation is used is the semiconductor injection laser. Here the following conditions obtain: low cavity Q, small length ($\sim$300 um), very wide gain curve and short relaxation times. The injected current is switched from an amount near that needed for oscillation into the operating range and even though the output vs. current is highly nonlinear this is acceptable for a digital system. Modulation rates of $\sim$50 MHz are achieved before high frequency components of the modulating current initiate relaxation oscillations. Some of the same laser properties that allow high speed direct modulation would mitigate against use of an external modulator. Recent advances indicate that the direct modulation range may be extended to 100's of MHz by injection of quasi-monochromatic radiation to help suppress relaxation oscillations [L-1].

The possibility of internally modulating a laser cavity must also be considered. Introduction of a variable loss in a laser cavity obviously has a large effect on the output and similarly an element whose phase length can be varied will change the resonance characteristics. Mechanical, acoustooptic and electrooptic elements are all used in such intra-cavity modulation to produce Q-switching, cavity dumping, mode locking, frequency and amplitude regulation, etc. These uses are clearly important and we mention them as important applications of modulating elements but cannot discuss them here because of the complex interdependence of modulator, gain medium, resonator, etc.

Most laser sources are far from being ideal single frequency oscillators: they may operate with a number of discrete modes oscillating simultaneously, the width od each mode may be many MHz, and the center frequency may be only poorly defined. All of these ills are potentially curable, but their effect is to make amplitude modulation much to be preferred over phase or frequency modulation for transmission of information. However, a suitable way to change optical attenuation directly has not been found. In the search for a variable loss mechanism, such phenomena as the electric field induced shift of a band edge (Franz-Keldysh Effec) and absorption by free carriers have been considerd. The former is a small effect and if one operates on the steepest part of the absorption edge in order to get large changes, the insertion loss is prohibitive. Free carrier absorption is limited bysuch things as recombination time, power required and the fact that it is relatively small in the visible. However, the λ^2 dependence makes it more attractive in the infrared, where, as we shall see, other effects become weaker. The dispersion associated with this absorption is also sizeable in the infrared and has been considered for device application [M-1].

Usable changes in refractive index can be produced by strain, magnetic fields and electric fields. All materials have a strain--optic effect. the relation being given through a dimensionless fourth rank tensor p_{ijkl}. This leads to acoustooptic effects. Magneto-optic behavior is described by a second rank tensor susceptibility whose elements depend on magnetization. Similarly, all materials show a quadratic dependence on electric field (Kerr effect) with the coefficients being fourth rank tensors. Only crystals without a center of symmetry have a linear electrooptic (LEO or Pockels) effect and this is described in turn by a third rank tensor. Physically, the electric field description would be better replaced by polarization.

Each type of modulator can also be classified according to whether it is a bulk modulator acting on an unguided optical beam or a device using a wave guided in one or both dimensions. Waveguide dimensions of the order of a wavelength lead to very low modulating powers, since only this volume needs to be driven. A comprehensive list of references is given in [T-1]. Bulk modulators are diffraction limited and for a cylinder of diameter d and length l we have:

$$\frac{d^2}{l} = S^2 \times 4\lambda/n\pi \geq 4\lambda/n\pi,$$

which defines the safety factor S [K-1] and where n is refractive index and λ free space optical wavelength.

Acoustooptic modulators depend ondiffraction of light by an acoustic wave. The acoustic standing wave or traveling wave acts as a phase grating and, while the individual orders are phase modulated sidebands, the most useful results come from their angular separation. Deepest modulation is obtained when operating in the so-called Bragg regime in which the interaction length $l \gg \Lambda^2/\lambda$, with Λ the acoustic wavelength.

The acoustic interaction arises through changes in refractive index which can be found from

$$\Delta(1/n^2)_{ij} = p_{ij,kl}\,\varepsilon_{kl},$$

where ε_{kl} is the strain and the strain optic tensor is symmetric in ij and can usually be taken as symmetric in kl. The acoustic strain is generally produced by a piezoelectric transducer bonded to the active medium, but in some cases the generating medium itself is used. Acoustic power, P_a, crossing an area, A, will produce a change in refractive index given by [P-1] :

$$\Delta n = \sqrt{n^6 p^2 P_a/2\ \ V^3 A} = \sqrt{M_2 P_a/2A}$$

where p is the appropriate (combination of) strain optic coefficients, V is the sound velocity, M_2 a figure of merit, and ρ the density. Many materials have values of p as large as .2 or .3, but there are large variations in the other material parameters. For example, TeO_2 has large n and remarkably small shear wave velocity, with a resultant very large M_2. Water would be a very good material because of small V and ρ, but for many purposes its large, frequency dependent acoustic attenuation rules it out.

In a bulk modulator, let us take a $\hat{y}$ directed acoustic wave with a cross section of z = l and x = h, where the wave may have been generated by an external piezoelectric element. An optical wave propagating at an angle θ_B to z and with cross section y = t, x = h passes through the same region. In general, the beams will have been focused and some attempt made to maximize overlap in the region of the beam waists. The acoustic wave constitutes a thick phase grating with $\Delta\phi = 2\pi\Delta n l/\lambda$. Under these conditions only one diffracted (first) order beam at an angle $2\theta_B$ to the incident wave adds up in phase. The angle θ_B is given by the well known Bragg condition sin $\theta_B = \lambda/2\Lambda$ and the fractional scattered intensity is $\sin^2(\Delta\phi/2)$.

One limit on bandwidth is clearly the transit time for the acoustic beam across the optical beam waist. A second limitation arises from the dependence of the Bragg angle on acoustic wavelength, since the modulation depth is reduced if the entrance angle θ_B is not appropriate. The optimum condition occurs when the diffraction angles of the light and sound waves are the same [G-1] . This leads to the condition

$$P_a/\Delta f = (\Delta\phi)^2 \lambda^3 f_a h/(2\pi^2 M_2 n V^2).$$

If we also take into account the fact that a minimum value is placed on t^2/l by diffraction and make use of the safety factor, S, defined earlier, we have [H-1] for extinction of the zero order wave:

$$P_a/\Delta f = 125\ \lambda^3 S^2/nVM_2\ \text{mW/MHz}.$$

The burden then is entirely on material properties apart from one's ability to minimize S. Pactical results show numbers as good as ~ 1 mW/MHz.

There have been a number of theoretical and experimental reports on interactions of optical waves guided in one dimension with surface acoustic waves. The situation has been summarized by Hammer [H-1]. In general one can say that gains in efficiency over the bulk modulator performance have not been large because of poor overlap between the optical and acoustic wave volumes. Additionally, it appears that efficient acoustic wave generators are narrow band, so information handling capacity cannot be large. The question of transducer efficiency was also neglected in the discussion of bulk modulators.

Magnetic modulators do not seem to have won wide acceptance to date in spite of the fact that they are predicted to be quite efficient and the two experimental models to be discussed here would seem to verify this. In addition, one of these modulators introduces the notion of periodic variation of coupling <u>or</u> phase constant. We will discuss some of the present limitations that have limited the use of these modulators.

Magnetooptic modulators are based on the Faraday effect which usually exhibits itself as a rotation of the plane of optical polarization proportional to magnetization (M_z) along the direction of propagation. In diamagnetic materials and in paramagnetic materials at room temperature these effects are too small to be of practical interest. Restricting the discussion to ferri (or ferro) magnetic substances we next note that one must operate at wavelengths where the effect is small in order to avoid large losses. For magnetization along $\hat{z}$ we have: [T-2]

$$\begin{vmatrix} D_x \\ D_y \\ D_z \end{vmatrix} = \varepsilon_o \begin{vmatrix} n_x^2 & -i\delta & 0 \\ i\delta & n_y^2 & 0 \\ 0 & 0 & n_z^2 \end{vmatrix} \begin{vmatrix} E_x \\ E_y \\ E_x \end{vmatrix}$$

and with the assumptions mentioned, the elements of the magnetooptic tensor (δ and n^2) are real and $\delta \ll n^2$. The sign of δ depends on the sense of the magnetization. If $n_x = n_y = n$ the normal modes for propagation in the z direction are $\pm$ circularly polarized waves with indices $n^{\pm} = n \pm \delta/2n$, and a linear polarization is rotated by an amount $\theta = \pi \ell \delta / n\lambda$ in a length ℓ.

The most suitable materials are the cubic Yttrium Iron Garnet (YIG) and derivatives using other elements in place of some of the iron or yttrium. YIG is quite transparent in the range 1.15 to 4.5 μm, although doping tends to restrict this window. The fact that there is no comparable material in the visible is one reason for lack of interest. Since the direction or magnitude of magnetization must vary in a modulator the small anisotropy of these materials is a feature as is the very small ferrimagnetic damping.

A bulk modulator using Gallium doped YIG was built and operated at 1.52 μm by LeCraw [C-1]. In this structure polarized light travels along the axis (z) of a cylinder. A biasing (x) field normal to the axis is provided to saturate the magnetization and is adjusted so the effective internal field, H_i, puts the ferrimagnetic resonance frequency well above the highest modulating frequency. Under these conditions, when a modulating field h_z is provided the (saturated) magnetization M_s follows the resultant H field essentially adiabatically and a varying z-component of magnetization results in a varying Faraday rotation. The figure of merit for this modulator in mW/MHz can be found from:

$$\frac{P}{\Delta f} = \frac{10^2}{8\pi^2} \; S^2 \; H_i \, 4\pi M_s n \lambda^3 \theta^2 / \delta_s^2 \text{ mW/MHz},$$

where δ_s is the value δ would assume for propagation along M_s and cgs units are used. This indicates that 18.5 S^2 mW/MHz will give 100% modulation with a bandwidth of 200 MHz. The sample was oriented to take advantage of the anisotropy energy and the gallium substitution reduces M_s without a major reduction in δ_s. Reducing M_s also reduces H_i. Larger bandwidth requires larger H_i.

One dimensional waveguide modulators using, e.g., a film of a Ga and Sc doped YIG grown epitaxially on a nonmagnetic GdG garnet have been built and analyzed by Tien [T-3]. In addition to the requirements of small anisotropy and large δ_s these films must have a refractive index greater than the substrate and as the lattice constants must match very closely to make good wave-guides. Although Tien has shown ferrimagnetic damping to be of small importance here, there is again the requirement that higher operating frequencies require a higher biasing field and correspondingly larger driving field. The switching current follows a serpentine path such that each time it crosses the path of the guided wave it provides a driving h_z field of alternating sign. The period, P, of this alternation is chosen so $2\pi/P$ equals the mismatch, $\Delta\beta$, in TE and TM phase constants. This periodic variation in the coupling term δ permits complete TE→TM conversion even though $\Delta\beta \neq 0$. The use of this modulator is restricted to $\lambda \gtrsim 1.15\,\mu m$ by losses.

It is interesting to note that transverse magnetic fields can be used to modulate TM modes in this waveguide. The effect is not due to the higher order Cotton-Mouton or Voigt effect but is rather due to the fact that the guide asymmetry causes a net circular polarization of electric field components normal to the modulating field.

The linear and quadratic electrooptic effects produce a field dependent change in refractive index leading naturally to phase modulation. Thus, amplitude modulation requires interference phenomena. The LEO exists only in crystals lacking a center of symmetry whereas all materials have a Kerr effect. In practice the Kerr effect would be used with a biasing field which removes the center of symmetry and gives a linear dependence on an added time varying field. The LEO in certain ferroelectrics seems to be best described in terms of such a biasing by the sponaneous polarization. Limiting discussion to the LEO, the effect of an applied modulating field on optical polarizability is due to a direct effect on the electronic polarizability plus a change in this polarizability which is brought about by lattice displacements within the unit cell. The first mechanism persists for frequencies into the optical range and is related to the optical nonlinear coefficient. The lattice term frequency limit is set by the optical phonon resonances ($\sim 10^{13}$ Hz), so the high frequency performance of the LEO is going to be determined by circuit problems. The transition from the optical nonlinear to LEO regions has been observed [F-1] experimentally. The results are well described by a phenomenological theory [G-1] which uses a simple anharmonic potential. Far from resonances one result of this theory is:

$$n^4 r = (\pi - n^2)(n^2-1)^2 \delta^C + (n^2-1)^3 \delta^D .$$

Here, r is a LEO coefficient, π is the dielectric constant, the δ's are frequency independent, the first term on the right depends on lattice motion and the second is solely electronic. This equation also shows why polarization rather than electric field is physically important (as was recognized by Pockels). Various viewpoints on the physical effects are well summarized by Kaminow [K-2]. On the low frequency side the LEO persists to dc. Acoustic resonances excited through the piezoelectric effect can cause problems unless the crystal is clamped.

Electrooptic coefficients are formally defined by the tensor relation

$$\Delta(1/n^2)_{ij} = r_{ij,k} E_k + s_{ij,kl} E_k E_l ,$$

where we shall focus on the LEO coefficient $r_{ij,k}$. One can adopt either of two viewpoints in discussing the LEO: first, one may use the traditional geometric construction of the indicatrix (or index ellipsoid) and find deformations caused by the LEO; second, one can find a polarization component induced through the LEO and treat it as a driving term. The second viewpoint is the natural one to use in coupled transmission line problems. In a principal axis system and with $\Delta n_{ij} \ll n_{ii}$ or n_{jj} (invariably the case), one has an induced polarization

$$P_i = -\varepsilon_o n_{ii}^2 n_{jj}^2 r_{ij,k} E_j E_k$$

with E_k the modulating field and the j and i fields at optical frequencies. Using the indicatrix construction we have

$$\sum_{i=1,3} \sum_{j=1,3} \left((1/n^2)_{ij} \, \delta_{ij} + r_{ij,k} E_k \right) x_i x_j = 1$$

in a principal axis system with subscipts enumerating the axes. It is customary to make use of the fact $r_{ij,k}$ is symmetric in ij to use a matrix notation where ij is contracted to a single index according to the scheme $11 \rightarrow 1$, $22 \rightarrow 2$, $33 \rightarrow 3$, $23 = 32 \rightarrow 4$, $13 = 31 \rightarrow 5$, $12 = 21 \rightarrow 6$. In general this leaves a 3 x 6 matrix with 18 independent elements, but crystal symmetry reduces this number substantially in most cases: e.g., 4 in $LiNbO_3$ and 1 in cubics.

The largest LEO effects are generally found in ferroelectrics such as $LiNbO_3$ and $LiTaO_3$ which are transparent in the range 0.4 to 5 μm. A comprehensive tabulation of coefficients is given elsewhere [K-1]. The size of the effect, which is presumed due to large K and asymmetry associated with the spontaneous polarization, has made these crystals important. One and two dimensional waveguides have also been constructed. Potassium dihydrogen phosphate (KDP) and its isomorphs were among the first materials used and remain important because of excellent optical quality, large size and a mode of operation using transmission on the optic axis. They transmit only to ~1.3 μm and are water soluble, however. Cubic crystals such as the III-V semiconductors are important because they transmit to over 10 μm (GaAs and CdTe, e.g.) and because they are optically isotropic, but most of all because they can be incorporated as part of an optical integrated circuit.

Consideration of particular modulator structures may help to show how the indicatrix is used. If a field is applied along x_3 (c-axis or polar axis) of $LiTaO_3$ we need only two of the four LEO coefficients and have

$$(1/n_1^2 + r_{13} E_3)(x_1^2 + x_2^2) + (1/n_3^2 + r_{33} E_3) x_3^2 = 1$$

and the crystal remains uniaxial since x_1, x_2 are interchangeable. If the wave propagates along x_2 (transverse operation) the refractive index for polarization along x_3 is found by setting $x_1 = x_2 = 0$ and so $x_3 = n_3 = 1/2\, n_3^3 r_{33} E_3$ is the new index. Similarly, for polarization along x_1, $x_1 = n_1 - 1/2\, n_1^3 r_{13} E_3$. The x_3 polarization is phase modulated with modulation index $2\pi \Delta n_3\, \ell/\lambda$. If the initial polarization is at 45° to x_3 and x_1, a retardation $(2\pi\ell/\lambda)(\Delta n_3 - \Delta n_1)$ of π can cause the polarization to be rotated 90° if the natural birefringence is suitably compensated. This is a common modulator configuration and values of $P/\Delta f = 1.1\, S^2$ mW/MHz with a 1.3 GHz bandwidth have been demonstrated [D-1]. The calculated performance has the form $P/\Delta f = S^2 4 \varepsilon_o K_3\, \lambda^3 10^9 / n^7 (r_{33} - r_{13})$ mW/MHz, where K_3 is the relative dielectric constant and we have taken $n_1 \approx n_3 \approx n$. The values of r and λ are in MKS units. For the tetragonal KDP type crystals with the field E_3 along the optic axis the indicatrix equation is

$$(1/n_1^2)(x_1^2 + x_2^2) + (1/n_3^2)x_3^2 + 2r_{63}E_3 x_1 x_2 = 1.$$

This can be diagonalized by rotating x_1, x_2 45° about the x_3 axis with a resultant retardation for a wave propagation along the optic axis becomes

$$(2\pi\ell/\lambda)(\Delta n_1^1 - \Delta n_2^1) \text{ where } \Delta n_2^1 = -\Delta n_1^1 = -n_1^3 r_{63} E_{3/2}.$$

When this becomes π we have a simple half wave plate. Since the applied field is along the direction of propagation (longitudinal operation) the cross sectional area can be made large to accomodate high power beam. A similar configuration can be used with cubic crystals where there is only one independent coefficient $r_{41} = r_{52} = r_{63}$. Cubic crystals can also be used for transverse operation with a maximum retardation of $(2\pi\ell/\lambda)(n^3 r_{41} E_1)$. Using cubics, or any crystal with a 3-fold axis, if light propagates along this axis a field in a plane normal to the axis causes a retardation $\sqrt{2/3}\, n^3 r_{41} E$ regardless of the direction of E in the plane. However, if E is rotated uniformly the birefringent axes rotate uniformly in the opposite sense. The resulting rotating wave plate can be used to shift frequency [K-1].

We will limit the discussion of electrooptic waveguide modulators to $LiNbO_3$ in order to keep this to a manageable length.

References to a number of papers on semiconductor waveguide modulators can be found in [H-1] and [T-1] as well as references to one dimensional guided wave structures on $LiTaO_3$ or $LiNbO_3$.

A phase modulator using a Ti diffused stripe (4.6x1 μm) on $LiNbO_3$ has been built [K-3] and shown to require only .0017 mW/MHz for a one radian modulation index with a 600 MHz bandwidth. Using somewhat different structures, two groups [U-1], [N-1] have demonstrated polarization rotation in a manner analogous to the bulk modulator described earlier. The same experiment on the Ti diffused stripe should require $P/\Delta f = (\pi r_{33}/(r_{33}-r_{13}))^2 \times .0017 \approx .03$ mW/MHz for extinction. A second way to use the phase modulator in conjunction with directional couplers is mentioned below.

We use the term directional coupler modulators to describe modulators involving the coupling of physically separated waveguides when they are brought close enough for significant overlap of their evanescent fields [M-1]. The coupling of energy from one guide to the other is well known to depend on the coupling coefficient K, the coupling length L and the difference $\Delta\beta$ in phase constants of the unperturbed guides. If R_o and S_o are the initial modal amplitudes on the two lines, the resultant amplitudes are given by

$$\begin{pmatrix} R \\ S \end{pmatrix} = \begin{pmatrix} A_1 & -jB_1 \\ -jB_1^+ & A_1^+ \end{pmatrix} \begin{pmatrix} R_o \\ \rho_o \end{pmatrix}.$$

Here,

$$A_1 = \cos L\sqrt{K^2+\Delta\beta^2/4} + j\,\frac{\Delta\beta \sin L\sqrt{K^2+\Delta\beta^2/4}}{2\sqrt{\Delta\beta^2/4+K^2}}$$

and $B_1 = K \sin L\sqrt{\Delta\beta^2/4+K^2} \Big/ \sqrt{K^2+\Delta\beta^2/4}$. If $R_o = 1$, $S_o = 0$ it is clear that when $L\sqrt{K^2+\Delta\beta^2/4}$ is a multiple of π all the power will be in the "R" guide whereas if $\Delta\beta = 0$ and LK is an odd multiple of $\pi/2$ the power will completely cross over to the "S" guide. Using Ti diffused in $LiNbO_3$ waveguides of about 2 μm in cross section and separated by 3 μm a length of 1 mm has been shown experimentally [P-2] to provide crossover. By inducing a $\Delta\beta$ via the linear electrooptic effect the return of the light to the R guide was also demostrated. The fabrication of guides with the precise value of LK for crossover is presently very difficult. An elegant solution known as alternating $\Delta\beta$ and requires basically only that LK $\geq \pi/2$ has been proposed [K-4] and demostrated [S-1]. When used as an amplitude modulation in a digital communication system only one output would be required to have 20 to 30 dB extinction so the real advantage of alternating $\Delta\beta$ may lie in its potential for greater optical bandwidth. An amplitude modulator using two directional couplers joined by a phase modulator has also been suggested [K-3]. Here, the directional couplers are required to divide power equally and this could be done by providing somewhat more coupling than needed and trimming $\Delta\beta$ electrooptically. This should require .017 mW/MHz for complete switching.

REFERENCES

+ Denotes reviews which themselves contain extensive references.

+C-1 F-S. Chen, "Modulators for Optical Communication", Proc. IEEE, Vol. 58, pp. 1440-1457, Oct., 1970.

D-1 R.T. Denton, F.S. Chen and A.A. Ballman, "Lithium Tantalate Light Modulators", J. Appl. Phys., Vol. 38, pp. 181-187, March 15, 1967.

F-1 W.L. Faust, C.H. Henry and R.H. Eick, "Dispersion in the Nonlinear Susceptibility of GaP Near the Reststrahl Band", Phys. Rev., Vol. 173, pp. 781-787, Sept., 1968.

+G-1 E.I. Gordon, "A Review of Acoustooptical Deflection and Modulation Devices", Proc. IEEE, Vol. 54, pp. 1391-1401, Oct., 1966.

H-1 J.M. Hammer, "Modulation and Switching of Light in Dielectric Waveguides", Integrated Optics, ed. T. Tamir, pp. 139-200, Springer Verlag, 1975.

K-1 I.P. Kaminow and E.H. Turner, "Electrooptic Light Modulators", Proc. IEEE, Vol. 54, pp. 1374-1390, Oct., 1966.

+K-2 I.P. Kaminow in An Introduction to Electrooptic Devices, pp. 56-70, Academic Press, 1974.

K-3 I.P. Kaminow, L.W. Stulz and E.H. Turner, "Efficient Strip-Waveguide Modulator", Appl. Phys. Lett., Vol. 27, pp. 555-557, Nov. 15, 1975.

K-4 H. Kogelnik and R.V. Schmidt, "Switched Directional Couplers with Alternating $\Delta\beta$", IEEE J. Quantum Electron.,

Vol. QE-12, pp. 396-401, July, 1976.

L-1 R. Lang and K. Kobayashi, "Suppression of the Relaxation Oscillation in the Modulated Output of Semiconductor Lasers", IEEE J. Quantum Electron., Vol. QE-12, pp. 194-199, March, 1976.

M-1 J. H. McFee, R. E. Nahory, M. A. Pollack and R. A. Logan, "Beam Deflection and Amplitude Modulation of 10.6 μm Guided Waves by Free-Carrier Injection in GaAs-AlGaAs Heterostructures", Appl. Phys. Lett., Vol. 23, pp. 571-573, Nov. 15, 1973.

M-2 E. A. J. Marcatili, "Dielectric Rectangular Waveguide and Directional Coupler for Integrated Optics", Bell Syst. Tech. J., Vol. 48, pp. 2071-2102, Sept. 1969.

N-1 J. Noda, N. Uchida, M. Minakata, T. Saku, S. Saito and Y. Ohmachi, "Electrooptic Intensity Modulation in $LiTaO_3$ Ridge Waveguide", Appl. Phys. Lett., Vol. 26, pp. 298-300, March 15, 1975.

P-1 D. A. Pinnow, "Guide Lines for the Selection of Acoustooptic Materials", IEEE J. Quantum Electron., Vol. QE-6, pp. 223-238, Apr., 1970.

P-2 M. Papuchon, Y. Combemale, X. Mathieu, D. B. Ostrowski, L. Reiber, A. M. Roy, B. Sejourne and M. Werner, "Electrically Switched Optical Directional Coupler: Cobra", Appl. Phys. Lett., Vol. 27, pp. 289-291, Sept. 1, 1975.

S-1 R. V. Schimdt and H. Kogelnik, "Electrooptically Switched Coupler with Stepped $\Delta\beta$ Reversal Using Ti-Diffused $LiNbO_3$ Waveguides", Appl. Phys. Lett., Vol. 28, pp. 503-508, May 1, 1976.

+T-1 P. K. Tien, "Integrated Optics and New Wave Phenomena in Optical Waveguides", Rev. Mod. Phys. (to be published).

+T-2 W. J. Tabor, "Magneto-optic Materials", Laser Handbook, ed. F. T. Arecchi and E. O. Schulz-DuBois, pp. 1009-1027, North Holland Publishing Co., 1972.

T-3 P. K. Tien, D. P. Schinke and S. L. Blank, "Magneto-optics and Motion of the Magnetization in a Film Waveguide Optical Switch", J. Appl. Phys., Vol. 45, pp. 3059-3068, July, 1974.

U-1 S. Uehara, K. Takamoto, S. Matsuo and Y. Yamauchi, "Optical Intensity Modulator with Three-Dimensional Waveguide", Appl. Phys. Lett., Vol. 26, pp. 296-298, March 15, 1975.

Coherant Optical Engineering, F.T. Arecchi and V. Degiorgio (eds.)

NEW WAVELENGTH STANDARDS WITH FREQUENCY STABILIZED GAS LASERS

Sergio Sartori
Istituto di Metrologia G.Colonnetti - CNR, Torino

1. INTRODUCTION

The unit of length, the metre, is at present defined in terms of the wavelength in vacuum of the $2p_{10} - 5d_5$ krypton 86 line.

The Engelhard type ^{86}Kr lamp is used for implementation of the metre and when used in accordance with the reccomandations of the Comité International des Poids et Mesures (CIPM) /1/ the total uncertainty of the emitted wavelength is $\pm\ 4\cdot 10^{-9}$; but some applications in the fields of metrology and geophysics require better uncertainty. For this reason, new frequency stabilization methods of gas lasers have been improved and recently they have made possible the realization of a wavelength standard superior to the existing one.

A wavelength standard must have some well defined characteristics: long term stability and reproducibility of wavelength should be obtained at the highest possible degree; it should be able to give interference fringes with an optical path difference as long as possible; it should have at least a power output in the 0,1 mW region, to allow accurate measurements of fringes and of fringe fraction; it should be easily constructed and should not dissipate power to any extent which would require substantial cooling.

Both ^{86}Kr source and gas lasers have the last two requirements to a good extent; for the second requirement, we must compare the coherence length l of the emitted line, with wavelength λ:

$$l = \lambda^2 / \Delta\lambda \tag{1}$$

where $\Delta\lambda$ is the line width.

^{86}Kr $\Delta\lambda$ is of the order of $5 \cdot 10^{-13}$ m, while with a gas laser it can be $2 \cdot 10^{-21}$ m. From this point of view gas lasers coherent sources are definitely better than the ^{86}Kr incoherent source. Wavelength long-term stability and reproducibility depend mainly on the physical properties of the radiation source. ^{86}Kr lamp works with spontaneous emission, consequently the emitted radiation wavelength depends on the position of the energy levels involved in emission and can change only if those levels are in some way perturbed. In a laser we have stimulated emission and we need a resonant system, namely a cavity. The cavity resonance frequencies $\nu_q = c/\lambda_q$ are given by

$$\nu_q = q \frac{c}{2nL} \tag{2}$$

Here q is an integer and gives the order of the resonance mode; n is the refractive index of the medium inside the cavity, L the mechanical cavity length and c the velocity of light.

If $\mathcal{L} = 2nL$ is defined as the optical length of the cavity, we can see from equation (2) that frequency stability is

$$\frac{d\nu_q}{\nu_q} = - \frac{d\mathcal{L}}{\mathcal{L}}$$

If we need to have a wavelength standard from a gas laser, to implement a new definition of the unit of length, we have:

(i) to select a single frequency among the several frequencies at which a laser may operate and which correspond to the cavity resonance modes with which the laser gain exceedes any losses present in the laser;

(ii) to minimize any changes in the optical length $\mathcal{L}$ and to maintain it constant by comparing the single frequency emitted by the laser with some stable reference and by correcting length for any deviation by means of an error-sensing mechanism;

(iii) to compare the single laser frequency locked to the reference frequency to that of the present standard of the metre, to ensure continuity in the unit definition.

The purpose of this lecture is to point out briefly the different possibilities for the choice of the stable reference, their characteristics and the techniques used to lock the frequency of the laser to them; finally I wish to give some recent experimental results.

2. THE LAMB DIP AS A REFERENCE FOR STABILIZATION

The decrease of output intensity of a single-mode oscillator when tuned to the centre of the line is the celebrated Lamb dip. It occurs because the gain saturates more rapidly at the center of the line as a result of interference of two travelling waves propagating in opposite directions along the cavity axis. The composition of the running waves gives the standing wave electric field existing inside the laser cavity.

If we have a single-mode laser with single isotope gas filling, the power output tuning curve shows at its centre a symmetric reduction, known as Lamb dip. In the visible helium-neon laser, the dip is about 200 MHz wide (15% of the gain bandwith at the half-power points) and is 5-10% deep.

We may use the Lamb dip centre as a frequency reference as it is more sharply defined than the peak of the gain curve, because the dip is narrower than the gain curve width.

To lock the laser to the Lamb dip central frequency, one end mirror of the cavity is vibrated to modulate the laser frequency over the power curve; the consequent variation in output intensity acts as error signal.

In 1966 and 1967 wavelength measurements were performed at the National Bureau of Standards (NBS), USA, at the National Physical Laboratory (NPL), U.K., and at the Physikalisch-Technische Bundesanstalt (PTB), Germany, on helium-neon lasers operating at 633 nm and stabilized on the Lamb dip /2/. The wavelength of the same laser (Spectra Physics model 119) was measured in these three laboratories; the agreement of the results was better than 5 parts in 10^9. The fact that this uncertainty is of the same order as the uncertainty of the ^{86}Kr wavelength demonstrates that the three independent techniques involved in this intercomparison were free from significant systematic errors; furthermore, all other wavelength measurements performed at the three laboratories are comparable. Many years of measurement practice confirm that different lasers from the same manufacturer or from different manufacturers emit wavelengths which may differ by about 1 part in 10^7; a single laser of this kind may present a significant decrease of emitted wavelength during its operating life, up to 1 part in 10^7.

In the past two years, some measurements were performed at the Istituto di Metrologia G.Colonnetti(IMGC) Italy, on two Spectra Physics lasers model 119 and on a Hewlett Packard model 5500 A ; variation in wavelength during the time and between the three lasers examined was of the same order as that reported during the international intercomparison.

We can now conclude that the helium-neon laser stabilized on the Lamb dip has not the fundamental quality to be used as a primary standard.

In 1969 E.J.G. Engelhard /3/, the father of the krypton lamp, showed the right metrological point of view. He emphazized that to establish a primary wavelength standard by a line produced by a wavelength stabilized laser, the new standard would be the reference line rather than the laser line itself. The atomic or molecular line used as a reference line must be produced under controllable conditions, especially under a controllable pressure. The internal conditions of a helium-neon laser tube are not controllable in a natural way; that is the reason for the poor reproducibility of lasers of the same kind. The gas pressure in a laser tube changes during the tube life time, and the natural emission lines (and the Lamb dip) are therefore shifted during laser operation.

3. SATURATED ABSORPTION FREQUENCY STABILIZATION

An absorption line - compared with an emission line produced, for instance, with gas discharge - has the important advantage as a reference line that it is not affected by the Stark effect.

Moreover, the pressure of a saturated molecular vapour is under controllable conditions, as it depends only on temperature, as is the case with the krypton lamp. Three systems have been mainly studied: the coincidence of the 633 nm He-Ne laser with the $^{127}I_2$ /4/ and $^{129}I_2$ /5/ molecular absorption spectra; of the 3,39 μm He-Ne laser with the CH_4 molecule /6/; of the 10,6 μm radiation from a CO_2 laser with the SF_6 molecule /7/.

I personally believe that at present the I_2 stabilized He-Ne 633 nm laser stands the best chance of metrological applications: it is in the visible; its cost is comparatively low; it is easy to set up and transport; it has been studied in many different laboratories and its characteristics are well known. For these reasons I shall illustrate only this laser, bearing in mind that the same problems are found with the other systems.

The laser I shall describe has been realized at IMGC /8/ (see figures 1 and 2); it is derived from a NPL design /9/. Inside the laser cavity, 40 cm long, is placed a tube containing a low-pressure $^{127}I_2$ vapour, controlled in temperature at 15°C. Seven absorption saturated lines having a width of about 5 MHz fall within the tunable range of the laser. Direct observation of the increase in laser power output corresponding to the dips produced by saturation of the absorption can be achieved in carefully controlled environments.

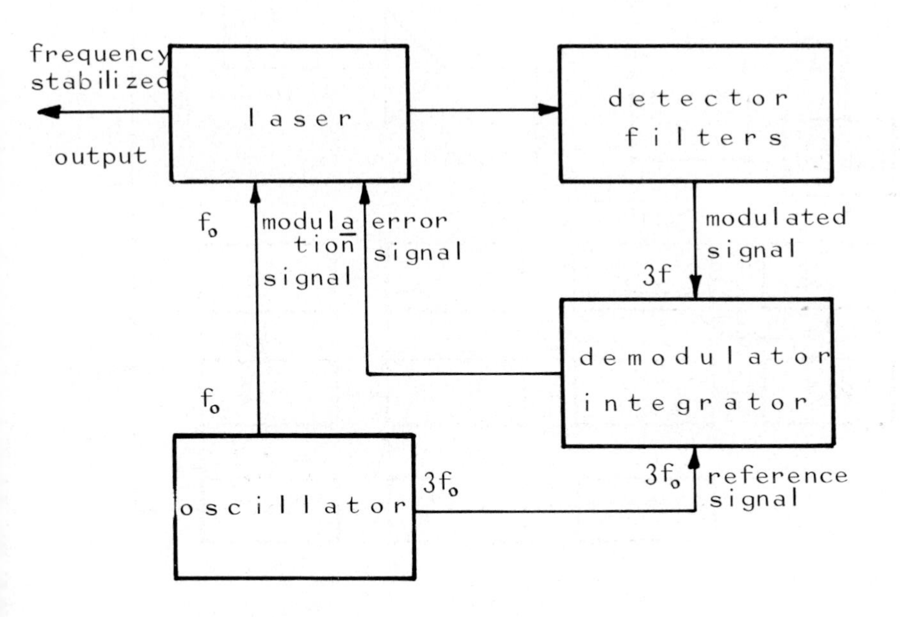

But phase sensitive detector (PSD) techniques may give better results. The length of the laser cavity must be modulated and the resulting modulation of power output is detected through a PSD as the laser is tuned over its operating range. At the oscilloscope there will appear a wide feature which reflects the curvature of the laser gain curve with the fine structure superimposed which is due to seven iodine lines falling within the tuning range.

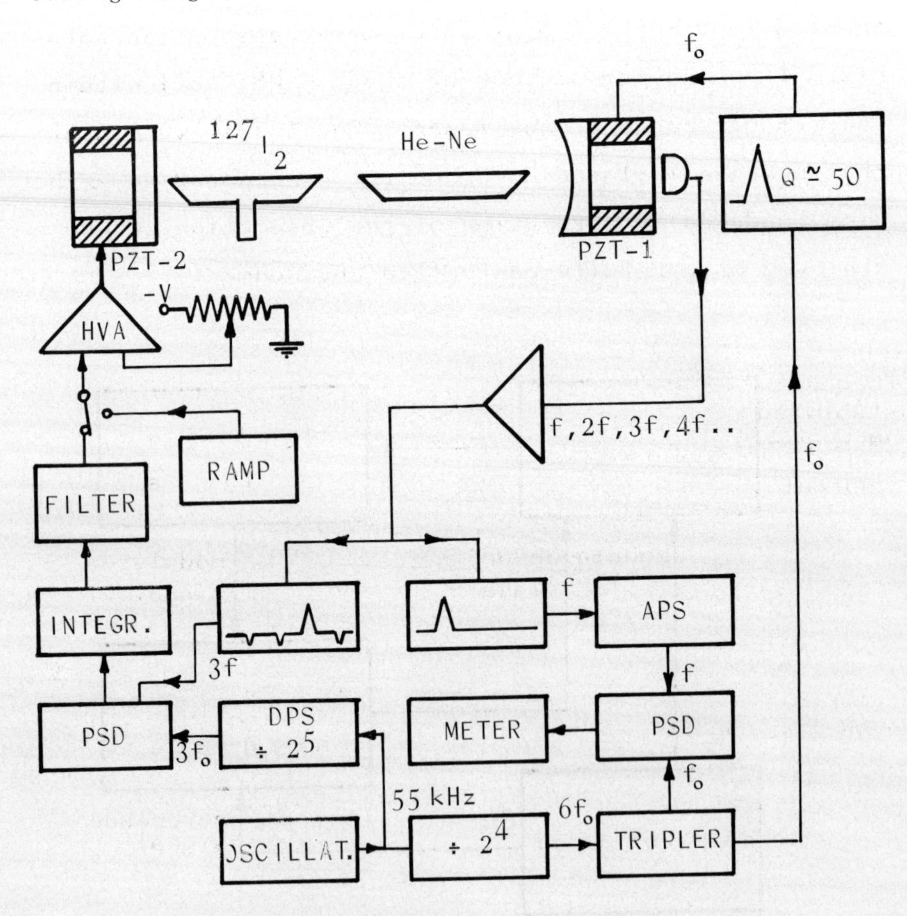

One of these line can be choosen as a reference line and the laser locked to it. But when this technique is used, it is difficult to identify the centre of each iodine line because the slope of the gain curve shifts the centre frequency. The problem can be overcome using the third harmonic technique, first proposed by A.J. Wallard /9/. The power output curve is differentiated three times, thereby ensuring the asymmetric discriminant shape necessary for stabilization and also eliminating the larger background slope. This is obtained by observing the third harmonic of the fundamental modulation frequency present in the laser output; it represents the rate of curvature change of the iodine peaks, so that, when it is used as an error signal, it stabilizes the laser frequency to the point of maximum rate of curvature change of the iodine absorption line. This point is practically unaffected by the general slowly varying shape of the curve of power output frequency.

Errors in stabilization on the centre of a saturated absorption line may occur for two reasons: third harmonic signal due to non linearity of circuit components (PZT, detector, amplifiers); residual signal with a frequency different from the third harmonic at the input of the PSD.

To discuss experimental results, we must first define what we mean as frequency stability. On this topic no really satisfactory agreement has been reached till now.

First of all, we must convince ourselves that the only parameter we can measure is the relative frequency stability between two similar lasers. We can collect information about frequency variations over short periods - up to 1 s - and over long periods - up to several hours. Frequency shifts of the beat frequency occuring to variations in the reference frequency environment of one laser (temperature, laser power, etc.) must be specified for an international standard. Frequency long-period stability of the beat frequency between two lasers realized in different countries, with different techniques, will virtually garantee the reference frequency stability.

A parameter, which has been used in the microwave frequency standard field for some years, seems to be very useful in defining frequency stability and getting comparable information of the type of noise that limits the performance of a laser; it is the Allan variance /10/.

The Allan variance is defined, in a beat experiment, as the averaged value of N pairs of readings of the beat frequency, when N is increased to infinite (see N very large):

$$\langle \sigma^2(2,\tau) \rangle = \frac{1}{N} \sum_{i=1}^{N} \left(\frac{f_{2i} - f_{(2i-1)}}{2} \right)^2$$

Here τ is the sample integration time, f is the beat measured frequency.

When plotting the logarithm of the Allan variance as a function of the integration time, one obtains information about flicker noise and white noise perturbing a laser over a particular time averaging period.

4. EXPERIMENTAL RESULTS ON IODINE STABILIZED LASERS

During the past month of March, at the Physikalisch-Technische Bundesanstalt (PTB) six He-^{20}Ne lasers stabilized on $^{127}I_2$ and one He-^{22}Ne laser stabilized on $^{129}I_2$ were compared together /11/. The lasers had been designed and constructed by the BIPM, the NPL and the PTB. The temporary results show reproducibility and long term stability of this type of lasers within $\pm$ 10 kHz ($\pm$ $2 \cdot 10^{-11}$).

At the end of June a He-^{20}Ne, $^{127}I_2$, laser from IMGC was compared in Paris with a similar laser from BIPM /12/.

The average difference in frequency between the two lasers, computed on six sets of measurements, is 2,3 kHz, with a 3,9 kHz standard deviation for each set. Variation of reference frequency due to the iodine temperature, and to the amplitude of the modulation were also measured.

Wavelength measurements were recently performed in many laboratories, by means of interferometric comparisons to the ^{86}Kr 606 nm primary standard line /13/. The satisfactory agreement in the results (within $3 \cdot 10^{-9}$, mainly due to the ^{86}Kr uncertainty) suggested these two resolutions to the 15th Conférence Générale des Poids et Mesures (CGPM 1975).

"1^ Resolution: taking into account that the radiations produced by the gas lasers stabilized on an absorption line are wavelength standards with accuracy and reproducibility better than those of the ^{86}Kr radiation defining the metre, the 15th CGPM nevertheless estimates premature to adopt a new definition of the metre; asks to the BIPM and to the National Laboratories to continue their researches on those radiations; asks to the CIPM to co-ordinate these researches.

2^ Resolution: taking into account the excellent agreement among the results of wavelength measurements on the radiations of lasers stabilized on a molecular absorption line in the visible or infrared spectrum, with an extimated uncertainty of $\pm 4 \cdot 10^{-9}$, corresponding to the metre implementation uncertainty, taking also into account the frequency measures of many of those radiations, the 15th CGPM recommends to use the value resulting for the velocity of light in vacuum: $c = 299\,792\,458$ $m\ s^{-1}$."

The recommended value for $^{127}I_2$, line i, radiation wavelength is 632 991,399 pm; and for CH_4, band ν_3, is 3 392 231 , 40 pm .

The value of the velocity of light has been mainly obtained from the experimental results of the NBS, Boulder /14/. The path to join the frequency standard to the wavelength standard is: caesium, klystrons, HCN laser, H_2O laser, CO_2 laser, CH_4 stabilized laser, I_2 stabilized laser, krypton; frequency against caesium and wavelength against krypton have been measured on the same CH_4 stabilized laser radiation.

We have now four values (λ ^{86}K, λ $^{127}I_2$, λ CH_4, c) and an agreement among them within $\pm$ $4 \cdot 10^{-9}$. To improve the mutual compatibility of standards, we have three possibilities.

First, we might maintain the present definition of the unit of length, but change the recommended standard; this will raise difficulties in the implementation of the future standard (i.e. the iodine laser) and astronomers and geophysic men will remain unsatisfied; they need not have to change the velocity of light each time a metrologist makes a new more accurate determination of it.

Second, we might change both definition and standard, going directly to the wavelength of a stabilized gas laser; but what laser will be the best? And the astronomers will still be unsatisfied.

Third, we might change length into distance, and define the metre as the distance travelled by an electromagnetic wave (with a defined wave front) in vacuum in a time of 1/299 792 458 s. Astronomers would appreciate this; but length measuring people would be less happy as they would have to start from the frequency of caesium to calibrate their end gauges.

Quite a number of papers have been written on this topic and many more will be in future. I may conclude this short review remembering that about two thousand year b.C. in China the unit of length was defined approximately in this way: it is the distance between two knots of a cane playing a definite note. The human history shows changement in technology, not in fundamental principles.

References

/1/ CIPM Procès Verbaux, october 1960

/2/ K.D.Mielenz, K.F.Nefflen, W.R.C.Rowley, D.C.Wilson, E.Engelhard - (1968) Reproducibility of Helium-Neon laser wavelength at 633 nm, Appl.Opt. 7, 289

/3/ E.J.G.Engelhard - (1969) The feasibility of establishing a new primary wavelength standard of length by a laser line, Electr.Techn. 2, 71

/4/ G.R.Hanes, C.E.Dahlstrom - (1969) Appl.Phys.Lett. 14, 362 - see also: G.R.Hanes, K.M.Baird - (1969) I_2 controlled He-Ne laser at 633 nm: preliminary wavelength, Metrologia 5, 32

/5/ J.D.Knox, Yoh-Han Pao - (1970) Appl.Phys.Lett. 16, 129

/6/ R.L.Barger, J.L.Hall - (1969) Phys.Rev.Lett. 22, 4

/7/ P.Rabinowitz, R.Keller, J.T.Latourrette - (1969) Appl.Phys.Lett. 14, 376

/8/ F.Bertinetto, B.I.Rebaglia, M.Liverani, S.Gualini (1975) Present development of iodine stabilized He-Ne lasers at the Istituto di Metrologia Gustavo Colonnetti, Alta Freq. 46, 569

/9/ A.J.Wallard - (1972) Frequency stabilization of the helium-neon laser by saturated absorption in iodine vapour, J.Phys.E 5, 926

/10/ D.W.Allan - (1966) Proc.Inst.Elect.Electron Engrs. 54, 221

/11/ J.M.Chartier - (1976) Internal report of BIPM.

/12/ J.M.Chartier, BIPM: private communication; F.Bertinetto, B.I.Rebaglia, IMGC: private communication.

/13/ W.G.Schweitzer, Jr., E.G.Kessler, Jr., R.D.Deslattes, H.P.Layer, J.R.Whetston - (1973) Description performance, and wavelength of iodine stabilized lasers, Appl.Opt. 12, 2927
see also: G.R.Hanes, K.M.Baird, J.De Remigis - (1973) Appl.Opt. 12, 1600
W.C.R.Rowley, A.J.Wallard - (1973) J.Phys.E 6, 647
Documents CCDM/75 - 5^ session, Paris 1973.

/14/ K.M.Evenson, J.S.Wells, F.R.Petersen, B.L.Danielson, G.W.Day, R.L.Barger, J.L.Hall - (1972) Phys.Rev.Letters 29, 1346.

Coherant Optical Engineering, F.T. Arecchi and V. Degiorgio (eds.)

DYE LASERS - A SURVEY

P. Burlamacchi
Laboratorio Elettronica Quantistica
Via Panciatichi 56/30, Firenze, Italy

1) Introduction

The present seminar is inteded to give a survey of the state of the art of dye lasers and a guideline to a selected literature.

For a general introduction to the physics and technology of dye lasers one may refer to well known books and review papers (1 - 3).

A host of efficient dyes are presently available, making dye lasers tunable over all the visible spectrum, and improved compunds can be foreseen due to intensive research in this field; however the main characterization of dye lasers does not lie in the type of active material but seems actually to be located in different classes of pumping methods.

Three main classes of dye lasers have been deeply studied and commercially expoited, namely:

a) Flashlamp pumped
b) Pulsed laser pumped
c) C.W. laser pumped

2) Flashlamp pumped dye lasers

Relatively high efficiency (4) (up to about 2%) and simple construction (5) are advantages of pulsed, flash pumped dye lasers.

High energy per pulse (400 j) (6,7) and high average power (~100 W) have been reported. Major problems to be solved are:

1) Thermally induced schlieren
2) Shock waves
3) Dye photodegradation
4) Flashlamp lifetime

- Thermally induced refraction gradients are the main drawback when very narrowband tuning has to be obtained, although in some cases this effect can be conveniently used to create tunable thermal waveguide lasers (9-10). Transverse rapid flow of the dye solution is an essential condition for high repetition rate of flash pulses but severe non uniformity of the pumping light may occur in this situation (11).

Shock waves are important effects in coaxial flashlamps or close coupled cavities, but can be virtually eliminated by using elliptical reflectors so that the flashlamp is sufficiently distant from the dye cell; (the shock wave travels at about 1 mm/μs). A solution of copper sulphate for flashlamp cooling acts as an U.V. absorbing filter and reduces heating of the dye prolonging the useful life of the dye solution by a factor of ten.

Photochemical degradation of the dye solution is a serious problem for high power dye lasers. This problem can be circumvented to some extent if one uses a sufficiently high ratio of reservoir volume to dye cell volume.Research on more stable dyes and additives to prevent degradation is in progress (12,13).

Flashlamp lifetime is probably the major problem for useful applications of dye lasers, expecially for industrial purposes, such as isotope separation and photochemistry. The best reported lifetime for commercial flashlamp is 10^6 shots at energy per shot below 10 J, using summering mode operation (5) i.e. by flowing a continous current of a few tens of mA through the flash electrodes. At higher energy and short pulses, the lifetime will be considerably lower. At the present time, a laser with output power higher than a few tens of watts, using commercially available flashlamps, would not survive one day.

Special flashlamps have been studied by groups of United Technology. Spark pumped dye lasers have been operated at very high repetition rate (14) and vortex stabilized flashlamps have been successfully operated at high average power (15) (16) (8). A very easy and convenient approach, expecially for laboratory use, is the use of ablating wall flashlamps which operate with low pressure air, by simply connecting a quartz tube to a vacuum pum (17).

Much effort has been devoted to the development of efficient high average power lasers. Up to now the use of flashlamp pumping seemed to be the only possible approach with the major problems of thermal effects, dye degradation, and lamps lifetime still to be solved. Very recently, the development of efficient, power scalable U.V. lasers, such as rare-gas halide T.E.A. lasers, might provide, in the near future, an ideal pumping source for production of high average power from tunable dye lasers (20).

3) Nitrogen Laser pumped dye laser

The nitrogen laser pumped dye laser offers, at the present time, the most versatile and rather simple way to generate coherent light over the broadest spectral interval, from the near U.V. to the near I.R.. Spectral width which can be achieved, is comparable with that obtained by flash pumping (~ 100 MHz); pulse duration is in the order of a few ns, repetition rates range from single pulse to 1,000 P.P.S. and peak powers of 100 kw or more are achieved. The dye solution does not need to be flowing at low repetition rate or can be magnetically stirred, greatly simplifying handling and replacing dyes for different spectral intervals. Thermal effects are not severe, expecially when the nitrogen laser emission is not exactly focused on the active medium.

The nitrogen laser owes its wide development to its use as a pumping source for dye lasers. Strictly speaking it is not a laser but a high gain amplifier which emits superradiant U.V. light in a narrow line at 3,371 Å, in a transverse discharge. Unfortunately the intrinsic characteristics of the transverse discharge in N_2 does not permit scaling the laser to large dimension and high energy per pulse (21). Recently compact atmospheric pressure nitrogen lasers have been reported, which can be used for simple inexpensive and reliable ecitation sources for very low power lasers (22, 23). The output of the trasverse discharge nitrogen laser has the form of a sheet of light which can be focused on a narrow strip 0,1 mm in width and a few cms in length on the dye cell. The gain achieved in this way is extremely high (30÷200 db/cm) (24) and the dye solution superradiates very easily, emitting a collimated beam of broad band radiation.

4) <u>C.W. Laser pumping</u>

The only continuos wave dye laser actually existing is operated with an argon or krypton laser beam, focused to a very narrow spot (Ø = 50 μm) which longitudinally pumps a narrow sheet of dye solution rapidly flowing in a transverse direction. The first C.W. dye laser was operated by Peterson (25) in a small cell with emispherical geometry, using Rhodamine 6G as active molecules. A more versatile and efficient configuration was developed by Dienes et al.(26).

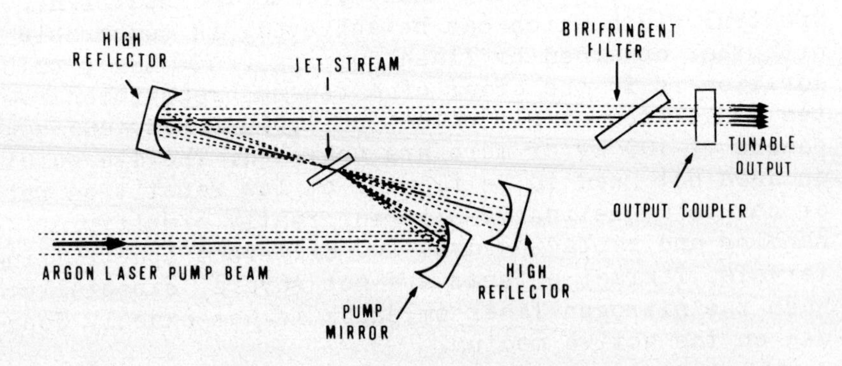

Fig. 1: Folded three mirrors astigmatically compensated cavity with non collinear pumping.

The actual arrangement for a C.W. dye laser, which has become generally accepted, is the non collinear pumping of a thin, optical quality laminar foil of dye emitted by a jet nozzle in a folded, three mirror astigmatically compensated cavity (27) (Fig.1).

A great number of dyes have given laser action, ranging from 400 nm to 800 nm pumped with various U.V. and visible lines. A list of about 200 different molecules have been given by Drexage (1) however new compounds are constantly coming into the market or tested in laboratoires.

5) Dye Laser amplifiers. Injection Locking

Spectral narrowing of dye laser amplifiers may be obtained relatively easily by injecting a narrowband radiation from a C.W. or low power pulsed oscillator into a high gain pulsed amplifier. In the past, the feasibility of such a laser scheme, was demonstrated by Erickson and Szabo (28). They injected the radiation from an argon laser into a dye cell pumped by a N_2 laser. The broad band emission from the pulsed oscillator was almost completely converted into the bandwidth of the oscillating C.W. line, giving a peak power of 4 KW, with 50 mw of injected power. Following this approach, Turner et al. (29) and Burlamacchi et al. (30), obtained tunable spectral narrowing in a flashlamp pumped amplifier with injection intensities of a few tens of mw, demonstrating the possibility of substantial average power gain by this method . Mode locking of high power laser oscillators, by injecting to a weak mode locked signal from a C.W. laser, was proposed and demonstrated by Moses et al. (31).

The method of injection locking appears to be of general interest because it combines the high performance of single mode low power highly stabilized narrowband lasers and the high peak and average power obtainable from pulsed lasers. For the dye laser the future of the method is linked to the effective reliability of high average power lasers, which is still a very urgent problem, as discussed in section 2).

6) Mode locking of dye lasers

Methods and technology of ultra short pulses production by mode locking of the dye laser oscillators, do not differ substantially from the methods used in gas and pulsed solid state lasers. The matter has been extensively reviewd in several articles (32-36). The main characteristics of dye laser in this case, is the broad bandwidth, which allows transform limited pulses of less than one picosecond in duration.

Flashpumped dye lasers can be easily mode locked using various saturable absorbers. The most widely used for Rh 6G is D O D C I (33 dietyloxadicarbocyanine) dissolved in ethanol (37).

C.W. dye lasers give the possibility of generation of a continuous train of wavelength tunable subpicosecond pulses, which allow the use of sophisticated and sensitive detection techniques, useful in spectroscopy and in metrology. Detailed methods for obtaining stable trains of pulses of a few tenths of a picosecond, have been described by Ruddock and Bradley (38) and by Ippen and Shank (39-40).

7) Tuning methods

Most of applications of dye lasers depend on their unique capability to be tuned continuously in wavelength over a wide range. Furthermore C.W. dye lasers, operated with a jet stream, appear to be the most suitable lasers for generation of highly stabilized narrow band light, perhaps setting new records for frequency and time standards. The dye laser head can in fact be constructed very compactly and the pumping system consists only of a narrow laser beam which introduces into the cavity a few watts of radiating power. The introduced heat is quickly removed by the jet stream so that the long term drift, due to thermal fluctuations, is strongly reduced.

Fluorescence of dye solutions is almost completely homogeneously broadened, so that, by simply using a frequency selective cavity for the laser oscillator, the emitted power can be concentrated in a single narrow line without substantial loss.

Coarse wavelenght selection of dye laser emission is possible by:

1) Changing dye molecule
2) Changing dye concentration
3) Changing dye cell length and almost any other parameter.

Fine tuning is obtained by using wavelength selective devices such as:

1) Prisms inside the cavity (41) or a grating in

place of one of the end mirrors (42)
2) Interferometric selectors such as Fox and Smith (43) (44), or Fabry Perot etalons inside the cavity (45)
3) Acusto-optical or electro-optical filters (46)
4) Distributed feedback (47)
5) Rotational dispersion (Birifringent tuning elements)
6) Pressure tuning (48)

The combination of gratings, prisms and Fabry Perot etalons of various free spectral ranges, and other tuning elements, is often used in order to obtain a narrow linewidth. The ultimate linewidth obtainable depends mostly on the type of laser. The worst situation appears to be for high energy flash pumped lasers where, due to thermal and acoustic effects, attainment of a bandwidth smaller than about 100 MHz might be an extremely difficult task, and operation might be very unreliable.

8) Tuning of N_2 pumped lasers

Tuning methods for pulsed N_2 pumped dye laser have been recently extensively reviewed by Wallenstein (49). For comprehensive literature on the subject one may refer to the references quoted by him.

Due to the very high single pass gain which is achieved easily in dye media pumped with very fast U.V. pulses, rather high losses can be tolerated in the laser cavity, allowing the use of highly dispersive gratings and high finesse Fabry Perot filters. The system, first developed by Hansch (50), and presently used in commercial lasers, is represented in Fig. 2.

The ultimate linewidth which can be achieved with the use of an expanding telescope and Fabry Perot etalon, is in the GHz range, corresponding to the inverse of the pulse length. A reduction to about 10 MHz can be achieved by the use of an external confocal resonator with a free spectral range of 2 GHz and a finesse of 200.

Although severe losses in power output are introdu-

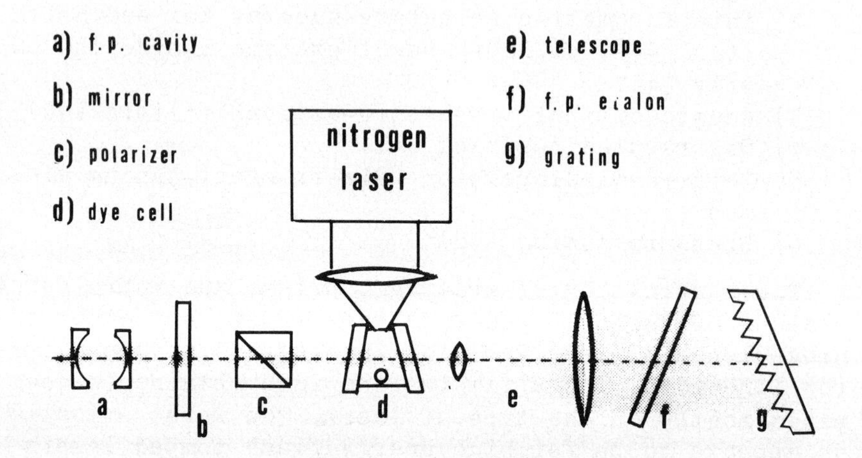

Fig. 2: Tuning components of a Nitrogen laser pumped dye laser (50).

ced by this arrangement, the technique remains very useful for high resolution spectroscopy.

9) C.W. dye laser tuning and stabilization

C.W. dye lasers certainly provide a superior monochromaticity and frequency stability as compared with pulsed lasers, although, for some applications, the high peak power and the short duration provided by pulsed lasers is more desirable.

C.W. laser systems with linewdth in the range of 10 MHz tunable from 500 to 700 nm, can be considered in a maturity stage of actual technology.

Tuning is generally accomplished by a prism or birifringent filter inside the cavity, and with the addition of low finess etalons. With this combination, when the tuning elements are properly matched, a single mode of the cavity can be selected.

Once the appropriate mode of oscillation of the dye laser cavity is achieved, fine tuning can be obtained by simply varying the cavity length, by means of a piezoelectric transducer. Servo systems are generally necessary to lock the resonances of the tuning elements to the oscillating mode. The natural istantaneous bandwidth of the single mode laser is in the range of a few KHz but free running jitter, even in the best evironment conditions, usually introduces a frequency spread up to 10 MHz.

Essential conditions to reduce jitter are:

a) prevent acoustic noise, vibration and thermal instability in the optical cavity;

b) the solvent must be kept at constant temperature;

c) air bubbles and particles must be removed by means of filters;

d) a gear pump for dye circulation is convenient together with a heavy surge tank to attenuate pressure fluctuations and vibrations from the pump.

To obtain the ultimate linewidth, frequency jitter must be eliminated by means of fast servo systems which lock the dye laser mode to any one of the resonances of a high finesse reference cavity, which usually lie ~300 MHz apart one from the other. Long term drift of the reference cavity may be virtually eliminated by locking a resonance to a hiperfine absorption line of an atomic transition by means of a fixed frequency laser. In order to obtain a continuously tunable radiation with controlled frequency a second dye laser is connected to the first one using frequency offset locking, which locks the frequency of the two lasers together with a preselected difference. The frequency offset must have a tuning range larger than the resonances of the reference cavity in order to garantee a complete frequency coverage. A schematic diagram of a possible system is presented in Fig. 3.

Following this approach, but using somewhat different laser and servo systems extremely high frequency stability, up to a few parts in 10^{-13}, with linewidth down to one KHz have been reported essentially by three groups: at M.I.T. (51) and the National Bureau of Standards (52) in the United States, and at the

University of Munchen, Garching in W.Germany (53).

In conclusion at the present time three main classes of dye lasers have obtained a deep attention in the research field and extensive commercial developement. The main characterisation resides in the different sources of pumping ligth: flashlamps, N_2 pulsed lasers and cw ion lasers. The possibility of acheaving high peak and average power, high efficiency, ultra-short pulses and stable, tunable, narrow bandwidth output have been discussed.

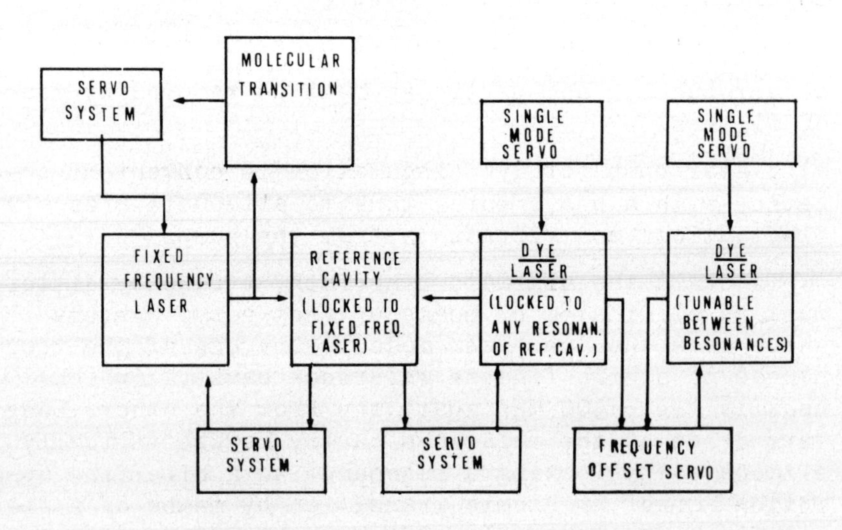

Fig. 3: Schematic diagram of a continuously tunable, stabilized dye laser.

References

1) Dye Lasers, Springer Verlag, Ed. F.P.Schäfer 1973

2) B.J.STEPANOV, A.N.RUBINOV, Sov. Phys. Uspeki 11 304 (1968)

3) B.B.SNAVELY, Proc. IEEE 57 1374 (1969)
C.W.SHANK, Rev. Mod. Phys. 47 649 (1975)

4) P.BURLAMACCHI, R.PRATESI, Appl. Phys. Lett. 23 475 (1973)

5) J.JETWA, F.P.SCHÄFER, Appl.Phys. 4 299 (1974)

6) V.A.ALKSEEV et al., Sov. J.Quantum Electr. 1 643 (1972)

7) F.N.BALTAKOV, B.A.BARIKHIN, L.V.SUKHANOV, JEPT Letters 19 174 (1974)

8) W.W.MOREY, W.H.GLENN, C.M.FERRAR, Technical Report R 75-921617 - 13 ARPA order n. 1806 AMEND n. 16 (1975)

9) P.BURLAMACCHI, R.PRATESI, L.RONCHI, Appl.Optics 14 79 (1975)

10) P.BURLAMACCHI, R.PRATESI, Appl. Phys. Lett. 28 124 (1976)

11) H.W.FRIEDMANN, R.G.MORTON, Appl. Opt. 15 1494 (1976)

12) D.BOSTING, D.OUW, F.P.SCHÄFER, Opt. Comm. in print.

13) F.P.SCHÄFER, Tunable Laser and Applications, Springer Verlag, Ed. by A.Mooradian, T.Jaeger and P. Stokseth 1976 p.50

14) C.M.FERRAR, Appl.Lett.Phys. 23 548 (1973)

15) M.E.MACK, Appl. Opt. 13 46 (1974)

16) W.W.MOREY, W.H.GLENN, Optica Acta, to be published (Nov. 1976)

17) C.M.FERRAR, Rev. Scient. Instr. 40 1436 (1969)

18) J.J.TURNER, E.I.MOSES, C.L.TANG, Appl. Phys. Lett. 27 440 (1975)

19) P.BURLAMACCHI, R.SALIMBENI, Opt. Comm. 17 6 (1976)

20) Laser Focus, The news in focus, p. 10 Dec. 1975

21) J.J.LEVATTER, SHAO-CHI LIN, Appl. Phys. Lett. 12 703 (1974)

22) V.HASSON, H.M.Von BERGMANN, D.PREUSSLER, Appl.Phys. Lett. 28 17 (1976)

23) H.STROHWALD, H.SALZMANN, Appl. Phys. Lett. 28 272 (1976)

24) T.W.HÄNSCH, F.VARSANYI, A.L.SCHAWLOW, Appl. Phys. Lett. 18 108 (1971)

25) P.G.PETERSON, S.A.TUCCIO, B.B.SNAVELY Appl. Phys. Lett. 17 245 (1970)

26) A.DIENES, E.P.IPPEN, C.V.SHANK, IEEE, Jour. of Q.E. pag. 338 March 1972

27) H.W.KOGELNIK, E.P.IPPEN, A.DIENES, C.V.SHANK, IEEE J. of Q.E., QE8 373 (1972)

28) L.E.ERICKSON, A.SZABO, Appl. Phys. Lett. 18 433 (1971)

29) J.J.TURNER, E.I.MOSES, C.L.TANG, Appl. Phys. Lett. 27 440 (1975)

30) P.BURLAMACCHI, R.SALIMEBENI, Opt. Comm. 17 6 (1976)

31) E.I.MOSES, J.J.TURNER, C.L.TANG, Appl. Phys. Lett. 28 258 (1976)

32) A.J.DE MARIA, W.H.GLENNjr., M.J.BRIENZA, M.E.MACK, P. IEEE 57 2 (1969)

33) P.W.SMITH, P.IEEE 58 1342 (1970)

34) A.DIENES, Opto Elect. 6 99 (1974)

35) D.J.BRADLEY, G.H.C.NEW, P.IEEE 62 313 (1974)

36) D.J.BRADLEY, Opto Elect. 6 25 (1974)

37) D.J.BRADLEY, F.O'NEILL, Opto Elect. 1 69 (1969)

38) I.S.RUDDOCK, D.J.BRADLEY, Appl. Phys. Lett. 29 296 (1976)

39) C.V.SHANK, E.I.IPPEN, Appl. Phys. Lett. 24 296 (1976)

40) E.P.IPPEN, C.V.SHANK, Appl. Phys. Lett. 27 488 (1975)
Laser Spectroscopy - Lecture Notes in Physics. Springer Verlag, P. 408 (1975)

41) F.P.SCHÄFER, H.MULLER, Opt. Comm. 2 407 (1971)
F.C.STROME, J.P.WEBB, Appl. Opt. 10 1348 (1971)

42) M.B.SOFFER, B.B.McFARLAND, Appl. Phys. Lett. 10 266 (1967)

43) P.W.SMITH, IEEE J. Q.E., QEQ 343 (1965)

44) R.E.GROVE, F.Y.WU, S.EZEKIEL, Opt. Engin. 13 531 (1974)

45) G.M.GALE, Optics Comm. 7 86 (1973)

46) D.J.TAYLOR, S.E.HARRIS, S.T.K.NIEH, T.W.HÄNSCH, Appl. Phys. Lett. 19 269 (1971)

47) Y.AOYAGI, T.AOYAGI, K.TOYADA, S.NAMBA, Appl. Phys. Lett. 27 687 (1975)

48) R.WALLENSTEIN, T.W.HÄNSCH, Appl. Opt. 13 1625 (1974)

49) R.WALLENSTEIN, Optica Acta, to be published Nov.76

50) T.W.HÄNSCH, Appl. Opt. 11 895 (1972)

51) R.E.GROVE, F.Y.WU, L.A.HACKEL, D.G.YOUMANS, S.EZE KIEL, Appl. Phys. Lett. 23 442 (1973)

F.Y.WU, R.E.GROVE, S.EZEKIEL, Appl. Phys. Lett. 25 73 (1974)

R.E.GROVE, F.Y.WU, S.E.EZEKIEL, Optical Engin. 13 531 (1974)

52) R.L.BARGER, J.H.HALL, Appl.Phys. Lett. 22 196 (1973)

R.L.BARGER, M.S.SOREM, J.L.HALL, Appl. Phys. Lett. (1973)

R.L.BARGER, J.B.WEST, T.C.ENGLISH, Appl. Phys. Lett 27 31 (1975)

J.L.HALL, S.A.LEE, Tunable Laser and Applications edited by A.Mooradian and al.n Springer Verlag (1976)

53) W.HARTIG, H.WALTHER, Appl.Phys. 1 171 (1973)

H.WALTHER, Physica Scripta 9 297 (1974)

M.STEINER, H.WALTHER, C.ZYGAN, IX International Conference on Quantum Electronics - paper A3 - Optics Comm. 18 July 1976.

Coherant Optical Engineering, F.T. Arecchi and V. Degiorgio (eds.)

SPECKLE INTERFEROMETRY

by A E Ennos

Division of Mechanical and Optical Metrology
National Physical Laboratory Teddington Middx England

I Laser Speckle Characteristics

Laser speckle is a phenomenon associated with the scattering of coherent light by a rough surface, ie one whose local irregularities in depth are greater than one quarter of a wavelength. The interference of the individual waves scattered over a wide range of angles is responsible for the bright and dark speckles, and thus the speckle field is not localised in space, but fills the complete volume occupied by the interfering waves. The detailed nature of a speckle pattern depends upon the angular extent over which scattered waves are received in the region under consideration. It is convenient (but not fundamental) to divide speckle patterns into two types:-

Fig 1a

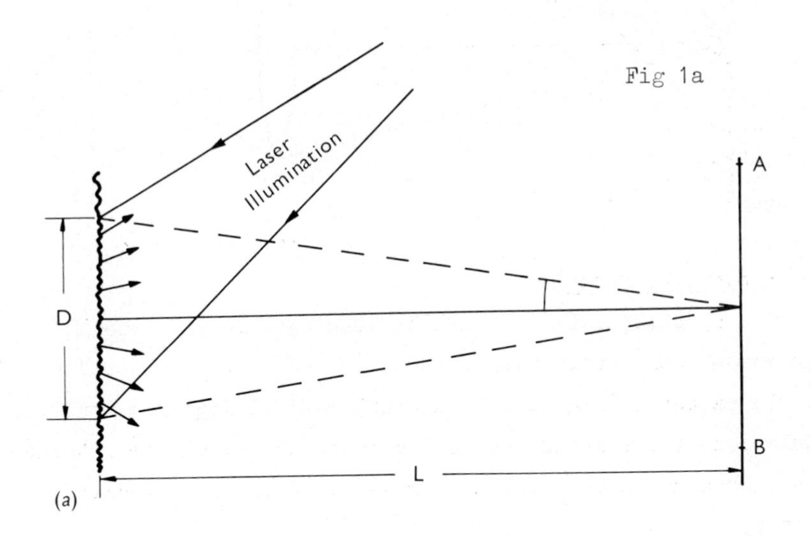

1) Objective Speckle

An objective speckle pattern is formed in the whole of the space in front of a rough surface illuminated by coherent light. A flat screen held at some distance from the scattering surface will have a section of the speckle pattern projected on to it, (Fig 1a). The average size of the speckle will depend on the angle subtended by the scattering area at the screen. For a circular illuminated patch of diameter D, the speckle diameter σ_o formed on a screen distant L, by light of wavelength λ, will be given by

$$\sigma_o \approx \frac{1.2 \lambda L}{D} \qquad \ldots (1)$$

Fig 1b

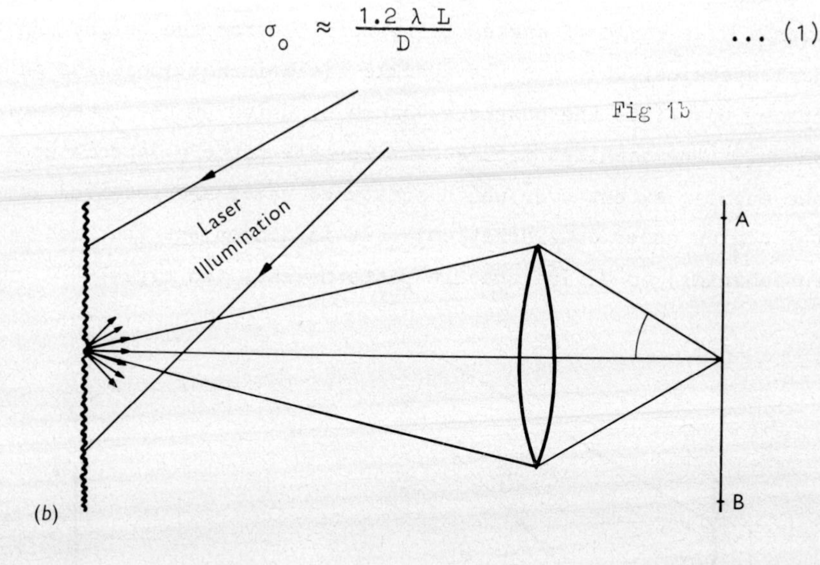

2) Subjective Speckle

The speckle pattern formed by imaging a rough surface, illuminated coherently with a lens, is termed the subjective speckle pattern, (Fig 1b). The pattern is a high-frequency modulation of the intensity of the image that would be obtained with incoherent light. The average diameter of a speckle is given by

$$\sigma_s \approx \frac{0.6 \lambda}{N A} = 1.2 (1 + M)\lambda F \qquad \ldots (2)$$

where N A is the numerical aperture of the image-forming lens, F its aperture ratio (f/number) and M the magnification of the image. The subjective speckle is the equivalent of the objective speckle that would be generated by the lens pupil considered as a scattering surface. A concept useful when considering interference of speckle patterns is to think of the subjective speckle pattern as being the image of a similar pattern, of different scale, lying in the object plane. The average diameter of these speckles is given by

$$\Sigma \approx 1.2\,(1 + M)\lambda F/M \qquad \ldots (3)$$

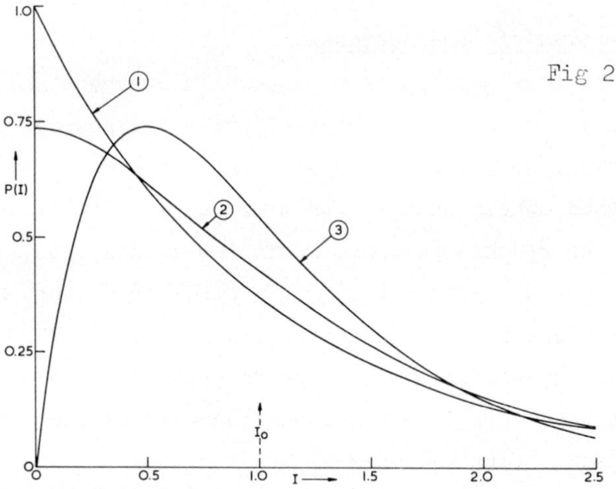

Fig 2

The brightness distribution of a speckle pattern depends upon whether it is fully-developed or not. Fully developed patterns are only obtained when all the waves contributing to a particular speckle are coherent with one another. This implies that the scatterer must not depolarise the light. The probability density function of the brightness distribution of fully-developed speckles is given in Fig 2, curve (1). The equation of this curve is

$$p(I) = \frac{1}{I_o} \exp\left(-\frac{I}{I_o}\right) \qquad \ldots (4)$$

The most probable brightness is zero, ie there are more completely dark speckles than those of any other brightness.

A speckle pattern derived from a depolarising scatterer can be considered to be the incoherent superposition of two independent speckle patterns, mutually polarised at right angles to one another. The probability density function of such a pattern is given by curve (3), Fig 2, with equation

$$p(I) = 4\,(I/I_o^{\,2})\,\exp\,(-2\,I/I_o) \qquad \ldots (5)$$

The most probable brightness is now no longer zero, but about one half the average brightness.

II Interference of Laser Speckle

The light forming a speckle pattern is coherent and is thus capable of interference. In general, when the sources of two speckle patterns do not have the same angular size or are not axially aligned, their interaction produces a modification to the size of the speckle pattern, or in the second case, a modulation of brightness within individual speckles. The interference conditions of most practical interest are

(i) the superposition of a speckle pattern with a uniform wave directed along the axis of the speckle-forming waves

(ii) the superposition of two speckle patterns derived from coaxial scatterers of the same angular aperture.

(i) (a) Coherent combination of speckle and uniform fields. The size and brightness distribution of the original pattern are both changed; size is approximately doubled, due to halving of the maximum angle between interfering rays; brightness distribution is changed depending upon the relative brightnesses of the speckle and uniform field. For equal average brightness, (Burch [1970])

$$p(I) = \frac{2}{I_o}\exp\left[-\left(1+\frac{2I}{I_o}\right)\right]\mathcal{J}_o\cdot 2\sqrt{2I/I_o} \qquad \ldots (6)$$

Curve (2) Fig 2 shows this function. It is not greatly different from curve (1) in that the most probable brightness of a speckle is zero.

(b) Incoherent combination of speckle and uniform field. In this case the size of the resultant speckles remains the same, but their contrast is strongly reduced; there will be no speckles of a brightness less than that of the uniform field.

(ii) (a) Coherent combination of two speckle fields. A third speckle field, differing in detail from either of its components, will result. The average size of a speckle and the brightness distribution of the speckles will be the same as that of a single speckle field, ie it will have a negative exponential probability density curve.

(b) Incoherent combination of two speckle fields. Superimposing two speckle fields incoherently produces a third speckle field, with similar average speckle size to the components, but with a modified speckle brightness distribution (Fig 2, curve (3)).

III Visual Speckle Interferometry

For visual speckle interferometry, interference of type (1) above is employed (Archbold, Ennos and Taylor [1970]); a subjective speckle pattern, derived from the rough surface of the object to be studied, is formed on the retina of the eye by a lens system, and a coherent 'reference' beam is directed along the same speckle-imaging axes. The speckle size is made large by reducing the aperture of the lens system. Depending upon the relative phase of the reference beam and a particular speckle, the resultant brightness will take on a well-defined value. If the surface giving rise to the speckle pattern moves towards or away from the observer, the brightness of the combination will vary cyclically from bright to dark and back again with every phase increment of π radians, or for a surface displacement in the line of sight of $\frac{\lambda}{2}$. Each speckle in the field undergoes this cyclic brightness variation independently, with random phase, since the phase of the light from the original speckles (without reference beam) varies randomly from speckle to speckle.

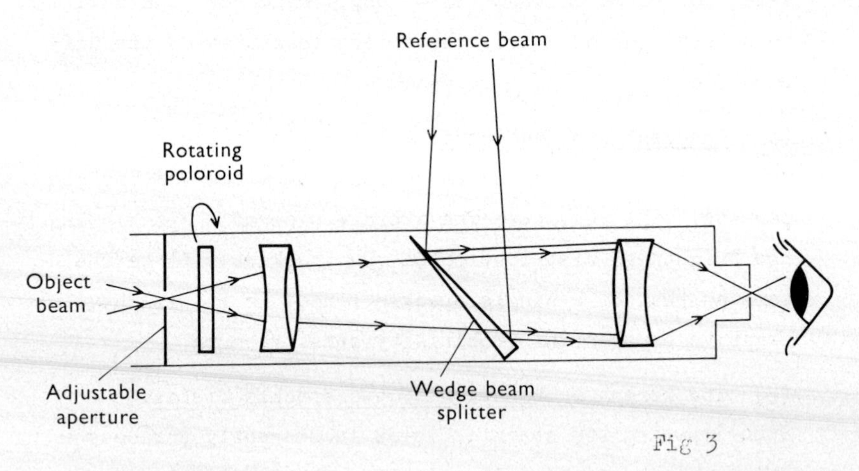

Fig 3

Figure 3 shows one form of visual speckle interferometer (Stetson [1970]). The laser-illuminated object is viewed through a telescope having a small entrance pupil (an iris diaphragm). Between the lenses of the telescope is mounted a wedged beam-splitter, by means of which a reference beam, observed from the same laser source, is introduced. To obtain the correct conditions for speckle interference, the reference beam source must appear to lie in the centre of the entrance pupil of the telescope. An aperture at the exit pupil intercepts one of the beams reflected by the wedged beam-splitter.

The visual speckle interferometer may be used for detecting very slow object drift or slight atmospheric turbulence; the speckles then appear to 'twinkle' over the area of field affected. It has greater use, however, in the direct determination of surface vibration modes. In places where a surface is vibrating in the line of sight by $\frac{\lambda}{4}$ or greater, the high contrast speckle will appear blurred out, but it will remain clear in the nodal regions. The interferometer is commonly used in conjunction with time-average holography to find the resonant frequencies of a vibrating body, and to decide upon the best means of its excitation.

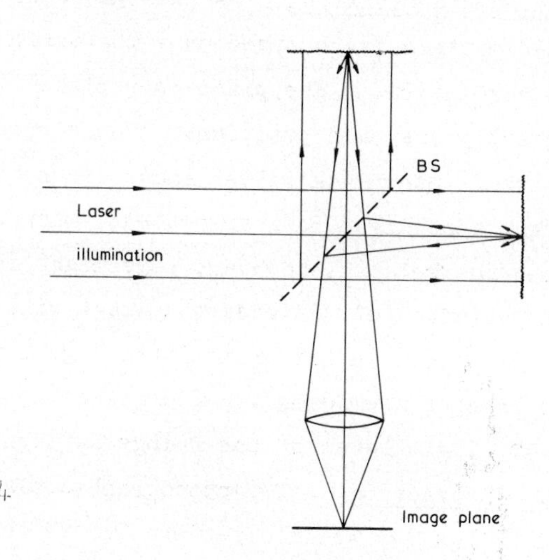

Fig 4

IV Interferometry based upon Speckle Correlation

Consider the interference of two speckle fields (case II (ii)). This may be visualised as occurring in a Michelson interferometer where the two mirrors are replaced by scattering surfaces (Fig 4). In the image plane two subjective speckle patterns $F_1(x,y)$ and $F_2(x,y)$ are generated by the two scattering surfaces. These combine coherently to form a third speckle pattern $F_3(x,y)$. If one of the scatterers moves in the direction of its normal, the combined speckle field will change from $F(x,y)_o$ to $F(x,y)_\delta$, where δ is the phase change in all parts of the field due to the displacement. When $\delta = 2n\pi$, every speckle will have returned to its original brightness, and the pattern as a whole will have become re-correlated with itself. When $\delta = (2n + 1)\pi$, the pattern will have maximum de-correlation with itself. If means can be found of distinguishing between correlated and uncorrelated patterns, the interferometer system can be used for measurement of displacement (or phase change). There are a number of different methods for performing the correlation process, (Leendertz [1970], Archbold, Burch and Ennos [1970]).

(i) Photographic Mask Correlation. A photograph of the combined speckle pattern is recorded on a photographic plate of adequate resolution. The plate is replaced in its holder in exactly the same position. Since bright speckles now fall on their own black silver images when the speckle is correlated, zero light is transmitted; when the speckle patterns are uncorrelated, however, some light is transmitted. A variation of δ over the image will then result in a variation in the transmitted light, and speckle correlation fringes are formed. The method is inefficient, due to the low light level of the fringe pattern.

(ii) Photographic Subtraction. Two photographic plates A and B are exposed to the two speckle patterns to be correlated. A positive contact print of plate B is made on another plate C. A and C are then placed in register and the resulting fringe pattern viewed by transmission through the pair.

(iii) Double Exposure with Non-linear Photographic Recording. Consider one speckle, formed by the superposition of two component speckles having amplitudes a_1 and a_2 and phase difference ε.

$$\text{Intensity } I = a_1^2 + a_2^2 + 2a_1 a_2 \cos \varepsilon \qquad \ldots (6)$$

Two recordings in which ε remains the same or changes by $2n\pi$ will result in a total intensity

$$I_1 = 2(a_1^2 + a_2^2 + 2a_1 a_2 \cos \varepsilon) \qquad \ldots (7)$$

If the phase changes by $(2n + 1)\pi$ between the two exposures, the total intensity will be

$$I_2 = a_1^2 + a_2^2 + 2a_1a_2\cos\varepsilon + a_1^2 + a_2^2 - 2a_1a_2\cos[(2n+1)\pi + \varepsilon]$$

or
$$I_2 = 2(a_1^2 + a_2^2) \qquad \ldots (8)$$

The second case is equivalent to the addition of two uncorrelated speckle patterns. Making use of the different brightness distribution characteristics of speckle patterns (Fig 2), a differential blackening of a photographic emulsion can be obtained for the two cases if a heavy exposure is given, due to the non-linear nature of the emulsion sensitivity characteristic.

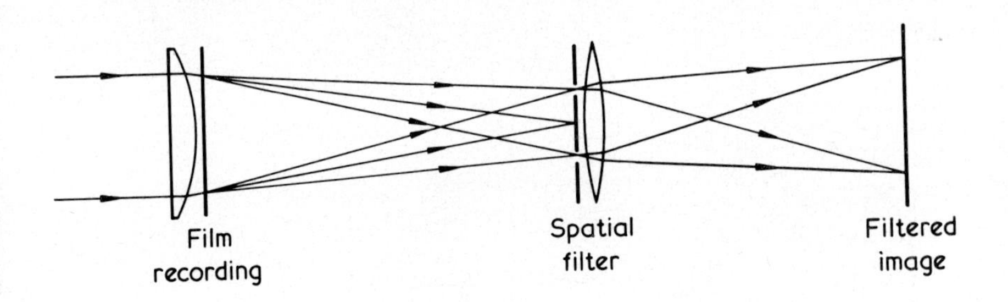

Fig 5

(iv) Double Exposure with Image Displacement followed by Spatial Filtering. A double-exposure is recorded with a small lateral displacement of the image between recordings. The negative, when placed in a diffractometer, will generate parallel equi-spaced fringes (Young's fringes) if the displaced speckle patterns are correlated (See Section VII); otherwise a diffraction halo only is generated. The correlated and uncorrelated areas are differentiated by spatial filtering (Fig 5). The double slot aperture allows through only the $\pm \frac{1}{2}$ order Young's fringe. High contrast fringes can be obtained in this way.

V Special Purpose Speckle Interferometers

A Michelson-type interferometer with scattering surfaces instead of mirrors (Fig 4) may be used, with suitable means of speckle correlation, for measuring the component of displacement of one of the scatters in the line of sight direction (in a similar manner to a conventional interferometer). This gives contours related to the change in height profile of the surface. The quality of the speckle correlation fringes is improved if only one of the mirrors is replaced by a scatterer; the 'reference' beam then becomes a uniform plane wave.

Using different configurations of speckle interferometer, components of displacement other than the normal direction can be measured.

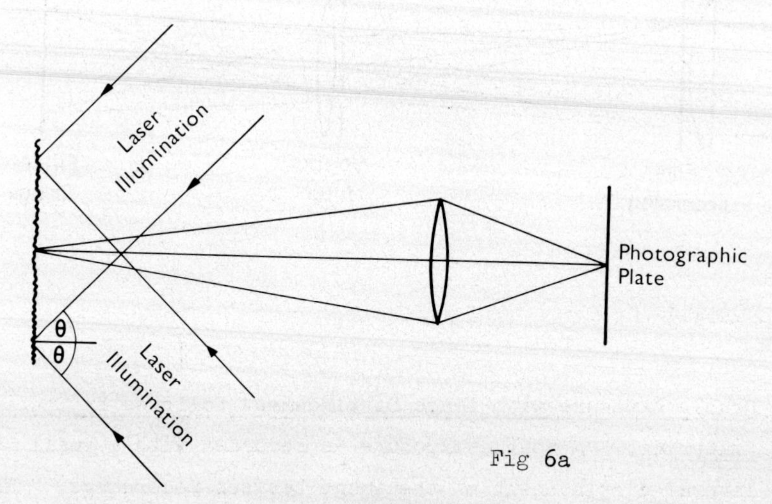

Fig 6a

(i) In-plane Displacement Interferometers. By illuminating a surface with two beams directed on to it obliquely at equal angles to the normal, an interferometer sensitive only to the in-plane component of displacement is obtained (Fig 6a), (Leendertz [1970]). The two speckle patterns formed by each illuminating beam acting alone can be thought of as the two interference fields. For an out of plane displacement d_z, both beams suffer a path difference

$d_z(1 + \cos\theta)$. For an in-plane displacement d_x, the path difference is $2d_x \sin\theta$. Thus re-correlation of the speckle pattern in the image plane occurs whenever

$$2d_x \sin\theta = n\lambda \qquad \dots (9)$$

By using a suitable method of correlation, a fringe pattern relating to the in-plane motion, with fringe spacing $\frac{\lambda}{2 \sin\theta}$, is obtained

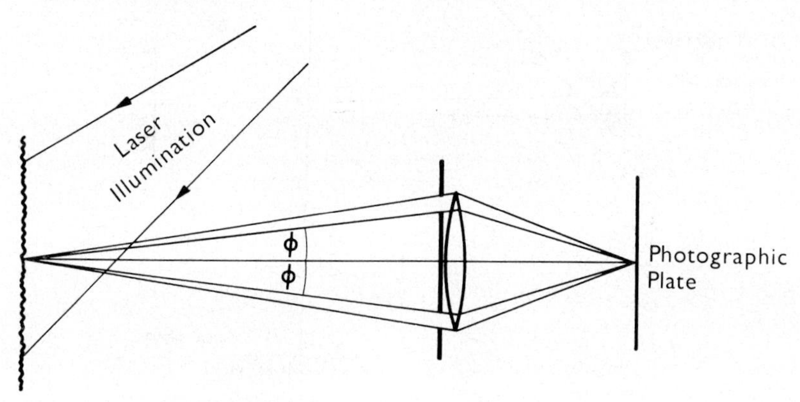

Fig 6b

An alternative form of interferometer uses single-beam illumination, but splits the imaging cone of rays into two parts by means of a double slot aperture in the imaging lens (Fig 6b), (Duffy [1972]). Each 'speckle' of the image is similar to that which would be formed by a single slot, but its brightness is modulated by a grating-like structure generated by the two interfering beams. Again a z - translation of the object causes no change, since the path lengths of the two beams are affected equally; a lateral translation will however alter the phase of the elementary gratings. For a double exposure, the gratings will reinforce one another whenever

$$2d_x \sin\phi = n\lambda \qquad \dots (10)$$

but for half integral values of wavelength they will cancel each other out. This technique is less sensitive than the double illumination method, since $\phi << \theta$. However, spacial filtering is carried out in a simple manner by viewing the negative in the direction of the strongest first-order diffracted light.

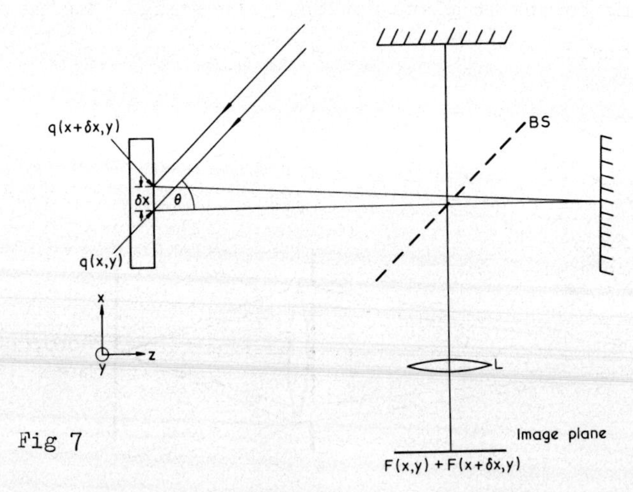

Fig 7

(ii) Speckle Shearing Interferometers. Shearing interferometers in conventional optics are used for measuring the slope of a wavefront. In speckle interferometry they can be used for measuring the change in slope (due, for example, to deformation) of a scattering surface. Shearing may be effected using a Michelson interferometer-type arrangement by tilting one of its mirrors (Fig 7) (Leendertz and Butters[1973]). For a point on the surface deforming by (u, v, w) in the three orthogonal directions, and for image shear δx, the speckle at the image will suffer a phase change of Δ given by

$$\Delta = \frac{2}{\lambda}\left[\sin\theta \frac{\partial u}{\partial x} + (1 + \cos\theta)\frac{\partial w}{\partial x}\right] \cdot \delta x \qquad \ldots (11)$$

For $\theta = 0$, the fringes obtained by correlating the speckle patterns will give contours of $\frac{\partial w}{\partial x}$, the slope change.

Alternative methods of effecting the shear use an optical system based upon the twin-slot aperture recording system of Fig 6(b). To effect the shear, either (a) optical micrometer plates are mounted in front of the two slots (Hung and Taylor [1973]), or (b) the image is recorded out of focus (Hung, Rowlands and Daniel [1975]).

(iii) Contouring by Speckle Interferometry. In an analogous way to holographic contouring, a speckle interferometer can be used for generating contours over a three-dimensional surface by changing the laser wavelength between the two recordings (Butters and Leendertz [1974]). For wavelengths λ_1 and λ_2 the contour interval is $\Delta h = \frac{\lambda_1 \lambda_2}{2(\lambda_1 \sim \lambda_2)}$; eg, for argon wavelengths 488.0 nm and 486.5 nm, $\Delta h = 14.2\ \mu m$.

Comparison of two complicated shapes (eg, turbine blades) has been achieved using speckle interferometry contouring (Denby, Quintanilla and Butters [1976]). However, to obtain good contrast fringes, it is necessary for the master shape to be specularly reflecting, and to be illuminated by a wave giving rise to single direction scattering; a hologram illuminator is used for this.

VI Electronic Speckle Interferometry (ESPI)

Speckle interferometry can be carried out with a detector having low spatial resolution, since speckle size can be matched to it by reducing the aperture of the imaging lens (Butters and Leendertz [1971]). Television (TV) techniques can thus be employed, if a vidicon camera is used with a lens of aperture f/11 or smaller. The advantages of such a system are

1) A real-time display on a bright, large-size monitor is obtained.
2) No lengthy photographic processing is required.
3) The TV signal from the vidicon camera can be electronically processed and stored if necessary.

4) Speckle correlation can be performed in real time by combining image storage and electronic image subtraction.

The disadvantages of ESPI are

1) Expense and complexity of the system.
2) The high light levels required when using a low aperture system.

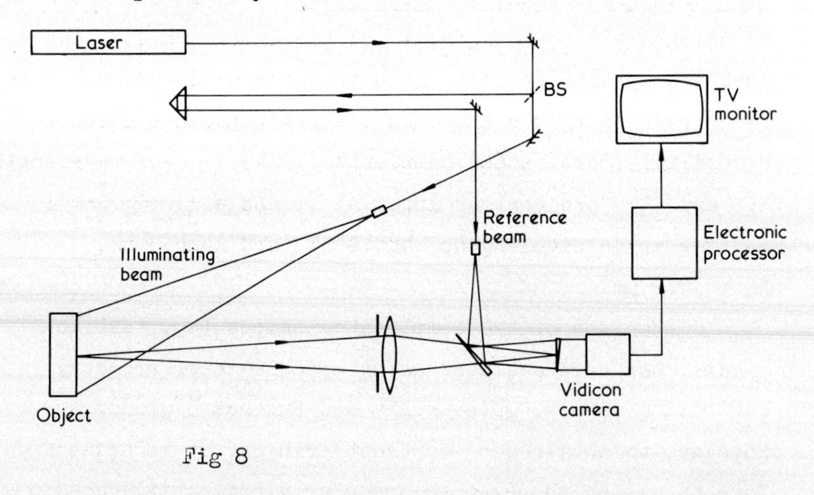

Fig 8

A system for ESPI is shown in Fig 8. The interferometer uses a uniform reference wave as in the visual speckle instrument. For vibration analysis, the signal is amplified and passed through a band-pass filter tuned to the speckle frequency; it is then full-wave rectified and displayed on the monitor. Not only are nodal areas shown up, but also fringes delineating the maxima of the $J_o^2\left(\frac{4\ \pi\ d_z}{\lambda}\right)$ function (where d_z is the vibration amplitude), as for time-averaged hologram interferometry. Transient vibration patterns may be viewed directly.

The ESPI is used for displacement measurement by storing successive frames of the TV picture of the undistorted object on magnetic tape or magnetic disc. This signal is subtracted synchronously from the signals from the vidicon camera obtained when the object is strained. The resultant fringe patterns on the monitor are equivalent to those of real-time holographic interferometry.

VII <u>Laser Speckle Photography</u>

The operation of a speckle interferometer requires that the scattering surface should not move laterally by more than one speckle diameter Σ. If in fact it does move more than Σ, the lateral displacement of a localised area can be measured by positional correlation of the pattern at that point, without using a reference beam (Archbold and Ennos [1972]). A double exposure photograph of the laser-illuminated surface is recorded on fine-grain film with a lens at moderate aperture (eg, f/4), before and after lateral movement of the object. The intensity recorded is

$$F(x,y) + F(x + d_x, y) \qquad \dots (12)$$

where d_x is the image movement. This equation can be written

$$F \otimes \left[\delta(x,y) + \delta(x + d_x, y)\right] \qquad \dots (13)$$

where $\otimes$ represents a convolution and $\delta(x,y)$ is the delta function at (x,y).

If the photographic film is processed to operate on the linear portion of the t-E curve, the amplitude transmittance t is given by

$$t = a - b\{F \otimes \left[\delta(x,y) + \delta(x + d_x, y)\right]\} \qquad \dots (14)$$

where a and b are constants. The film is then analysed by observing its diffraction spectrum in the back focal plane of a lens. The light distribution is given by the Fourier transform of t.

$$U(u,v) = a\,\delta(u,v) - b\,\{\tilde{F} \cdot \left[1 + \exp(ikvd_x)\right]\} \qquad \dots (15)$$

where $k = \frac{2\pi}{\lambda}$ and u, v are angular coordinates. The first term is the directly transmitted light. The intensity of the second term is

$$I \propto |\tilde{F}|^2 \cos^2 \left(\frac{kvd_x}{2}\right) \qquad \ldots (16)$$

$|\tilde{F}|^2$ represents the diffraction halo due to the speckle pattern, and the $\cos^2$ term its modulation (Young's fringes).

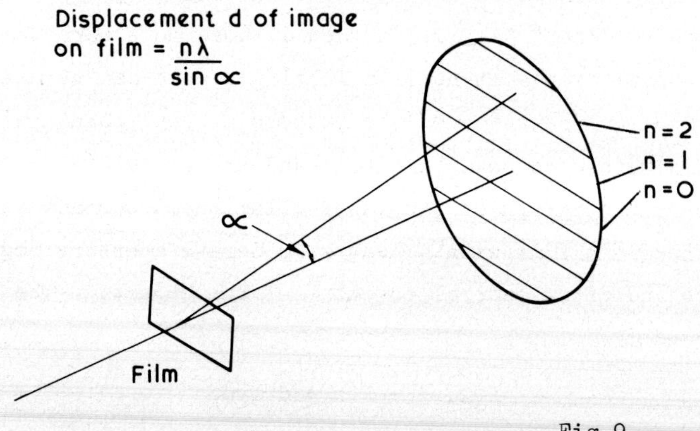

Fig 9

(i) Surface Displacement Measurement. The lateral displacement vector of each point of the recorded image may be evaluated from the Young's fringe pattern (Fig 9). Displacement magnitude $d = \frac{n\lambda}{\sin \beta}$; displacement direction is at right angles to the fringe pattern. Displacement of the object $d_o = \frac{d}{M}$. Measurement is unaffected by a small displacement in the line-of-sight direction.

An alternative method of analysis is to filter the negative image using an aperture stop in the back focal plane of a transform lens (Fig 10). The filtered image is modulated by contour fringes of the lateral displacement d* resolved in the azimuth direction of the aperture stop; contour interval corresponds to $\frac{\lambda}{\sin \beta_o}$, β_o being the altitude angle of the aperture stop.

(ii) Slope Change Measurement (Tilt). For a 'two-dimensional' rough surface, the objective speckle pattern rotates as a whole by angle $\delta\Psi$ according to

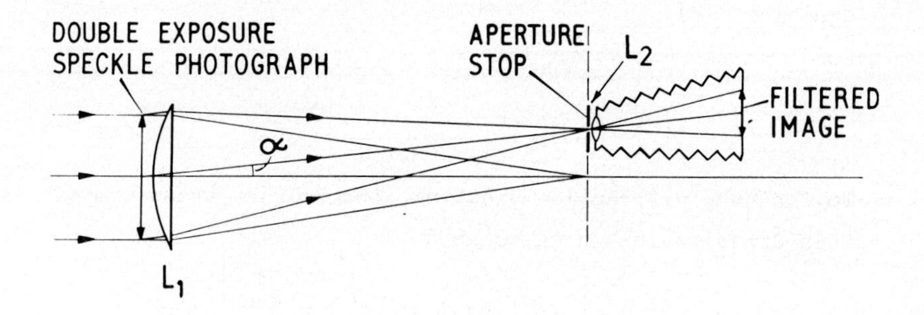

Fig 10 IMAGE FILTERING SYSTEM

$$\delta\Psi = (1 + \cos i/\cos r)\delta\Phi \qquad \ldots (17)$$

where $\delta\Phi$ is the angle of rotation, and i and r are the illuminating and viewing angles relative to the surface normal. For $i = r = o$, $\delta\Psi = 2\delta\Phi$. The speckle pattern rotates as if 'reflected' by the surface. Tilt can be measured by recording a double-exposure photograph with camera out of focus by distance δu (Archbold and Ennos [1972], Tiziani [1972]). For a speckle displacement d' on the film, the tilt angle

$$\delta\Phi = d'/2M'\ \delta U \qquad \ldots (18)$$

where M' = system magnification. Tilt can be measured independently of lateral displacement, by focusing on the plane lying behind the object which contains the 'mirror image' of the laser source, treating the surface as a mirror (Gregory [1976]).

(iii) Vibration Measurement. Vibration in the plane of a surface can be analysed by recording a time-averaged speckle photograph (Tiziani [1971]). Each speckle will be drawn out into a streak. Instead of the recorded speckle position being regarded as true delta functions $\delta(x,y)$ and

$\delta(x + d_x, y)$, as in displacement measurement, a position probability density function $W(x)$ is substituted in equation (14).

For a sinusoidal vibration $x = x_o \sin\omega t$,

$$W(x) = \frac{1}{\pi x_o} \frac{1}{\sqrt{1 - (x/x_o)^2}} \qquad \ldots (19)$$

To find the diffraction spectrum, the Fourier transform of this distribution is calculated.

$$\omega(\nu) = \frac{1}{\pi x_o} \int_{-x_o}^{x_o} \left[\exp (i\nu s)/\pi x_o \sqrt{1 - \frac{s^2}{x_o^2}} \right] \cdot ds$$

where $\nu = \frac{2 \sin \beta}{\lambda}$

Integrating, $\omega(\nu) = J_o(\nu x_o)$ ... (20)

The intensity distribution of the diffracted light is, therefore, that of the autocorrelation halo modulated by a Bessel function $J_o^2(\nu x_o)$, ie a modified Young's fringe pattern. Vibration can occur in two dimensions on the surface, eg circular motion, elliptical motion, etc. The diffraction spectra of particular trajectories (eg Lissajous' figures) are characteristic (Archbold and Ennos [1975]). The trajectory itself can be found by recording a double exposure speckle photograph, one exposure time-averaged with the surface vibrating, and one stationary, with a lateral displacement. The modulated Young's fringes can then be regarded as a Fourier hologram, from which is reconstructed two images of the trajectory (Lohmann and Weigelt [1975]).

VIII Comparative Advantages of Speckle Metrology Methods

Speckle interferometry and speckle photography provide non-contacting methods for measuring displacement and deformation in engineering structures. The choice of method to be used must take into account its limitations. All the speckle methods depend upon correlation of an imaged (subjective) speckle pattern, and they will fail if random decorrelation takes place. This will occur wherever:-

(1) The surface changes microscopically, eg due to oxidation or contamination.

(2) The surface moves in the line-of-sight direction by a distance greater than the Rayleigh depth of focus $(\pm 2[(1 + M)/M]^2 F^2\lambda)$.

(3) The surface tilts to such an extent that the objective speckle pattern formed at the lens entrance pupil is translated across the pupil by a distance comparable with the pupil diameter. A more stringent tolerance applies if the lens has aberrations, or if penetration of laser light into the object surface causes 'three-dimensional' scattering.

Speckle interferometry requires high stability of the apparatus and high coherence of the laser source. It is most suitable for measuring small deformations comparable in size to the speckle diameter, with an accuracy of $\sim \frac{\lambda}{5}$. By using an interferometer of low numerical aperture the range of measurement can be extended, but at the expense of the object tilt that can be tolerated. (limitation (3)).

Speckle photography requires only low mechanical stability and coherence of the laser source, since no reference beam is used. Sensitivity depends upon the demagnification of the system. For example, at 1:1, the uncertainty of measuring the speckle displacement is $\sim \frac{\lambda}{3}$. Progressive demagnification results in a proportional reduction in sensitivity. Errors in relative displacement, and therefore in strain measurement over the field of view can arise from distortion in the lens system, and from line-of-sight motion or tilt of the object. However, the object tilt limitation [(3) above] is less restricting for speckle photography than for interferometry, since a large lens aperture can be used.

References

General

'Laser Speckle and Related Phenomena' ed. J C Dainty Vol 9 of Topics in Applied Physics (Springer-Verlag, Berlin, Heidelberg, New York, 1975).

Chapter 6 Speckle Interferometry by A E Ennos

Chapter 5 Information Processing using Speckle Patterns by M Francon.

Archbold, E., Burch, J.M. and Ennos, A.E. (1970). 'Recording of in-plane surface displacement by double-exposure speckle photography'. Optica Acta 17, 883-898.

Archbold, E., Ennos, A.E. and Taylor, P.A. (1970). 'A laser interferometer for the detection of surface movements and vibration'. Optical Instruments and Techniques, ed. J. Home Dickson (Oriel Press, Newcastle-on-Tyne), 265-275.

Archbold, E. and Ennos, A.E. (1972). 'Displacement measurement from double-exposure laser photographs'. Optica Acta 19, 253-271.

Archbold, E. and Ennos, A.E. (1975). 'Two-dimensional vibrations analysed by speckle photography'. Opt. Laser Technol. 7, 17-21.

Burch, J.M. (1970). 'Interferometry with scattered light'. Optical Instruments and Techniques, ed. J. Home Dickson (Oriel Press, Newcastle-on-Tyne), 213-229.

Butters, J.N. and Leendertz, J.A. (1971). 'Holographic and video techniques applied to engineering measurement'. Measurement and Control, 4, 349-354.

Butters, J.N. and Leendertz, J.A. (1974). 'Component inspection using speckle pattern'. Proc. Electro-Optics International '74 Conference (Brighton, England), 43-50.

Denby, D., Quintanilla, G.E. and Butters, J.N. (1976). 'Contouring by electronic speckle pattern interferometry'. The Engineering Uses of Coherent Optics, ed. E.R. Robertson (Cambridge University Press), 171-197.

Duffy, D.E. (1972). 'Moiré gauging of in-plane displacement using double aperture imaging'. Appl. Opt. 11, 1778-1781.

Gregory, D.A. (1976). 'Speckle photography in engineering applications'. The Engineering Uses of Coherent Optics, ed. E.R. Robertson (Cambridge University Press), 263-282.

Hung, Y.Y. and Taylor, C.E. (1973). 'Speckle-shearing interferometric camera - a tool for measurement of derivatives of surface displacement'. Proc. Soc. Photo-Opt. Instrum. Eng. 41, 169-175.

Hung, Y.Y., Rowlands, R.E. and Daniel, I.M. (1975). 'Speckle-shearing interferometric technique: a full-field strain gauge'. Appl. Opt. 14, 618-622.

Leendertz, J.A. (1970). 'Interferometric displacement measurement on scattering surfaces utilizing speckle effect'. J. Phys. E. (Sci. Instrum.) 3, 214-218.

Leendertz, J.A. and Butters, J.N. (1973). 'An image-shearing speckle-pattern interferometer for measuring bending moments'. J. Phys. E. (Sci. Instrum.) 6, 1107-1110.

Lohmann, A.W. and Weigelt, G.P. (1975). 'Measurement of motion trajectories by speckle photography'. Opt. Commun. 14, 252-257.

Stetson, K.A. (1970). 'New design for laser image-speckle interferometer'. Opt. Laser Technol. 2, 179-181.

Tiziani, H. (1971). 'Application of speckling for in-plane vibration analysis'. Optica Acta 18, 891-902.

Tiziani, H. (1972). 'A study of the use of laser speckle to measure small tilts of optically rough surfaces accurately'. Opt. Commun. 5, 271-276.

Coherant Optical Engineering, F.T. Arecchi and V. Degiorgio (eds.)

LASER INTERFEROMETRY

A. Sona

CISE, Segrate, Milano, Italy

INTRODUCTORY REMARKS

Just after its discovery it was evident that laser was going to play a prioritary rôle in interferometry for the very obvious reason that the laser itself is an interferometer with an active medium. Laser emission is indeed a very close approximation to the ideal monochromatic wave used in all optical treatise to demonstrate the behaviour of various interferometers. On the other hand one could exploit, prior to the advent of lasers, only sources with an emission which is a much rougher approximation to the ideal monochromatic wave. Even the best spectral lamp is equivalent in electronic terms to a filtered noise generator. It is still with surprise that one has to conclude that all classical optics instrumentation was set up with "noise sources" used at best by extremely skilled opticists.

The basic differences between coherent and non coherent sources can be evidenced by giving the probability distribution function of the field in the Glauber's α plane or, classically, the corresponding distribution of the analytic signal in the complex plane (Fig. 1, 2). The above description however is not of much use as regards interferometry. By considering the general scheme of an interferometer one can conclude that the significant information as regards stationary interference is contained in the cross-correlation function of the interfering fields (Fig. 3). This function reduces to the autocorrelation function in the case of an ideal Michelson interferometer.

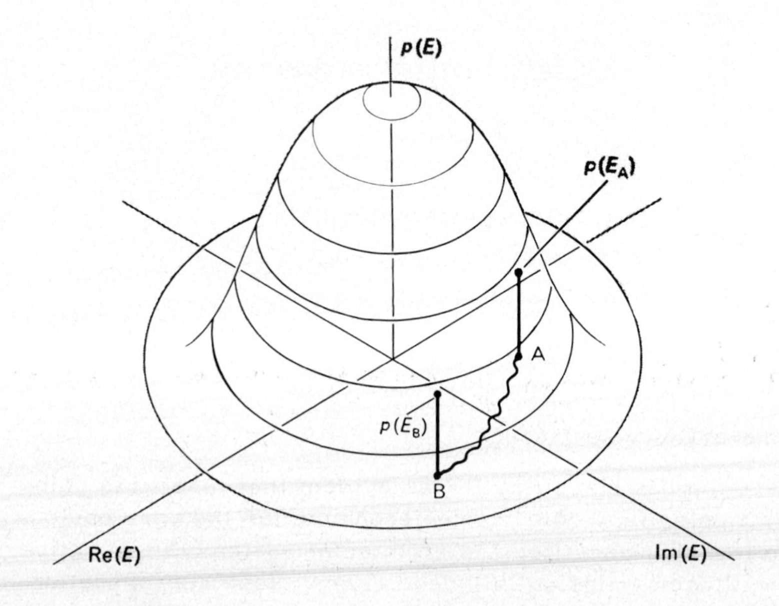

Fig. 1 – Probability distribution function of a Gaussian field from a non coherent light source.

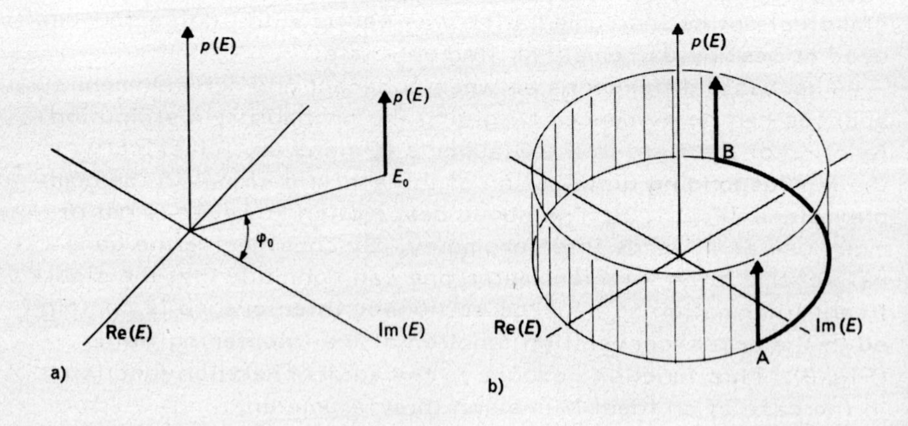

Fig. 2 – Probability distribution function of a coherent field from a laser source.

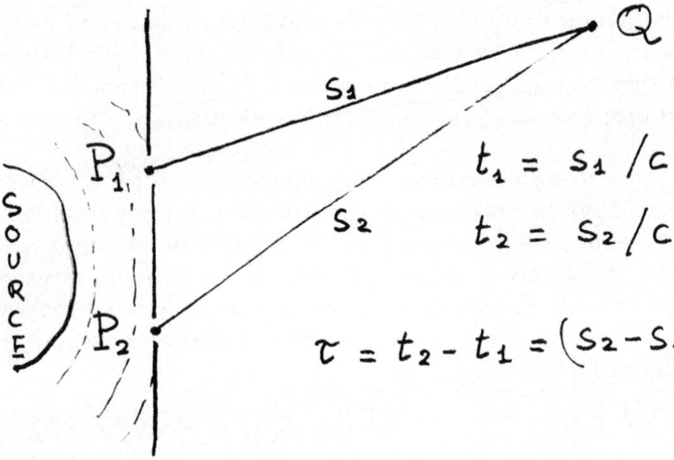

Q

s_1

P$_1$

$t_1 = s_1/c$

SOURCE

s_2

$t_2 = s_2/c$

P$_2$

$\tau = t_2 - t_1 = (s_2 - s_1)/c$

– Fields at P_1, P_2 :

$$V_1(t) = \tilde{A}_1(t) \exp i \{\phi_1(t) - 2\pi\nu t\}$$
$$V_2(t) = \tilde{A}_2(t) \exp i \{\phi_2(t) - 2\pi\nu t\}$$

– Resulting field at Q :
(one cartesian component only)

$$V(Q,t) = K_1 V_1(t - t_1) + K_2 V_2(t - t_2)$$

(K_1, K_2 in general both real or imaginary so that:

$$K_1 K_2^* = K_1^* K_2 = |K_1| \cdot |K_2|$$

– Intensity at Q :

$$I(Q) = \langle V(Q,t)\, V^*(Q,t) \rangle_{T_D}$$

T_D – Detector averaging time.

Fig. 3 - General scheme of an interferometer

LASER INTERFEROMETERS

A broad variety of interferometers and interferometric experiments with lasers have been studied in fifteen years of research and developments. The discussion will be here restricted to the applications to the field of measurements of distances, displacements and vibrations which is the most significant for practical purposes.[(°)]

The distance to be measured is compared with the wavelength of a reference source, for example a spectral lamp or a frequency stabilized laser. The comparison is usually performed with a Michelson interferometer, where the phase of the e.m. wave reflected by mirror M1 (see Fig. 4). The phase difference between the two waves results from the difference in their propagation times. It is given by :

$$\phi = 2\tau c/\lambda_{vac} = 2\pi\left[L_2 \langle n\rangle_2 - L_1\langle n\rangle_1\right]/\lambda_{vac}$$

$$= 2\pi d/\lambda_{vac}$$

where L_1 and L_2 are the geometrical lengths of the two arms of the interferometer, $\langle n\rangle_{1,2}$ is the average value of the phase refractive index along optical path 1 or 2 respectively and λ_{vac} is the vacuum wavelength of the reference source. Interference between plane waves gives rise to a rectilinear fringe pattern on the screen whereas interference between spherical waves results in a fringe pattern consisting of concentric rings. Both patterns are used in practice. The intensity at a point of the fringe pattern is obtained by superposition of two sinusoidal fields $E_1(t) = A_1 \sin\omega t$, $E_2(t) = A_2 \sin(\omega t + \phi)$ and subsequent squaring and time averaging,

$$I = I_1 + I_2 + 2(I_1 I_2)^{\frac{1}{2}} \cos 2\pi d/\lambda_{vac},$$

where I_1 and I_2 are the intensities of the interfering beams. The change in the optical path difference $d=2\left[L_2\langle n\rangle_2 - L_1\langle n\rangle_1\right]$

(°) Other applications of laser interferometry of practical interest such as holographic interferometry, Doppler velocimetry, laser gyroscopes, thickness monitoring, surface roughness evaluation will not be discussed here.

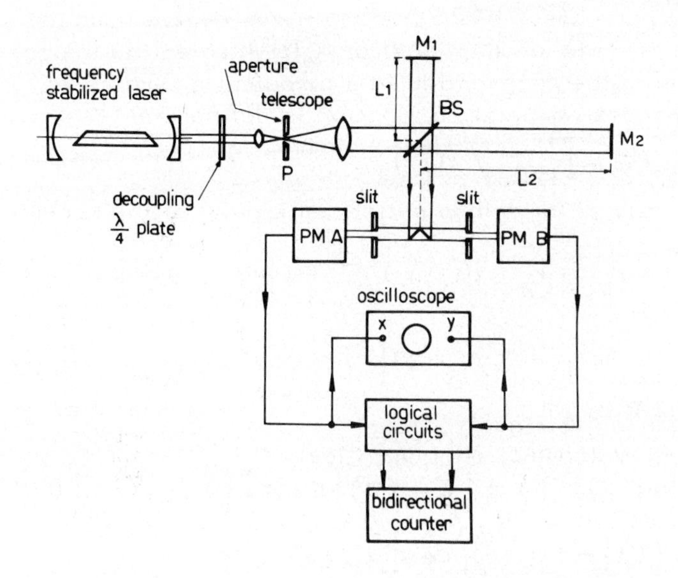

Fig. 4 - Interferometric set up for distance measurements

between the two interferometer arms can be measured by counting the number of fringes passing through a fixed point on the screen where the fringe system is observed. If the optical path difference is is changed by moving one mirror, for instance M2, by a distance ΔL_2, the corresponding number of fringes counted is $\Delta N = 2\Delta L_2 \cdot \langle n \rangle / \lambda_{vac}$, where $\langle n \rangle$ is the average phase refractive index of the medium in the region where the displacement ΔL_2 takes place. If the interfering fields are not sinusoidal but given by a statistical superposition of wave-trains of average duration τ_c, as in the example of spontaneous emission by a spectral lamp, then a proper description of the interference process is obtained by representing the two fields by their analytical signals $\tilde{A}_1(t)$, $\tilde{A}_2(t)$ (Gabor 1946, Born and Wolf 1964). The cross correlation function of the two fields is the complex quantity :

$$\Gamma_{12}(\tau) = \langle \tilde{A}_1(t)\ \tilde{A}_2^*(t+\tau) \rangle$$

and the autocorrelation functions of both fields separately is

$$\Gamma_{11}(\tau) = \langle \tilde{A}_1(t)\,\tilde{A}_1^*(t+\tau) \rangle \quad \text{and} \quad \Gamma_{22}(\tau) = \langle \tilde{A}_2(t)\,\tilde{A}_2^*(t+\tau) \rangle .$$

Here $\Gamma_{11}(0) = I_1$ and $\Gamma_{22}(0) = I_2$ are the intensities of the two fields. In our case, since the two fields derive from the same source, $\tilde{A}_1(t)$ is proportional to $\tilde{A}_2(t)$ (neglecting the perturbations due to the propagation in a turbulent atmosphere). The normalized cross correlation function is defined as

$$\gamma_{12}(\tau) = \Gamma_{12}(\tau) / (I_1 \; I_2)^{1/2}$$

The intensity of the fringe pattern at a point of the screen is then given by

$$I = I_1 + I_2 + 2(I_1 \; I_2)^{1/2} \; \mathrm{Re}\left[\gamma_{12}(\tau)\right] .$$

Here :

$$\mathrm{Re}\left[\gamma_{12}(\tau)\right] = \left|\gamma_{12}(\tau)\right| \cos\left\{2\pi d/\lambda_{vac} + \alpha_{12}(\tau)\right\}$$

where $\left\{2\pi d/\lambda_{vac} + \alpha_{12}(\tau)\right.$ is the argument of $\gamma_{12}(\tau)$ (Born and Wolf 1964, Glauber 1964).

By writing $\gamma_{12}(\tau) = \left|\gamma_{12}(\tau)\right| \exp i\left[\alpha_{12}(\tau) - 2\pi\bar{\nu}\tau\right]$

being by definition $\alpha_{12}(\tau) = 2\pi\bar{\nu}\tau + \arg\gamma_{12}(\tau)$

$$\mathrm{Re}\left[\gamma_{12}(\tau)\right] = \left|\gamma_{12}(\tau)\right| \cos\left[\alpha_{12}(\tau) - \delta\right]$$

where : $\tau = (S_2 - S_1)/c$ $\quad \delta = 2\pi\bar{\nu}\tau = 2\pi(S_2 - S_1)/\bar{\lambda}$

$\alpha_{12}(\tau)$ is the phase difference between vibrations at P_1 and P_2. The fringe visibility is usually defined by

$$V = \frac{I_{max} - I_{min}}{I_{max} + I_{min}} = \frac{2(I_1 I_2)^{1/2}}{I_1 + I_2} \left|\gamma_{12}(\tau)\right|$$

It is maximum when the two beams have the same intensity in which case $V = \left|\gamma_{12}(\tau)\right|$. This provides a direct means of measuring the function $\left|\gamma_{12}(\tau)\right|$ which varies from 1 at $\tau = 0$ to zero when τ is much larger than the coherence time τ_c of the source. The coherence time τ_c, the spectral bandwidth and the coherence length l_c of a spectral lamp are related by (Born and Wolf 1964):

$$\tau_c \simeq 1/4\pi\Delta\nu; \quad l_c = c\tau_c .$$

For most spectral lamps of practical interest the coherence time is in the nanosecond range and correspondingly the coherence length is in the order of 30 cm. Interference effects cannot be obtained with an optical path difference appreciably larger than the coherence length (see also Mielenz 1968). Frequency stabilized lasers have a coherence length in the range of a thousand kilometers. The practical limit to the usable optical path differences in laser illuminated interferometers is then set by perturbations associated with atmospheric propagation, for example, and not by coherence length. A typical setup for interferometric length measurement is shown in Fig. 4. The laser beams is sent to a beam expanding telescope which reduces the beam divergence (as the result of a wavefront curvature correction). To obtain a uniform and symmetrical intensity distribution in the output beam, a circular aperture is usually inserted in the focal plane P which performs spatial filtering of the input beam. Mirror M2 is often replaced by a corner cube reflector or by a "cat's eye" system+. Either system provides reflection in the direction of the incident beams and thus simplifies the alignment of the interferometer. Due to their compactness, stability and ruggedness, corner cube reflectors are often preferred. The fringe system is observed by a couple of photomultipliers. Through suitably positioned slits each one collects light from regions of the fringe pattern where the phase difference of the interfering beams differs by $\pi/2$. Since this $\pi/2$ difference is independent of the absolute value of ϕ, the resulting electric signal displayed on an x-y oscilloscope give rise to a Lissajous circle. The oscilloscope spot moves a full circle for a $\lambda/2$ displacement of one of the mirrors, the direction of rotation depending on the direction of the mirror displacement. A logical circuit coupled with a reversible counter processes the signals from the photomultiplier so that a count is added for each $\lambda/2$ increase (or subtracted for each $\lambda/2$ decrease) in the optical path difference.

A sufficiently fast reversible counter gives the correct indication of the displacement and is insensitive to zero average displacement such as mechanical vibrations or zero average index of refraction changes due to turbolence or sound waves. A resolution of $\lambda/8$ can be easily achieved by a suitable logical circuit (Arecchi et al. 1964a).

The use of fringe interpolators (Vali et al. 1965) allows the detection of displacements in the $\frac{\lambda}{100}$ range provided there is an adequate signal to noise ratio at the photodetectors. The

----

+ A "cat's eye" system consists of a focusing element, such as a positive lens, with a plane mirror in its focal plane. Spherical or parabolical converging mirrors may also be used in place of the positive lens.

interferometer of Fig. 4 may give rise to feedback effects on the laser source whose emission may be affected by the reflected light. (The injected signal provides a sort of frequency locking or pulling due to the coupling with the external interferometer). This effect can be exploited by detecting the variation of the laser intensity due to the feedback effect (Clunie et al. 1964). In other case, source perturbation is reduced by a decoupling system which consists of a polarizer and a quarter wave plate inserted into the laser beam. However, to get completely rid of feedback effects a nonreacting interferometer with separate beam splitting and beam recombining elements has to be used. One of various possible arrangements is shown in Fig. 5. The figure also shows the presence of two separate but related fringe systems F_1 and F_2 (Rowley 1966). The two fringe systems with lossless beam splitters are 180° out of phase (as required by energy conservation). A suitable choice of the reflectivity and transmission of a lossy beam splitter, for example a layer of aluminum, results in a phase difference between the two fringe systems of less than 180° down to 90° (Peck 1953, Rowley 1966). The two light outputs can then be collected on two photomultipliers and this allows a more efficient use of source intensity.

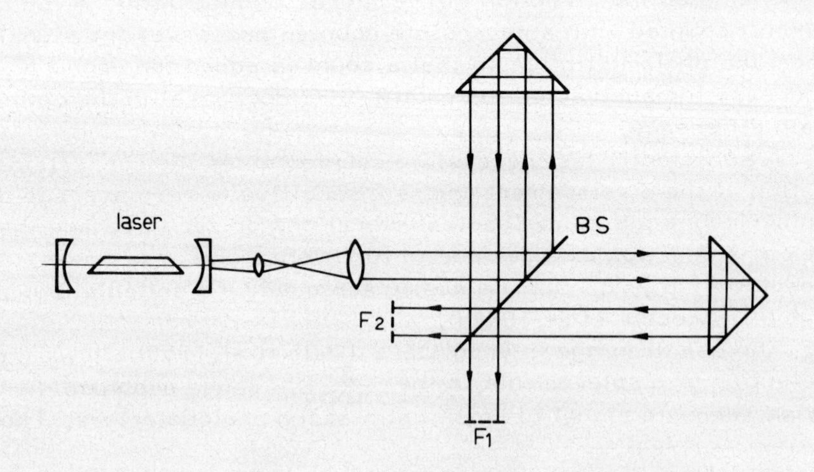

Fig. 5 - A non reacting two beam interferometer with a double fringe system.

Another interesting example of a nonreacting interferometer with two quadrature output fringe systems is shown In Fig. 6a (Minkowitz et al. 1966). A circularly polarized beam is split by BS1 into two beams each propagation in one interferometer arm. Due to the change in the sense of circular polarization upon reflection, the two beam recombined by BS2 have opposite sense of circular polarization. Notice that three reflections are required to reverse the propagation direction of a beam with a corner cube. The resulting field is plane polarized. The angle of the polarization plane is determined by the phase difference ϕ between the two beams. The plane of polarization rotates by 360° for a $\lambda/2$ displacement of one of the mirrors. The direction of rotation depends on the direction of the displacement. Two photodetectors are placed at the two interferometer outputs. Fringe detection and phase adjustment of the two signals is accomplished by the polarizers shown.

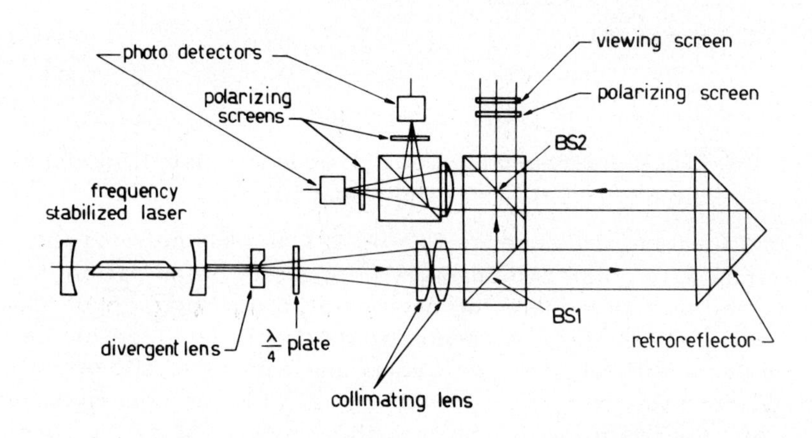

Fig. 6 - A non reacting interferometer with double fringe system operating with circularly polarized light. (Perkin Elmer Corporation).

An alternative type of interferometer exploiting a dual frequency He-Ne Zeeman laser (see Polanyi et al. 1965) was recently developed using the 1.15 μm transition (De Lang et al. 1969) or the 0.6328 μm transition (Dukes et al. 1970). The schematic diagram of this Hewlett & Packard interferometer is shown in Fig. 7 . A detailed description of this instrument will be given in the paper by Terentieff.

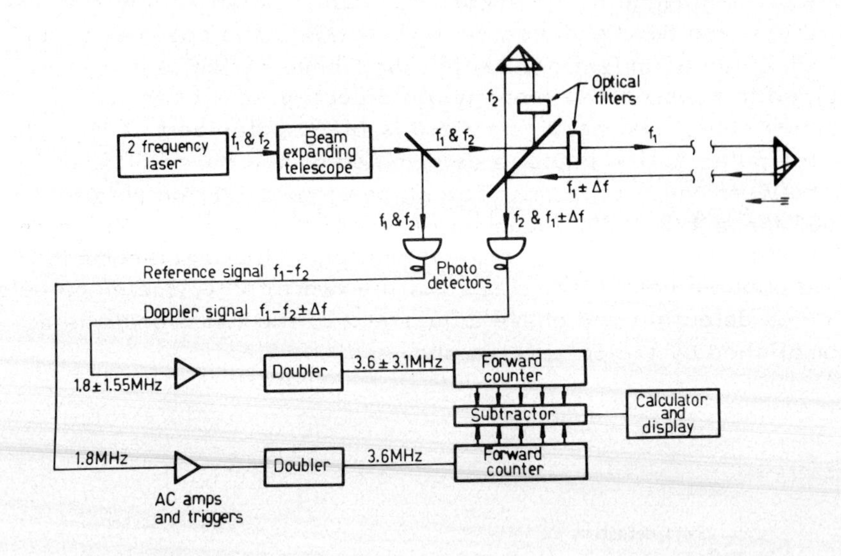

Fig. 7 - Block diagram of the dual frequency laser interferometer Hewlett-Packard Mod. 5525 A.

The frequency difference $f_1 - f_2 \approx 1.8$ MHz between the two circularly polarized beams with opposite polarization is monitored by a photodetector connected to the "add" input of a bidirectional counter. By means of a suitably polarization sensitive beam splitter, the two beams are separated and only the one with frequency f_1 is allowed to reach the far corner reflector. Any change in position Δs with a velocity v of the corner reflector will cause a Doppler shift $\Delta f_1 = f_1 \pm v/c$ in the returned frequency. The resulting beat frequency $f_1 + \Delta f_1 - f_2$ is detected by a second photodetector, where both beams are recombined, and sento to the subtract input of the counter. The net count is thus proprortional to the actual displacement Δs of the reflector and can be converted automatically from units into metric or English units. The direction of a movement is now detected simply by the increase or the decrease of the beat frequency. This method known as the frequency offset method is an alternative to the one previously discussed.

This interferometer has an exceptionally low sensitivity to atmospheric turbulence and/or attenuation. Correct operation is possible even when the returning beam is attenuated down to 5% of the emitted one.

The interferometer actually operates under conditions similar to a heterodyne arrangement which does not require, like other interferometers, nearly balanced intensities of the two recombining beams. The usable range is up to 60 m with a resolution of $\approx$ 0.01 mm. Interferometer measurements up to 400 m were performed to detect and analyse vibrations of buildings. This interferometer can be used also for measuring angles, flatness, alignment, straightness and orthogonality as described in the seminar of dr. Terentieff.

Interferometric distance measurements are mainly used in the following fields:

(1) Metrology (Length standard calibration);

(2) Mechanical tooling (Measurement of the displacement of mechanical parts in the workshop);

(3) Vibration detection and measurement;

(4) Geodesy and seismology (Detection of earth strains induced by earth or sea tides, seismic waves, continental drift, etc.)

Application(1) requires operation of the interferometer and of the connected calibration bench on an antishock mounted platform in a thermally controlled environment with acoustic isolation: the optical path is protected against turbulence effects but usually not evacuated. To achieve an accuracy of a few parts in 10^8 which is compatible with the accuracy of the present length standards, air refractivity corrections are required. One uses, for instance, the data from Edlèn reported in ref. (Edlèn 1966).

Application (2) requires accuracies in the range of a few parts in 10^6. Even in this case correction for air refractivity is required since the index of refraction of the air at ambient temperature and pressure in the range of 1.0003. The refractivity $\eta = n-1 = 300 \times 10^{-6}$ depends linearly on the air concentration (for a fixed air composition) according to the Lorenz - Lorentz law. A pressure change of 2.5 torr or a temperature change of 1 °C results in a 1×10^{-6} change in the index of refraction. In most commercial systems a small computer corrects automatically for pressure and temperature variations. It also converts from λ units to metric or English units. Note that interferometric techniques allow the measurement of reflector displacement i.e., only "incremental" length measurement. No "absolute" measurement of position is possible with an interferometer at least in its basic form.

Vibration detection can be accomplished by using Michelson type arrangements without special differences with respect to schemes already described provided :

a) the detector response time is faster than the period of the vibration to be analyzed (which is the case for most mechanical applications);

b) a corner reflector can be placed on the vibrating part.

The interferometer acts in this case as a length transducer providing a digital or analog output signal which can be recorded and analyzed subsequently.

If vibrations are to be measured on a diffusing surface than alternative arrangements similar to those used in Doppler Velocimetry must be considered. A typical example is shown in Fig.8 (P. Buckhave 75). The maximum range of this kind of interferometric instrumentation is of the order of 100 m the limit being set by the turbulence in the optical path. See also (R. A. Bruce 75).

ACCURACY, RESOLUTION AND RANGE

The accuracy in the measurement of a displacement is limited by the following factors. A displacement ΔL_2 given to mirror M2 provides a phase change in the optical path difference given by :

$$\Delta q = 2\pi \, 2 \, \Delta L_2 \langle n \rangle_{L_2} / \lambda_{vac}$$

under the assumption that the optical paths L_1, L_2 and related value of $\langle n \rangle_1$ and $\langle n \rangle_2$ have not undergone any change during the displacement. The corresponding digital output of the interferometer is given by the number of counts :

$$\Delta N = 2 \, \Delta L_2 \langle n \rangle_{L_2} / \lambda_{vac.}$$

The fractional accuracy in evaluating L_2 is limited by three main causes :

i) the accuracy in the laser wavelength which is of the order of 10^{-7} with Lamb Dip Stabilization

10^{-11} with saturated reference cell

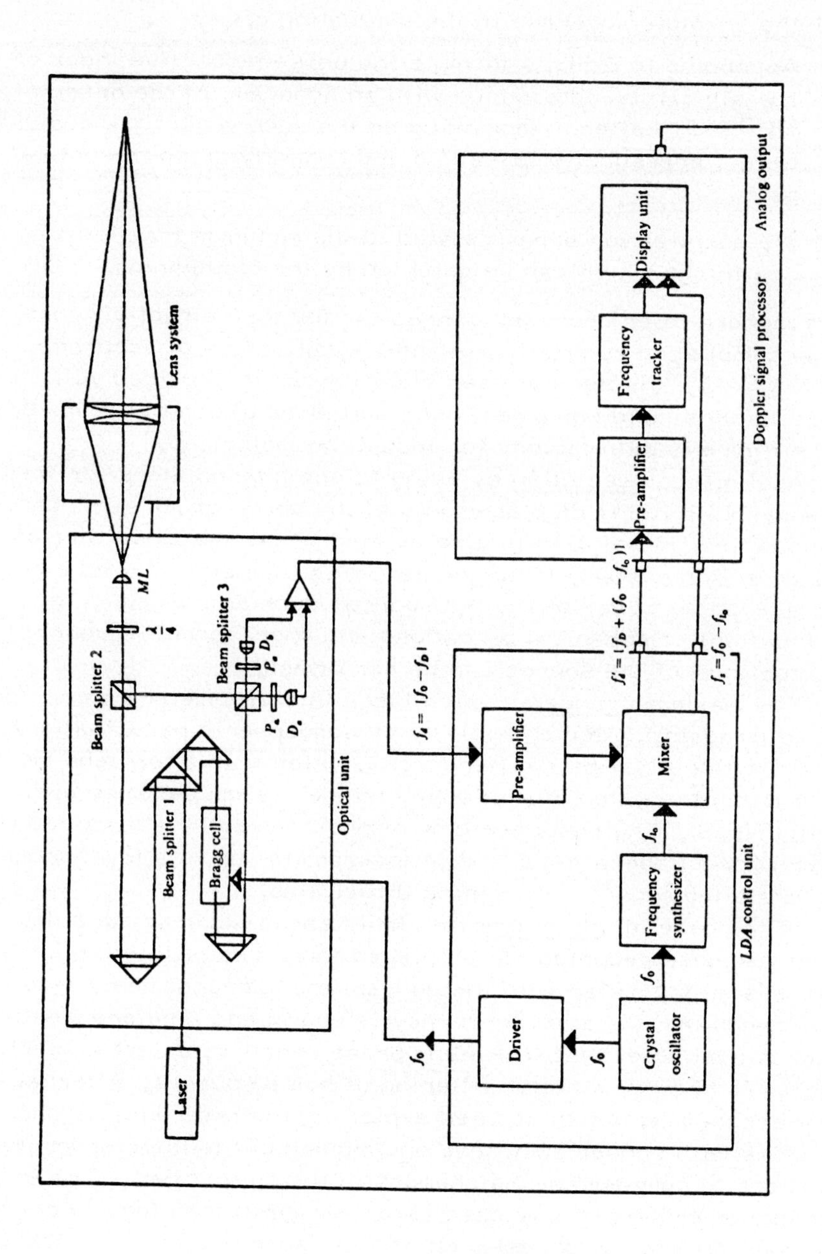

Fig. 8 - Block diagram of a Laser Doppler vibration meter

ii) the absolute accuracy in the evaluation of $\langle n \rangle_{L_2}$.

According to Edlèn's formula the phase refractive index of the air can be determined with an accuracy of the order of 10^{-8} or better depending on the accuracy of the evaluation of pressure temperature and humidity along the optical path

iii) the quantization error related to the minimum fraction of one fringe which can be detected by the electronics.

In a modern interferometer correction for air refractivity can be accomplished automatically and typical values of accuracy of $\pm$ 0,5 10^{-6} limited by laser stability can be obtained with reasonably simple equipment over distances of some tens meters. This figure is satisfactory for industrial metrology.

As regards resolution by using fringe interpolators (or frequency multipliers) displacements of the order of .01 μm can be detected. Resolution is also related to the response time of the detection system. If longer response time are allowed a better averaging of various noise sources such as atmospheric fluctuations, mechanical vibrations and short term frequency fluctuations of the source can be performed.

The maximum range at which laser interferometer can be operated is again limited essentially by atmospheric perturbations. Again with optical path differences greater than a few tens meters the observation of stationary fringe system becomes difficult. (Fig. 9). The response time of photodetectors allows however correct operation of the interferometer up to a few hundred meters although with increasing difficulties.

In the case of vibrations the limits set to accuracy are the same as those reported for displacements. The resolution achievable is also similar with the conventional arrangement. However when the vibration frequency is known and a reference signal is available one can exploit phase sensitive detection techniques to provide a better filtering of noise sources. Alternative arrangements can be used exploiting the reference signal to drive with proper amplitude one mirror of the interferometer in order to compensate the changes in the optical path. The interferometer acts in this case as a "zero monitor" for the optical path difference (a research on this technique is in progress at CISE by M. Corti and coworkers).

The above mentioned limits in resolution can be overcome by using an alternative arrangement shown in Fig. 10 (Boersch et al. 1967)

- Longitudinal phase fluctuations:

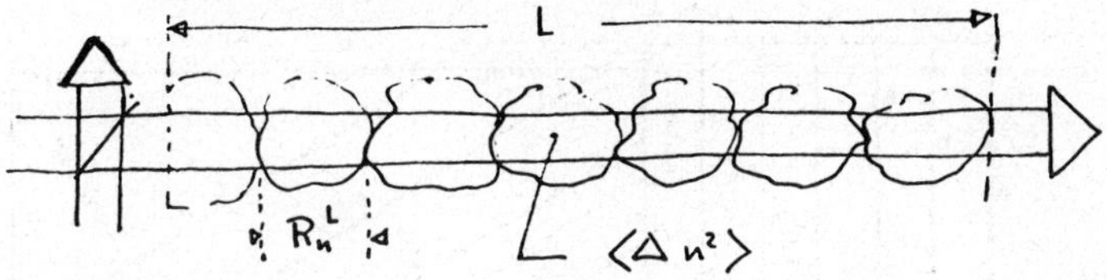

L

R_n^L

$\langle \Delta n^2 \rangle$

$$\langle \Delta L^2 \rangle / \lambda^2 = 4 \langle \Delta n^2 \rangle L \sqrt{\pi} \, R_n^L / \lambda^2 = \langle \Delta N^2 \rangle$$

- Spectral distribution of phase fluctuations:

$$\Delta N^2(f) = \frac{2}{f_o^L} \langle \Delta N^2 \rangle \sqrt{\frac{\ln 2}{\pi}} \exp - \left\{ \ln 2 \left(f / f_o^L \right)^2 \right\}$$

where $f_o^L = V_n^L / R_n^L$; V_n^L = velocity along optical path

Fig. 9a - Phase fluctuations due to atmospheric turbulence (bubble model). R_n^L is the bubble size depending on turbulence. For medium turbulence: $R_n^L \simeq 10$ m; $\langle \Delta n^2 \rangle \simeq 10^{-16}$ $\Delta L^2 / \lambda^2 \simeq 1$ at $R \simeq 50$ m

– Transverse coherence length r_o.

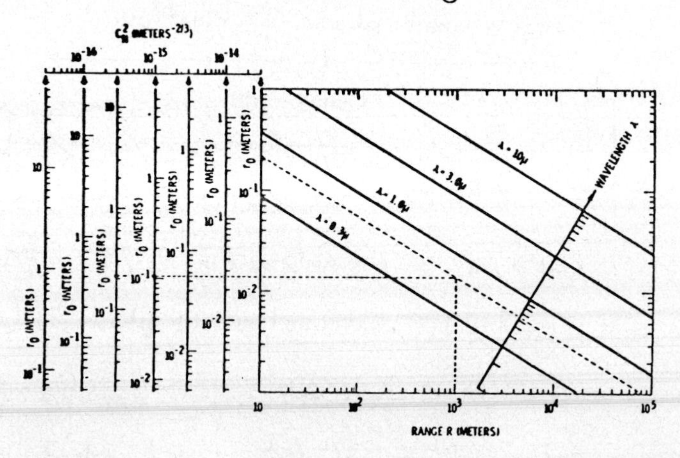

Fig. 9b – The dependence of r_o upon wavelength λ, path length R, and strength of turbulence as measured by the refractive index structure constant for horizontal propagation. The C_N^2 scale in the upper left covers the range of values normally encountered within several tens of meters of the ground.
(from D. L. Fried PIEEE 55, p. 67, 1967)

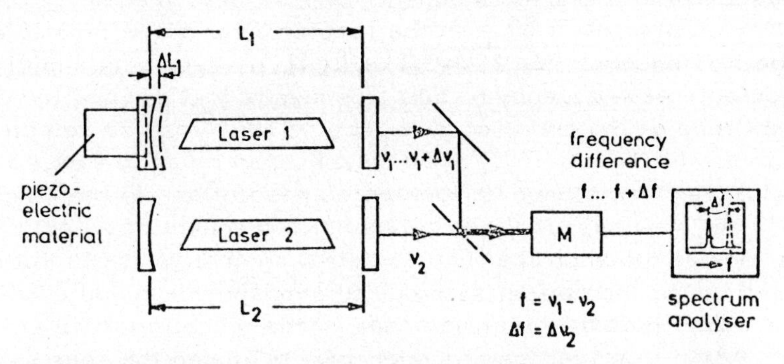

a - Laser interferometer for measuring small changes of length ΔL_1.

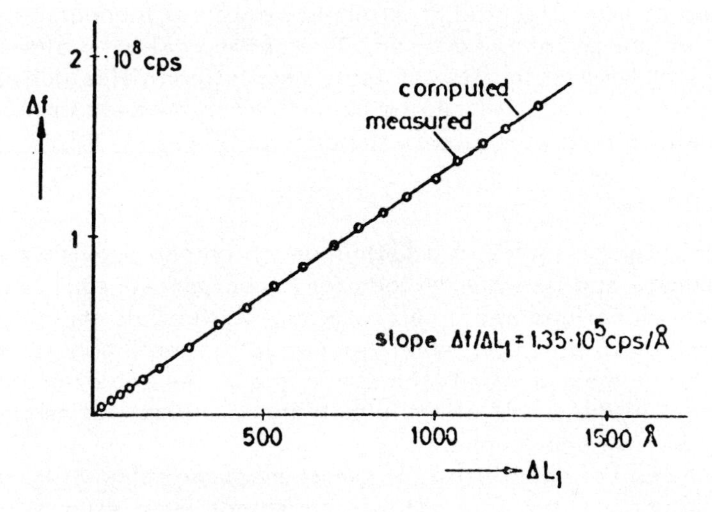

b - Frequency difference Δf versus shift of mirror separation ΔL_1 (L_1 = 35 cm)

Fig. 10 - Measurements of vibration by optical heterodyning

The vibration to be measured is applied to the mirror of laser L_1. The induced variation in the laser frequenc is measured by heterodining the laser frequency with a fixed frequency laser L_2. Any displacement ΔL_1 in the laser mirror gives rise to a change in frequency $\Delta \nu' \simeq -\Delta L_1/L_1 \nu_0$. By assuming a minimum detectable frequency of 5KHz this means that a vibration of an amplitude of the order of $\Delta L_1/L_1 = 10^{-14}$ can be detected with an He-Ne laser. If $L_1 = 50$ cm, ΔL_1 can be as low as .05 Å. Actual experiments have demonstrated resolutions in the range of 10^{-13}m. As regards the problem not infrequent of detecting vibrations at distance the limit is set by the range at which the fluctuations in the optical path due to atmospheric turbulence have an amplitude of the same order of the vibration to be analyzed. Again this limit can be overcome provided the spectral distributions of vibrations to be measured is significantly different from the one of atmospheric perturbations. A favourable case is the one in which a reference signal allows phase detection and an effective filtering of unwanted noise. If a loss in resolution can be tolerated an alternative way of increasing the range is to use an infrared laser. The range scales up almost linearly with wavelength and similarly resolution scales down. Interferometry at 10 μm can be performed in free air up to a few km depending on turbulence conditions (Figs. 9a, 9b).

CONCLUSIONS

Laser interferometry is a technique which can provide extremely accurate and flexible devices for direct measurements of distances, vibrations and displacements. With small changes in the optical setup planeity, straightness, alignment, angular deviation can be measured. All the above quantities are of practical interest in the fields of basic metrology, industrial metrology, geodesy, seismography.

In addition the possibility of remote measurements with non contacting probes is also a definite advantage over alternative methods.

A further advantage which will be possibly exploited in the future is related to the possibility of extracting this kind of information (or measuring any other physical variable which can be converted in a displacement or in a vibration) by equipment in environment without an easy access. Typical examples are electrical equipment in the high voltage regions of an electrical power plant. Another possible application is the use of laser interferometry for remote monitoring of nuclear reactors. In general this technique is a basic tool for a broad variety of electrooptical measuring systems.

REFERENCES

Arecchi, F. T. and A. Sona (1964a), Symposium on Quasi-Optics, Polytechnic Institute of Brooklyn.

Arecchi, F. T. and A. Sona (1964b), Nuovo Cimento 32, 1117.

Buckhave, P., DISA Information 18, 15 Sept. 1975.

Born, M. and E. Wolf (1964), Principles of Optics, 2nd ed. (Oxford).

De Lang, H. and G. Bouwhuis (1969), Philips Technical Review 30, 149.

Dukes, J. N. and G. B. Gordon, Hewlett-Packard Journal August 1970.

Edlèn, B. (1966), Metrologia 2, 81.

Gabor, D. (1946), J. Inst. Electr. Eng. III 93, 429 (London).

Gronchi, M. and A. Sona, (1970), Rivista del Nuovo Cimento 2, 219 (in English).

Herrick, B. R. and J. R. Meyer-Arendt (1966), Appl. Opt. 5, 981.

Hodara, H. (1968), Proc. IEEE 56, 2130.

Mielenz, K. D., K. F. Nefflen, W. R. C. Rowley, D. C. Wilson and E. Engelhard (1968a), Appl. Optics 7, 289.

Mielenz, K. D. (1968b), Advances in Electronics and Electron Physics, suppl. 4 (Academic Press).

Minkowitz, S. and W. Reid Smith Vanir (1967), Proceed. 1st Congress on Laser Applications (Paris) and J. Quantum Electronics 3, 237.

Peck, E. R. and S. W. Obetz (1953), J. Opt. Soc. Amer. 43, 505.

Polanyi, T. G., M. L. Skolnick, Tobias and R. A. Wallace (1965), Appl. Phys. Letts. 6, 198.

Rowley, W. R. C. (1966), IEEE Trans. on Instr. and Meas. 15, 146.

Tatarski, V. I. (1961), Wave Propagation in Turbulent Medium (McGraw Hill; N. Y.).

Vali, V., R. S. Krogstad and R. W. Moss (1965), Rev. Sci. Instrum. 36, 1352.

Bruce, R. A. and G. L. Fitzpatrick (1975), Appl. Opt. 7, 1621.

Boersch, H. et al. (1976), Proc. 1st Congress on Laser Application (Paris).

Coherant Optical Engineering, F.T. Arecchi and V. Degiorgio (eds.)

SANDWICH HOLOGRAPHY-PRACTICAL APPLICATIONS

Nils Abramson
Div. Production Engineering
Royal Institute of Technology
100 44 Stockholm 70
Sweden

INTRODUCTION

Optical interferometry has for a long time been used to compare the topography of two surfaces. Usually, one or both surfaces are substituted by their mirror images as in conventional interferometry or by their holographic images as in hologram interferometry.

In real-time hologram interferometry, a holographic image is compared with the object itself. The interference fringes do not exist in the holographic image but are formed when the object is studied through the hologram (live fringes).

In double-exposure hologram interferometry, one holographic plate is exposed twice using an unchanged holographic setup. In this case the two images that are compared cannot be changed during reconstruction. The interference fringes are then called frozen fringes.

It is also possible to expose the two images on a single plate using two different reference beams. In that case the comparison of the two holographic images can be manipulated by changing the direction and divergence of one of the reconstruction beams.

Finally, it is possible to combine two holographic images from two separate hologram plates. Gates has compared two in line holograms to study gas flow. His method was based on focusing one holographic image into another using high quality lenses. J. E. O'Hare introduced a method also limited to interferomeric study of gases, using two hologram plates, one in front of the other. Neither of the last three methods has become popular, probably because of the practical difficulties in keeping the necessary interferometric precision during reconstruction.

In the following a new method is described and discussed that makes possible a comparison of holographic images of diffusely reflecting objects from different hologram plates. No lenses are used. Instead the two hologram plates are simply placed one in front of the other during reconstruction with their emulsions towards the object. To make the plate positions optically and mechanically identical during reconstruction and exposure the same plate-holder is used all the time.

By making the exposures of a double exposed hologram on different photographic plates which before reconstruction are sandwiched together, with their emulsions separated, the back plate being the one which was exposed first, the following advantages are gained:

1. - If many exposures have been made it is possible to compare any two hologram plates to study interferometrically any changes in object shape that have taken place between the corresponding two exposures.

2. - If the sandwich-hologram during reconstruction is tilted in the same direction as the object tilted between the exposures, the object appears fringe-free. Thus fringes caused by unwanted object tilt can be eliminated.

3. - The tilt direction and sign of the object is easy to find because it is equivalent to the sandwich-hologram tilt direction and sign needed for fringe-free object reconstruction.

4. - The tilt amplitude of the object is easy to find because it is analogous to the sandwich-hologram tilt amplitude needed for fringe-free object reconstruction.

5. - The measurement of object tilt is simplified by the fact that the sandwich-hologram tilt needed for fringe-free object reconstruction is much greater (e. g. 1000 times greater) than the object tilt.

6. - By tilting the sandwich-hologram so that the studied interference fringes are rotated 90° the gradient of fringe spacing is transformed into fringe angle. Thus maximum strain and stress caused by bending is revealed directly without differentiation.

7. - Much larger displacements can be studied than possible using conventional hologram interferometry. Some 1000 fringes can be eliminated by simple sandwich-hologram tilt.

8. - Contouring fringes representing the intersection of the object by a set of equidistant parallel plates that are normal to

the line of sight can be produced by a simple transversal displacement of the illumination source between the two exposures.
9. - The contouring fringes can be tilted up to 180° by tilting the sandwich-hologram during reconstruction.
10. - It can be arranged so that object slope angle and sign are analogous to the sandwich-hologram tilt angle and sign needed for fringe-free object reconstruction.

PRACTICAL RESULTS

One holographic plate was placed in front of another in the same plate holder (Fig. 1).

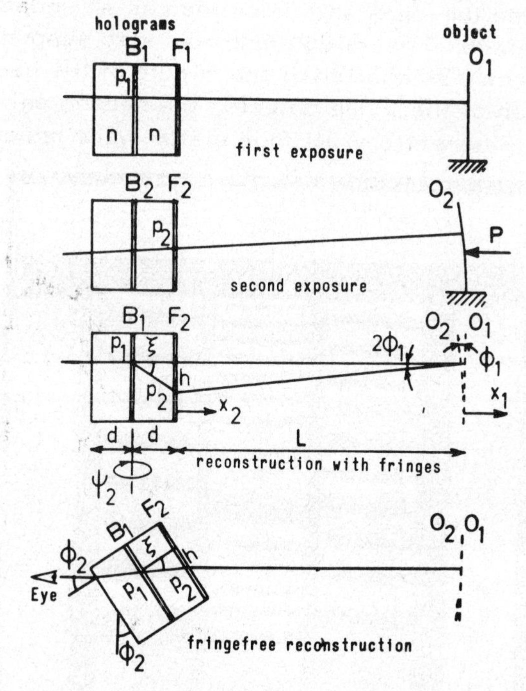

Fig. 1 - The top of the object 0 is tilted at angle Φ_1 by the force P. Therefore, a speckle ray from one object point is moved the vertical distance h from p_1 to p_2. B and F are the emulsions of back and front hologram plates, respectively. Glass plate thickness is d, and refractive index is n. Φ_2 and Ψ_2 represent sandwich rotation around a horizontal and vertical axis, respectively.
Object translation is x_1. Corresponding sandwich translation is x_2. The identical reference and reconstruction beams are excluded in the figure.

The two plates made contact with each other and had their emulsions toward the object. The emulsions were therefore separated by the glass base of the front plate. When the plates were exposed simultaneously both the object beam and the reference beam reaching the back plate passed through the front plate, which was not covered by an antihalo layer. The two exposed plates were taken away, and two new plates were placed in the plate holder.

The object consisted of three vertical steel bars (Fig. 2) that were fixed by screws at their lower ends to a rigid frame that surrounded the bars and functioned as an undeformed fixed reference surface. The middle bar was also supported at the top end. A force was applied at the middle of the bars. After the deformation of the three objects, the second pair of plates was exposed. Thereafter, all four plates were processed.

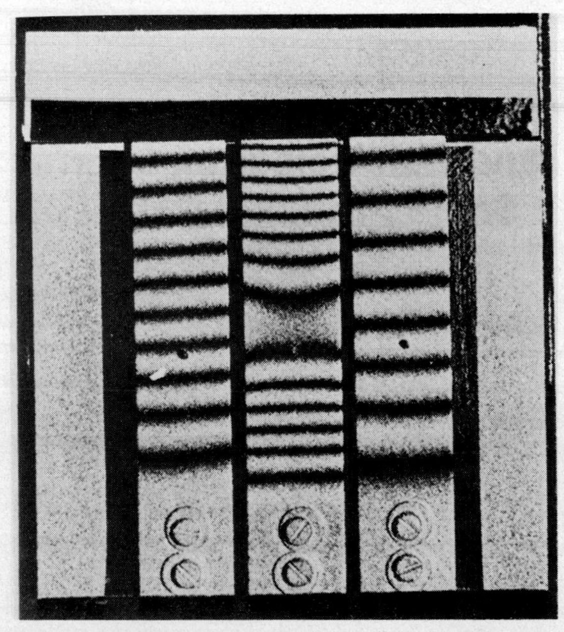

Fig. 2 - The lower ends of three bars are screwed to a stable fixed frame. The middle bar is also supported at the top end. Forces are applied at the middle of the bars between the two exposures. The right bar is deformed away from the observer. The other two bars are deformed toward the observer.

The back plate of the first exposure (B_1 of Fig. 1) was repositioned in the plate holder behind the front plate of the second exposure (F_2). The two plates were bonded together to form a sandwich hologram. 0_1 and 0_2 represent the two positions of one object, which between the two exposures was bent at angle ϕ_1 toward the holograms. L is the distance between the front hologram plate and the object (L = 1 m); L is also the distance between the point source of illumination and the object; D is the distance between the point source of the reference beam and the hologram plates (D = 2L); d is the thickness of one hologram plate (d = 1, 3 mm); n is the refractive index of the plates (n = 1, 5).

When the two plates were properly repositioned in the fixed plate-holder, no fringes were seen on the reconstructed image of the undeformed reference frame (Fig. 2). If however, the sandwich was tilted backward at angle ϕ_2 (Fig. 1) horizontal fringes were found on the frame, while the studied object became fringe free (Fig. 3)

Fig. 3 - The sandwich hologram has been tilted at angle ϕ_2 so that the top of the left bar (which was tilted at angle ϕ_1) is fringe free.

If the sandwich were instead rotated around a vertical axis at angle ψ_2, a set of vertical fringes were formed on the frame (Fig. 4).

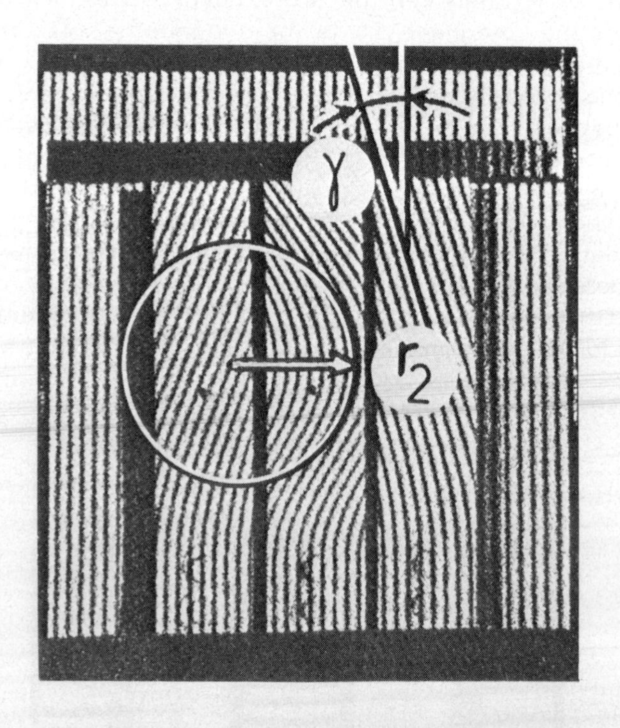

Fig. 4 - The sandwich hologram has been rotated at angle ψ_2 around a vertical axis. The angle (γ) between the fringes and a vertical line is a measure of object tilt angle (ϕ_1). The radius of curvature of the fringes (r_2) is a measure of object bending radius (r_1), from which the bending moment and maximal object strain can be calculated.

The fringes on the reconstructed deformed objects were in this case inclined at angle γ from the vertical direction. They were also bent, the radius of curvature being r_2. The radius of curvature of the deformed object was r_1.

We have found the following four dimensionless equations relating sandwich hologram functions to object displacements :

$$\phi_1 = \frac{d}{2Ln} \cdot \phi_2 \quad (1)$$

$$\phi_1 = \frac{d}{2Ln} \cdot \Psi_2 \cdot \mathrm{tg}\, \gamma \quad (2)$$

$$r_1 = \frac{2Ln}{d} \cdot \frac{1}{\Psi_2} \cdot r_2 \quad (3)$$

$$x_1 = \frac{d}{2Ln} \cdot 0,5 \; x_2 \quad (4)$$

MEASUREMENTS ON A MACHINE TOOL

In the following, a practical example is described of an industrial measurement which would have been quite difficult to make without the use of the new method of sandwich-holography[1, 2, 3]. The stability of a milling machine was tested holographically. The machine was 2 m high and weighted some 2 tons. The distance between the machine and the sandwich was 10 m. The milling machine, standing directly on the floor of the workshop, was painted white and a dark tent of black plastic film was built around it to shut out the daylight and to prevent the unwanted air turbulence.

The reference beam was reflected by a mirror which was placed close to the milling machine. To make possible a holographic imaging of the total object in spite of the fact that the coherence length of the laser was only about 15 cm, we used the holo-diagram for the planning of the holographic setup. The object was illuminated by a Spectra Physics He-Ne laser Model 125 with an output of some 60 mw. The exposure was 2x10 sec on Agfa-Gevaert Scientia 10E75. A pneumatic membrane placed between tool and workpiece produced a stimulated horizontal cutting force.

First one pair of holograms was exposed with the machine at rest. Thereafter air-pressure was applied to the membrane producing a force representing low cutting force. Another pair of holograms was exposed whereafter one plate from each exposure was put together to form a sandwich-hologram. The result is seen in Fig. 5. On the floor to the right of the machine was placed a heavy piece of steel to be used as a fixed reference surface against which the motion of the machine could be defined. When the sandwhich-hologram was tilted during reconstruction the eliminated fringe pattern could be moved from any part of the milling machine to this reference surface.

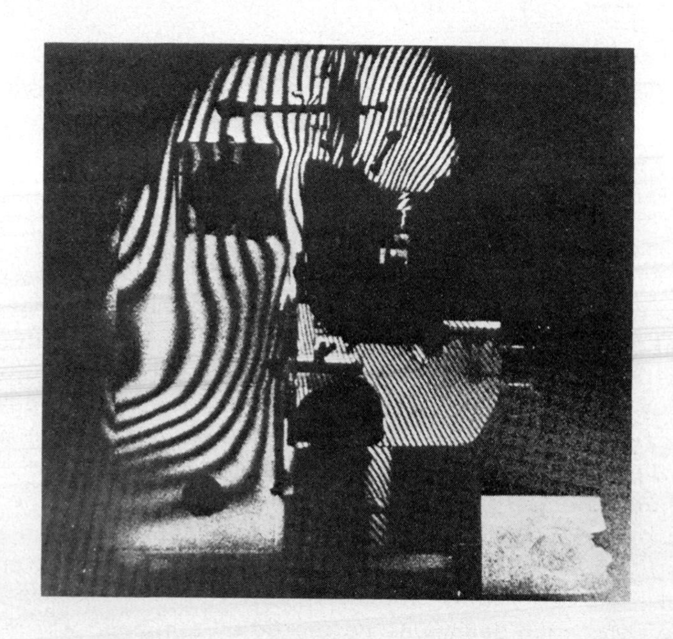

Fig. 5 - A milling machine has been loaded by a static force representing the cutting force. Every fringe represents a displacement of circa 0, 3 µm normal to the plane of the photo. Straight lines represent a tilt around an axis parallel to the lines. Curved lines represent deformation. Displacements forwards or backwards can not be distinguished directly.

Fig. 5 displays the motions of the milling machine in reference to the floor. Because of the large distances, illumination and observation were made almost normal to the total studied surfaces and therefore each fringe represents a displacement of about 0, 32 µm along the line of sight. Areas that are fringe-free have not moved in this direction (the base of the machine, the reference surface). Areas covered by straight parallel lines have tilted rigidly without deformation (the table has mainly tilted).

Areas covered by curved fringes have been deformed (the main body has been deformed by the force acting between tool and table). From Fig. 5 it is difficult to judge the deformation of the head of the milling machine because its fringe pattern is mostly caused by the large torsion of the main body.

After tilting the sandwich-hologram until the number of fringes crossing the head got as low as possible, it was however easy to study the deformation of the head, without any disturbances from the motions of the rest of the machine (Fig. 6). From the angles the sandwich-hologram had to be tilted to change the picture from that of Fig. 1 to that of Fig. 6, the tilt of the machine head could be evaluated with regard to magnitude, direction and sign.

The deformation of the table of the milling machine could, however, not be studied because the force used was too low.

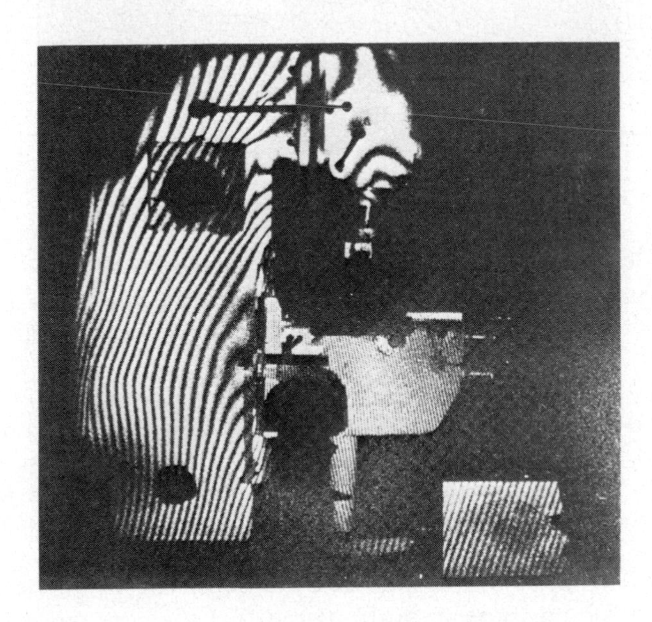

Fig. 6 - The sandwich-hologram has been tilted so that the deformation of the machine head can be studied without any influence on the fringe pattern from the deformation of the total machine.

Thus the force was increased until it was some four times larger and the hologram of Fig. 7 was made.

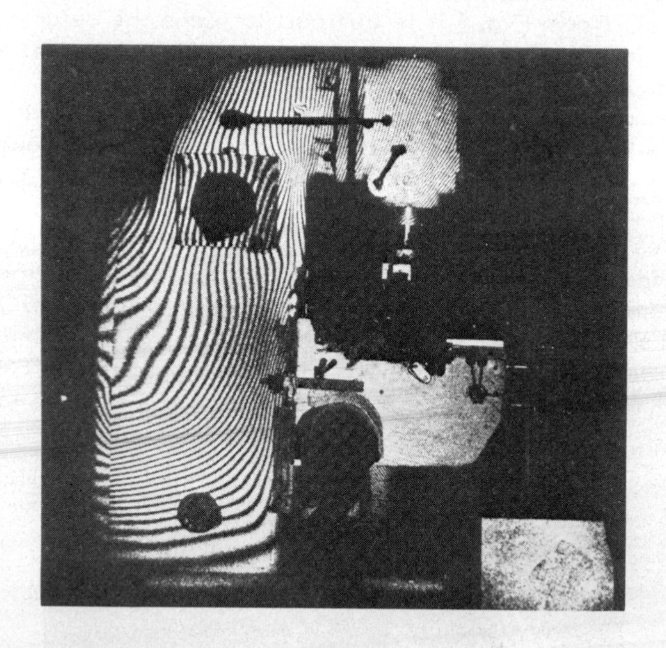

Fig. 7 - The load has been increased so that it is four times larger than in Fig. 1. The motion of the machine table became so large that the fringes can not be properly resolved.

The tilt of the table was so large that the fringes are not quite resolved. After the tilt of the sandwich hologram during reconstruction fringes appeared on the table and after some further tilt the number of fringes crossing it was set at a minimum (Fig. 8).

Now it is possible to measure the deformation of the table of the milling machine in spite of the fact that the deformation was so small that it had only caused circa eight of the about one hundred fringes that originally existed on the table. From the angles that the sandwich hologram had to be tilted to transform Fig. 7 into Fig. 8 it is also possible to calculate the magnitude, direction and sign of the table tilt, in spite of the fact that the fringes originally were so closely spaced that they could not be counted.

Fig. 8 - The same hologram as in Fig. 3 but the sandwich-hologram was tilted so that the deformation of the machine table is easy to study. From the tilt of the hologram the magnitude, direction and sign of table tilt can be evaluated.

Up till now we have found that with sandwich holography it is possible to measure at least ten times larger deformations than when ordinary holography is used. In the described experiment we increased the force still more and managed to eliminate some 500 fringes. In that case we used an extra glass plate to separate the hologram plates so that the distance between the emulsions was some 9 millimeters.

REFERENCES

1. N. Abramson, Appl. Opt. 13, 2019 (1974)
2. N. Abramson, Appl. Opt. 14, 981 (1975)
3. N. Abramson, Proceedings from : Engineering uses of coherent optics, Strathclyde, April 1975.

Coherant Optical Engineering, F.T. Arecchi and V. Degiorgio (eds.)
©North-Holland Publishing Company, 1977

PRINCIPLES AND APPLICATIONS OF INTENSITY CORRELATION SPECTROSCOPY

Vittorio Degiorgio

C. I. S. E., Segrate (Milano), Italy

INTRODUCTION

The term intensity correlation spectroscopy (ICS) designates the technique by which relevant properties of a scattering medium illuminated by a light beam are derived from the measurement of the correlation function of the intensity of the scattered light. In the last few years ICS has found many areas of applications in physics, chemistry, biology, and engineering. Its understanding requires an acquaintance with the statistics of light fields, theories of light scattering, and some knowledge of optical and electronic instrumentation. Detailed treatments of the ICS technique and of its must significant applications can be found in Refs. 1 and 2 and articles quoted therein. We shall discuss here a few important points, with the aim of making clear the basic principles of ICS.

PROPERTIES OF CORRELATION FUNCTIONS [3]

The statistical properties of any random process (such as an optical field) can be characterized by joint probability distributions or (and) correlation functions of any order. In practical cases, relevant information on optical fields is obtained by measuring only the lowest order correlation functions $G^{(1)}$ and $G^{(2)}$, defined as follows;

$$G^{(1)}(\vec{r}_1, \vec{r}_2, \tau) = \langle E^{+}(\vec{r}_1, t)\, E(\vec{r}_2, t+\tau) \rangle \qquad (1)$$

$$G^{(2)}(\vec{r}_1, \vec{r}_2, \tau) = \langle I(\vec{r}_1, t)\, I(\vec{r}_2, t+\tau) \rangle \qquad (2)$$

where E and $I = |E|^2$ are respectively the electric field and the intensity of the optical beam.

Quite generally it can be said that the spatial properties of cor-

relation functions reflect merely the geometry of the source (the scattering volume in a light scattering experiment). We are here more interested in the time dependence of $G^{(1)}$ and $G^{(2)}$ which contains information about the dynamics of source fluctuations. We put therefore $\vec{r}_1 \equiv \vec{r}_2$. Furtermore we consider only stationary fields, so that $G^{(1)}$ and $G^{(2)}$ depend only on the time delay τ.
Properties of $G^{(1)}(\tau)$:

$$G^{(1)}(0) = \langle I \rangle ; \left| G^{(1)}(\tau) \right| \leq G^{(1)}(0) \quad ; \lim_{\tau \to \infty} G^{(1)}(\tau) = 0$$

Properties of $G^{(2)}(\tau)$

$$G^{(2)}(0) = \langle I^2 \rangle ; \lim_{\tau \to \infty} G^{(2)}(\tau) = \langle I \rangle^2 ; \left| G^{(2)}(\tau) - \langle I \rangle^2 \right| \leq G^{(2)}(0) - \langle I \rangle^2$$

We recall also the definition of the optical spectrum $S^{(1)}(\omega)$

$$S^{(1)}(\omega) = \int G^{(1)}(\tau)\, e^{i\omega\tau}\, d\tau \tag{3}$$

If $S^{(1)}(\omega)$ is a symmetric function with respect to the central frequency ω_o, and we write the field as

$$E(t) = E_o(t)\, e^{-i[\omega_o t + \varphi(t)]} \quad (E_o(t) \text{ real})$$

the correlation functions can be expressed as

$$G^{(1)}(\tau) = \langle I \rangle\, e^{i\omega_o \tau} f(\tau) \tag{4}$$

$$G^{(2)}(\tau) = \langle I \rangle^2 (1 + g(\tau)) \tag{5}$$

where $f(\tau)$ and $g(\tau)$ are real.

The following relation holds for gaussian fields

$$g(\tau) = f^2(\tau) \tag{6}$$

Note that the knowledge of $G^{(2)}$ does not give completely $G^{(1)}$ even for gaussian fields. The information about the central frequency ω_o is lost.

LIGHT SCATTERING: GENERALITIES [4]

A schematic light scattering experiment is sketched in fig. 1. A monocromatic plane wave, linearly polarized, is incident upon a perfectly uniform transparent medium.
An optical detector in the position P reveals the presence of a nonzero light intensity, generally weak, propagating in directions other than that of the reflected and refracted beam. This is what is called scattered light.

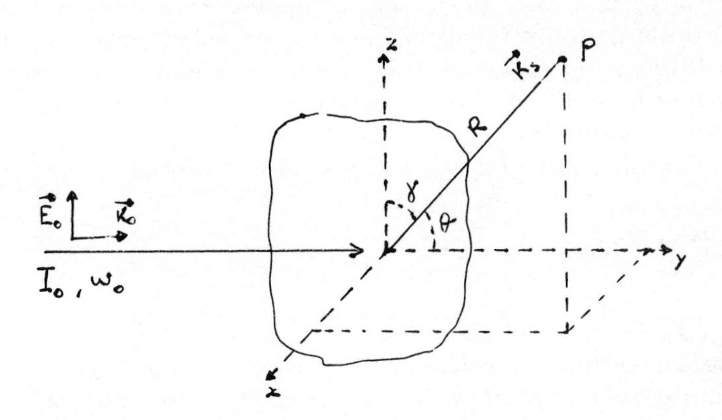

Fig. 1 - Schematic light scattering experiment. The incident beam propagates along the y axis and is linearly polarized along the z axis; ω_o and I_o are respectively its angular frequency and its intensity. The scattered light is observed at the point P having polar coordinates (R, θ, γ). θ is the angle between $\vec{k}_o$ and $\vec{k}_s$. The effective volume V which contributes to the scattered field collected at P does not include the entire sample, but it is rather defined by the cross section of the incident beam and the detection optics. The distance R is taken to be much larger than the linear size of the scattering volume V.

The physical origin of the scattering process can be understood in the following way. The illuminated medium interacts with the incident electric field at optical frequency through an electric polarizability per unit volume $\chi(\vec{r}, t)$. For sake of simplicity the medium is assumed to be optically isotropic (χ is a scalar quantity) and linear (χ independent of the amplitude of the incident field).
The polarization induced in each element of the illuminated volume is oscillating at the same frequency of the incident field. The field radiated by each volume element follows the well-known dipole radiation pattern. The field collected by a detector placed in the position R is the sum, with appropriate phases, of the contributions from each volume element. It is easy to show that, if $\chi(\vec{r}, t)$ is independent of $\vec{r}$, we get destructive interference in any direction, apart from that of the refracted beam. If, however, χ is a fluctuating function of $\vec{r}$, all the elementary contributions to the scattered field will not completely cancel out, and we do expect a nonzero scattered intensity.

The polarizability χ can always be written as

$$\chi(\vec{r}, t) = \langle \chi \rangle + \delta\chi(\vec{r}, t)$$

where $\langle \chi \rangle$ is the average part, independent of $\vec{r}$ and t for a homogeneous medium in stationary conditions, and $\delta\chi(\vec{r}, t)$ is the fluctuating part, which has zero average. From the intuitive considerations given above, it is clear that scattering is produced by $\delta\chi(\vec{r}, t)$.

The theoretical computation gives the following expression for the scattered field $\vec{E}_s(\vec{R}, t)$:

$$\vec{E}_s(\vec{R}, t) = \frac{1}{\langle \varepsilon \rangle} \vec{k}_s \times (\vec{k}_s \times \vec{E}_o) \frac{V e^{i(\vec{k}_s \cdot \vec{R} - \omega_o t)}}{4 \pi R} \delta\varepsilon(\vec{k}, t) \tag{7}$$

where $\varepsilon = 1 + \chi$ is the relative dielectric constant of the medium, V is the volume, $\vec{k}_s$ is the wave vector of the scattered field, and $\delta\varepsilon(\vec{k}, t)$ is defined by the Fourier transformation:

$$\delta\varepsilon(\vec{k}, t) = \frac{1}{V} \int_V \delta\varepsilon(\vec{r}, t)\, e^{-i(\vec{k} \cdot \vec{r})}\, d^3r \tag{8}$$

The vector $\vec{k}$, as shown in fig. 2, is defined by:

$$\vec{k} = \vec{k}_s - \vec{k}_o \tag{9}$$

Eqs. (7-9) indicate that of all the Fourier components of the fluctuation in dielectric constant only that particular component whose wave vector is the difference between tha wave vectors of the scattered and incident light is responsible for scattering in the direction of observation. It is interesting to observe that this is completely equivalent to say that the scattering process has to satisfy the Bragg condition for the reflection of the incident beam by a tridimensional grating characterized by a reciprocal vector $\vec{k}$.

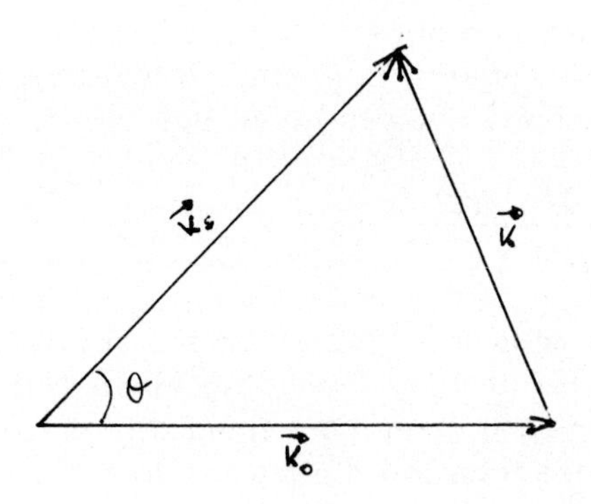

Fig. 2 - Triangle of wave vectors. If $k_s \simeq k_o$, the triangle is practically isosceles.

Eqs. (7) has been derived by using a perturbation approach which takes $\delta\chi(\vec{r}, t)$ to be small compared to $\langle\chi\rangle$ and which assumes the attenuation of the incident field to be negligible over the whole length of the scattering volume. Multiple scattering effects are, therefore, assumed to be very weak.

The scattered field $\vec{E}_s(\vec{R}, t)$ is a random function of position and time with a zero average. The time-dependent fluctuations of E_s exactly mirror the fluctuations in dielectric constant of wave vector $\vec{k}$. Fluctuations in dielectric constant are generally much slower than an optical period, that is, the energy associated with an elementary excitation in the medium is much smaller than the energy of the incident optical photon. As a consequence, the energy conservation theorem tells us that the energy of the scattered photon is practically the same as that of the incident one. By putting $|\vec{k}_s| = |\vec{k}_o| = k_o$, the momentum conservation relation (9) gives $k = 2k_o \sin\vartheta/2$ where ϑ is the angle between $\vec{k}_o$ and $\vec{k}_s$ (see fig. 2).

The space-dependent fluctuations of E_s depend only, in usual cases, on the geometry of the experiment. Indeed, if we compare $\vec{E}_s(\vec{R}, t)$ with $\vec{E}_s(\vec{R}+\delta\vec{R}, t)$, where $\delta\vec{R}$ is a displacement on the sphere of radius R centered at the origin of the coordinate system, we find that both amplitude and phase of E_s are different since the relative phases of the scattered fields from each volume element change by moving from $\vec{R}$ to $\vec{R}+\delta\vec{R}$. The coherence area A_c of the scattered field is qualitatively defined as the area (on the sphere of radius R) over which the scattered field is appreciably uniform in amplitude and phase. A more precise definition would imply the use of a spatial correlation function for the scattered field. The coherence area is given by:

$$A_c \simeq \frac{\lambda^2 \quad R^2}{A_s(\vartheta,\gamma)} \tag{10}$$

where $A_s(\vartheta,\gamma)$ is the area intersected on the scattering volume by a plane perpendicular to R and passing through the center of the scattering volume. It is evident from this definition that A_c depends upon the observation direction, that is, the angles ϑ and γ. The ratio λ^2/A_s is called the coherence solid angle. Eq. (10) can be interpreted as an extension of the well known result for a one-dimensional grating of size a, which gives a diffraction angle λ/a and, therefore, a spot size $\lambda R/a$ at a distance R.

The scattered field $E_s(R, t)$ at a given point is a random function of time. A complete characterization of it is given by the set of correlation functions:

$$G_m(t_1, \ldots, t_{2m}) = E_s^+(t_1)\ldots$$

$$\ldots E_s^+(t_m)\, E_s(t_{m+1})\ldots E_s(t_{2m})$$

where m runs from 1 to infinity. By using Eq. (7) a one-to-one correspondence between the correlation functions of the field and those of the dielectric constant can be established.

LIGHT SCATTERING: MACROMOLECULAR SOLUTIONS [5]

In a macroscopic system in thermal equilibrium at a temperature T, local fluctuations of the dielectric constant are caused by fluctuations in the thermodynamic parameters describing the state of the system. For instance, in a pure fluid the value of $\delta\varepsilon(\vec{r}, t)$ is mainly determined by local density fluctuations. We will discuss now a specific example, that is light scattering from a suspension of non-interacting macromolecules, small compared with the wavelength of light. For such a system, containing N_s macromolecules in the scattering volume

$$\delta\varepsilon(\vec{r}, t) = \Delta\varepsilon_r \sum_{i=1}^{N_s} \delta\left[\vec{r} - \vec{r}_i(t)\right] \tag{11}$$

where $\Delta\varepsilon_r$ is the difference in relative dielectric constant between a macromolecule and the solvent, and $\vec{r}_i(t)$ is the position of the macromolecule i at time t. Equation 7 then becomes (taking $\vec{k}_s \perp \vec{E}_o$)

$$E_s(\vec{R}, t) = \frac{E_o K_o^2}{4\pi \langle\varepsilon\rangle R} e^{i(K_s R - \omega_o t)} \Delta\varepsilon_r \sum_{i=1}^{N_s} \exp\left[i \vec{K} \cdot \vec{r}_i(t)\right] \tag{12}$$

This result can also be obtained directly by regarding each macromolecule as an elementary dipole radiator. The electric field correlation function is given by

$$G^{(1)}(\tau) = \frac{|E_o|^2 K_o^4 \Delta\varepsilon_r^2}{16\pi^2 \langle\varepsilon\rangle^2 R^2} e^{i\omega_o\tau} \sum_i \sum_j \left\langle \exp\left\{ i\vec{K} \cdot \left[\vec{r}_i(t) - \vec{r}_j(t+\tau)\right]\right\}\right\rangle \tag{13}$$

For independent scatterers, the position of particle i will at all times be uncorrelated with the position of particelle j. Thus only the terms for i = j will contribute to the double sum of Eq. (13). For random walk diffusion under the influence of Brownian motion it is easy to show that

$$G^{(1)}(\tau) = \langle I_s \rangle \exp(i\omega_o \tau) \exp(-Dk^2\tau) \tag{14}$$

where $\langle I_s \rangle$ is the average scattered intensity and D is the translational diffusion coefficient of the macromolecule. An intuitive justification of this result comes from the fact that $(Dk^2)^{-1}$ is roughly the time taken by a macromolecule to diffuse a distance $1/k$.

If N_s is not too small, the scattered field, being the superposition of many statistically independent contributions, is gaussian. The intensity correlation function is therefore immediately derived from Eqs. (14)and(6), and reads

$$G^{(2)}(\tau) = G^{(1)}(0)^2 + |G^{(1)}(\tau)|^2 = \langle I_s \rangle^2 \left[1+\exp(-2Dk^2\tau)\right] \tag{15}$$

THE INTENSITY CORRELATION TECHNIQUE

Let us evaluate in a typical case the correlation time $\tau_c = (Dk^2)^{-1}$. For a spherical particle, D is given by the Einstein-Stokes relation

$$D = \frac{K_B \quad T}{6\pi\eta r} \tag{16}$$

where K_B is the Boltzmann constant, T the absolute temperature, η the viscosity of the solvent, and r the radius of the particle. By considering a room temperature aqueous solution of spherical macromolecules with $r \simeq 20$ Å, we find $D \simeq 10^{-6}$ cm^2/sec. If the optical source is a He-Ne laser (λ=6328 Å) and the scattering angle is $\theta = 60°$, the correlation time comes out to be around 10^{-4} sec. We could evaluate alternatively the linewidth $\Delta\nu_c$ of the optical spectrum of the scattered light. From Eqs. (3) and (14) we obtain a Lorentzian spectrum, with $\Delta\nu_c = \frac{1}{2\pi\tau_c} \simeq 2$ KHz.

Besides the problem of the required frequency stability of the laser source, it is clear that such a narrow linewidth cannot be measured with the available optical instrumentation. We recall in fact that the smallest obtainable instrumental width of a Fabry--Perot interferometer is around 1 MHz. The measurement of $G^{(2)}(\tau)$ is simply performed by sending the scattered beam to a photodetector and by electronically processing the electric current at the output of the detector. Weak light beams are generally detected by high-gain photomultiplier tubes which yield a current pulse for each photon absorbed by the photodensitive surface. If the optical signal to be analyzed is so intense that many photons are absorbed within the response time of the photodetector, the output electric current i(t) can be considered as an analog signal proportional to the intensity of the light beam. If we assume that i(t) is a stationary random variable, its autocorrelation function $R(\tau)=\langle i(t)i(t+\tau)\rangle$ depends only on the delay τ and can be evaluated by a time average. In practice, $R(\tau)$ is measured by sampling i(t) periodically and performing the appropriate multiplications among the samples. Since it is easier and faster from an electronic point of view to perform multiplications among digital signals, the result of each sampling operation is quantized through an analog-to-digital converter. Several correlators based on this technique are now commercially available. They generally work at a maximum sampling frequency of 4 MHz, but real time operation is possible only at sampling frequencies lower than a few kilohertz.

If, however, the optical signal is so weak that it is very unlikely to detect more than one photon within the response time of the photodetector, the output electric current consists of a random train of nonoverlapping pulses. In this case the efficiency of correlators utilizing analog-to-digital conversion is very low, and it is more convenient to exploit the fact that the signal is already in digital form. Several photon correlators have been built in the last few years. A detailed description of their design and operation can be found in Refs. 6 and 7.

It should be recalled that the same information obtained from the intensity correlation function can be gathered by measuring the photocurrent power spectrum which is the Fourier transform of the correlation function. This was indeed the technique employed in the original applications of electronic spectroscopy to light scattering. Spectral analysis is still widely used for velocimetry applications (fluid mechanics and electrophoresis) because it gives directly the velocity distribution of the scatterers [2].

In usual light scattering experiments the scattered field has gaussian statistical properties, so that the time behavior of $G^{(1)}$ can be derived from the measured $G^{(2)}(\tau)$. It should be mentioned, however, that a direct measurement of $G^{(1)}$ can be performed also by the intensity correlation technique. Indeed if the scattered light under investigation is mixed on the photodetector surface with some of the unscattered light which acts as a local oscillator, and the intensity of the local oscillator is made much larger than the average scattered-intensity, it can be shown that the time dependent part of the intensity correlation is proportional to cos $(\omega_o - \omega_s)$ $|G^{(1)}(\tau)|$, where ω_o and ω_s are respectively the frequency of the incident beam and the central frequency of the scattered field. The reference-beam technique is particularly useful when the scattered field is not gaussian or is frequency shifted with respect to the incident field.

Since ICS introduces electronic instead of optical delays, there is practically no upper limit to the available time delays. The response time of the photodetector puts however a lower limit to the delay. A very interesting feature of ICS is that requirements on the temporal coherence of the optical source are much less severe than in the case of interferometric measurements. If Eq. (7) is written as

$$E_s(t) = E_o(t)\, a(t), \tag{17}$$

where a(t) is proportional to the amplitude of the appropriate Fourier component of the fluctuating polarizability of the scattering medium, and if we take into account that $E_o(t)$ and a(t) are statistically independent, the correlation functions of $E_s(t)$ are expressed by

$$G_s^{(1)}(\tau) = G_o^{(1)}(\tau)\; G_a^{(1)}(\tau) \quad \text{and} \tag{18}$$

$$G_s^{(2)}(\tau) = G_o^{(2)}(\tau)\; G_a^{(2)}(\tau), \tag{19}$$

where $G_o^{(1)}(\tau) = \langle E_o^+(t)\, E_o(t+\tau)\rangle$, $G_a^{(1)}(\tau) = \langle a^*(t)\, a(t+\tau)\rangle$, and so on. The function $|G_o^{(1)}(\tau)|$ is constant if τ is much shorter than the coherence time τ_{co} of the source, and goes to zero for $\tau \gg \tau_{co}$. Therefore $G_a^{(1)}(\tau)$ can be derived from the measured $G_s^{(1)}(\tau)$ only when its decay time is $\lesssim \tau_{co}$. Quite different is the situation for the measurement of $G^{(2)}$, because $G_o^{(2)}(\tau)$ becomes asymptotically a constant for large delays. Hence, provided the source is intensity stabilized, no matter how broad is its optical spectrum, $G_s^{(2)}$ is proportional to $G_a^{(2)}$. The only limitation to the spectral width of the source comes from the fact that Eq. (7) is valid only if the scattering volume is smaller than the coherence volume of the source.

APPLICATIONS

We list below tha main applications of ICS.

a) Fluids in thermal equilibrium:

Physical chemistry of macromolecules in solution: polymers, biological macromolecules, micelles. Information on size, shape, hydration, molecular weight, interaction, association-dissociation equilibria, motility of living microorganisms (bacteria, sperm cells).

Dense fluids: single and multicomponent systems. Measurement of thermal diffusitivity and mass diffusion coefficients. Study of phase transitions 8 .

Fluid interfaces: velocity of propagating surface waves (riplons). Information on surface tension and shear viscosity.

Plasmas and gases: velocity distribution of ions, electrons, atoms. Information on temperature and other parameters.

b) Flowing fluids (Laser Doppler Velocimetry)

Hydrodynamics : laminar and turbolent flows, convective instabilities.

Aerodynamics: wind tunnels

Combustion and flames

Blood flow

Electrophoresis: macromolecular motion under the action of an electric field.

Applications b) are reviewed in Corti's lectures in this volume. We only mention here the principle on which Laser Doppler Velocimetry is based. If we consider Eq. (12) and assume that the N_s particles are moving with a constant velocity $\vec{v}$, the position $\vec{r}_j(t)$ can be written as $\vec{r}_j(t) = \vec{r}_{oj} + \vec{v}\,t$. As a consequence the scattered field is frequency shifted with respect to the incident field of a quantity $\Delta\nu_D = \frac{\vec{k} \cdot \vec{v}}{2\pi}$.

We have not here the space to give even a sketchy treatment of all applications a). The following short discussion is confined to macromolecular solutions.

Macromolecules with a linear size much smaller than $1/k$ are essentially point scatterers, and only motion of their centres of mass will contribute to the time-dependence of scattered light fluctuations. If we further assume the scatterers to be non-interacting and identical, $G^{(1)}(\tau)$ is exponential (sec. Eq. 14) with a decay time $\tau_c=(DK^2)^{-1}$. The translational diffusion coefficient D is a parameter of some importance for the following reasons. First of all it can be used to determine the molecular weight of the macromolecule through the Svedberg equation 5 , once the sedimentation coefficient is known. Furthermore D provides a direct measure of the friction coefficient f_o through the relation $D=(K_B T/f_o)$. For spherical particles f_o is, in turn, given by the Stokes relation (cf. Eq. 16)

$$f_o = 6\pi\eta r_H \tag{20}$$

where r_H is the radius of the particle in solution, the so-called hydrodynamic radius. Note that when the macromolecule is hydrated r_H does not coincide with the radius of the dry macromolecule, and the comparison of r_H withe the dry radius gives information on the degree of solvation. For non-spherical particles a form factor has to be included into Eq. (20). Therefore configurational changes in the macromolecule produced by any variation of the environment, such as temperature or pH, can be detected by measuring changes of D. A typical application is the study of the reversible denaturation of small proteins.

The treatment becomes more complicated if macromolecular polydispersity and (or) interactions are taken into account. We refer to the specialized literature for a full discussion of these points. We just mention here that, from the experimental point of view, polydispersity appears as a departure of $G^{(1)}(\tau)$ from the single exponential behavior and interactions give rise to a concentration dependence of the diffusion coefficient D.

A large variety of macromolecules have been studied in the last few years by ICS. The list includes naturally occurring and man-made polymers in water and other organic solvents, many macromolecules of biological interest, such as properties, viruses, phages, microorganisms such as bacteria and sperm cells. Very recently ICS has been applied to micelles which are colloidal aggregates formed by surfactant molecules [9].

REFERENCES

1) - Photon Correlation and Light Beating Spectroscopy, H. Z. Cummins and E. R. Pike eds., (Plenum, New York, 1974).
2) - Photon Correlation Spectroscopy and Velocimetry, H. Z. Cummins and E. R. Pike eds., (Plenum, New York, 1977).
3) - F. T. Arecchi and V. Degiorgio in Laser Handbook, F. T. Arecchi and E. O. Schulz-Dubois eds., (North-Holland, Amsterdam, 1972), Vol. 1, p. 191.
4) - B. Chu, Laser Light Scattering (Academic, New York, 1974).
5) - H. Z. Cummins and P. N. Pusey in Ref. 1.
6) - E. Jakeman, and C. J. Oliver in Ref. 1; V. Degiorgio in Ref. 2.
7) - M. Corti, A. De Agostini and V. Degiorgio, Rev. Sci. Instrum. 45, 888 (1974).
8) - G. B. Benedek in Statistical Physics, M. Chretien, E. P. Gross, S. Deser eds. (Gordon and Breach, New York, 1968), Vol. 2, p. 1.
9) - M. Corti and V. Degiorgio in Ref. 2.

Coherant Optical Engineering, F.T. Arecchi and V. Degiorgio (eds.)

A MODERN LASER INTERFEROMETER: REALISATION AND APPLICATION

S. Terentieff

HEWLETT PACKARD
Meyrin, Switzerland

INTRODUCTION

The materialisation of length standards has gone a long way. First it was a body part - the foot, or the thumb - name of the inch in many languages. Then, with the introduction of the metric system, it was for a long time a specific piece of platinum carefully preserved in Sèvres (Paris) at the Bureau International des Poids et Mesures, until Metrologists agreed to have a more independent and reproduceable standard, as the wave lengths obtained interferometrically by a well defined transition line of a Krypton 86 lamp. Now the legal definition of the meter is 1.650.763,73 such wave lengths. Similar to this trend of dematerialising the standard, dimensional measurements have also evolved drastically. Most organisations use also now, beside classical means as gauge blocks or standard rule, Laser based Interferometers. From a linear measuring device, they have now evolved to major problem-solver in dimensional metrology with multi-measurement capability. Rather than only measuring length, as conventional interferometers did, the modern laser--based system measures now angle (pitch and yaw), flatness, alignment, straightness, and orthogonality (squareness).

This is a major contribution to the field of dimensional metrology. One of its most significant applications, in addition to the metrology laboratory, is in machine tool testing, calibration and operation. With laser-based systems, it is possible to test and define most of the geometrical characteristics of a machine tool. Also, since most of the measurements can be performed under dynamic conditions, laser-based systems are used as a transducer for adaptive control of machine tools.

PRINCIPLE OF THE TWO-FREQUENCY INTERFEROMETER

It was the emergence of the two-frequency laser and interferometer system which allowed for considerable optical and electronic flexibility. Systems have become less susceptible to environmental changes and electronic signal manipulation is simplified and more reliable. It is so much easier to handle and process the AC signals obtained with this technology, as opposed to the DC signals used with the classical laser inteferometry, and requiring DC Amplifier-drift problems!-with delicate treshold settings. The two-frequency interferometer system schematic is shown in Figure 1.

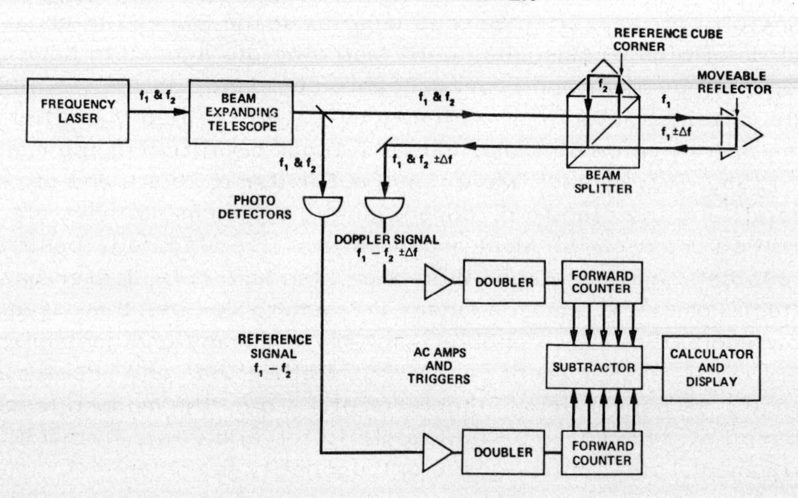

The light source for the interferometer is a single mode HeNe laser using Zeeman splitting to obtain two frequencies f_1 and f_2. f_2 is used in the reference arm of the interferometer and f_1 in the measurement arm. Any movement of the reflector causes a Doppler shift Δf in f_1. The two frequencies, f_2 and $f_1 + \Delta f$ after optical inteference at the beam splitter are electrooptically heterodyned to form the measurement signal. This is then compared to the reference signal $f_1 - f_2$ to obtain Δf which is proportional to the distance the reflector has travelled. By applying the appropriate conversion factor the distance is then displayed, or is otherwise available, either in the English or Metric system. Another essential advantage of the two-frequency laser of opposite circular polarization is the absence of Brewster window. That means that the tube is a solid, reliable glass envelope resulting in an extremely long life. The oldest Laser Tubes are over 6 years old, and still lasing strong.

ACCURACY

The accuracy of Laser Interferometers has been extensively tested and checked by all important standard organizations in Europe and America (BIPM, NPL, PTB, NBS, Istituto Colonetti, etc.). Let's say in a simplistic way that it is built in the properties of the matter, in our case in the neon atoms used in the Lasing Tube.

Accuracies of a few parts in 10^8 have often been reported for the equipment discussed here, within an adequate environment. In other words, the Laser Interferometer gives the possibility to have in house a secondary standard which compares very favourable with any absolute standard used throughout the world (Krypton 86 source or master rule).

COMPENSATION

During the conversion from wave lengths to the Metric or English system, which is in fact a multiplication, it is very easy to add the compensation for environmental changes. This can be done by either dialling in a suitable compensation number or by an automatic compensator which monitors the air parameters (temperature, pressure, humidity) and continuously updates the measurement information.

THERMAL EXPANSION

Usually, far more important (by a factor 10!) than the air influence is the actual, absolute temperature of the workpiece. The reference temperature is normally 20°C or 68°F. This is also compensated for very easily by means of multiplication, either dialling manually the adequate factor of expansion, or monitoring continuously the piece temperature with the automatic compensator.

GLOBAL METROLOGY SYSTEM

Multi-measurement capability became practical only after the development of the remote interferometer. This passive opto-mechanical device removes the optics from laser and electronics, makes them accessible, thus permitting their re-arrangement. A benefit of remote interferometry is the reduction or elimination of thermal problems, dead path errors, Abbé errors and mounting problems.

Linear Displacement

This is the classic application described above on which all subsequent systems are based.(Figure 1 detail)

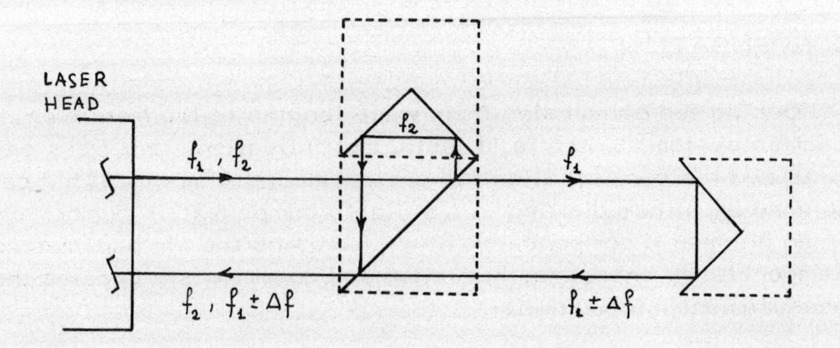

Angles

One typical example is the angular measurement configuration of the remote interferometer as shown in Figure 2. In this application the reference and measurement arm are used to make a differential measurement. This is accomplished by deflecting the reference beam parallel to the measurement beam and using as targets two cube-corner reflectors which are accurately spaced in a mounting structure. The differential reading of the two beams is directly proportional to the angular deviation from the datum plane of the mounting structure. This results in a measurements capability of 0. 1 arc-second resolution, high accuracy and wide dynamic range (Figure 2).

ANGULAR/FLATNESS MEASUREMENT

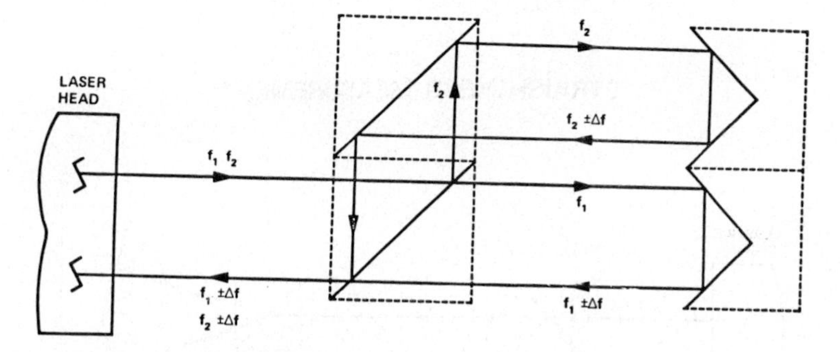

Pitch & Yaw

For this measurement; the mount holding both reflectors is fixed in the spindle. Therefore, during the motion of the machine head, its angular motion in both planes, vertical (pitch) and horizontal (yaw) can be measured.

Straightness

Another important geometric measurement is straightness (or alignment). Straightness is a translational measurement as compared to the above described angular (or rotational) measurement. This measurement can also be accomplished interferometrically.

One way of doing it is to use a polarizing prism, also called Wollaston prism, to split laser light in its two frequency components f_1 and f_2 (Figure 3). Each of the beams is then reflected back to the polarizing prisms by fixed plane mirrors, where they interferometrically combine. Any lateral translation of the polarizing prism as it travels along the axis of the laser beam is indicated as a deviation of a straight line.

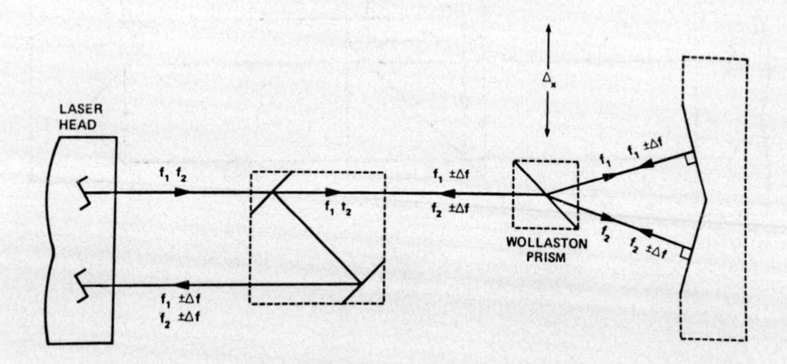

Figure 3

Squareness

Using the same set-up but bending the laser beam by exactly 90°, the squareness (orthogonality) between two surfaces can be determined. The square reference will be in this case a high precision pentaprism (Figure 4).

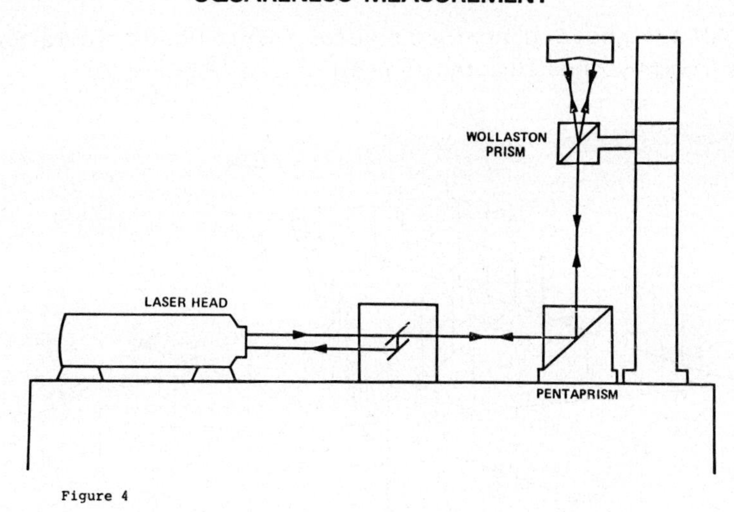

Figure 4

Parallelism

Combining the straightness capability and the pentaprism property, it is possible to relate to each other the straightness of several components of a machine tool. As an example, the parallelism of the head stock motion to the bed of a lathe can be measured.

Flatness

Additional capabilities of the Laser Measurement System include flatness measurement. Using the angle measurement configuration and accepted data reduction techniques, the topography of machine beds (as well as other reference surfaces) can be mapped out.

Calculating Capabilities

All the above-described measurements need calculations, even lengthy ones like the isometric plot (Figure 5).

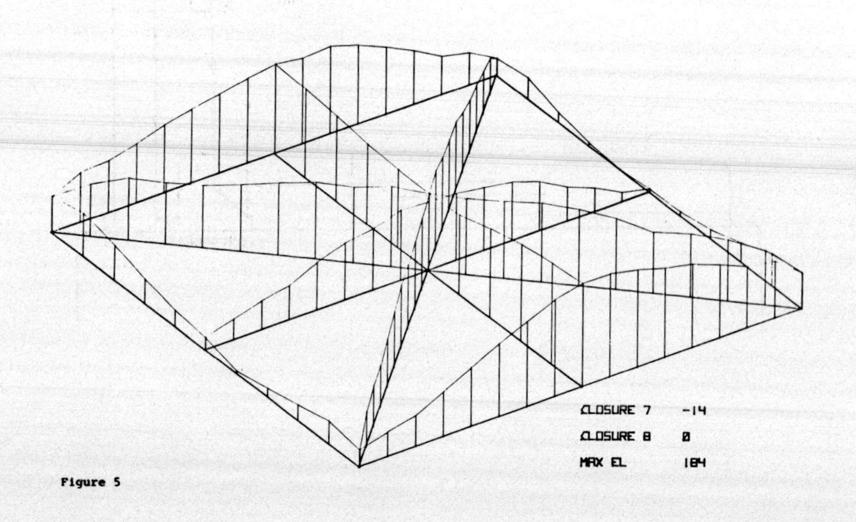

Figure 5

It is very easy to connect a Laser Interferometer to a powerful desk top Calculator. Measurements Data are directly transferred and stored for futher processing.

The results are clearly printed or plotted, presented in a protocoll-like fashion.

Numerous metrology programs are available (like NMTBA or VDI 3254 for N/C machines) - and can also be very easily developed for specific task because the desk-top calculator language is so direct and relatively simple. Statistics are very important to realistic and efficient metrology. Measurement cycles can be easily automated and repeated often enough without great expense so statistical rules and probabilities can be gained.

The extended use of Calculators have almost eliminated human errors due to excessive repetition and fatigue as often happens in metrology.

LASER TRANSDUCER

The idea to build in the machine itself the Testing device, making it the monitoring device, was not remote at all. Using the same elements, proven through years of experience (like the Laser Tube), a very modular measuring system was created (Figure 6). Using a single Laser source, up to 6 Axis of motion can be measured simultaneously, each Axis having its own interferometer and receiver.

For processing the electronic signals, many possibilities are open with various interface cards:

- Desk top Calculator (for Measuring Machine)
- Mini Computer for CNC
- Output signals compatible with normal, hardwired N/C

The simple preparation (no machined surface for rules or scales) and easy mounting, the big range without increased costs, of course the absence of calibration need, not to speak of the high accuracy and precision, make the Laser Transducer not only attractive but also competitive.

Many companies from the integrated circuit field use it: Step and repeat cameras (Elektromask, David Mann), Pattern Generator (Gerber), Electrobeam Machine (Cambridge Instruments), Measuring Machine (Leitz), etc.

Also Machine Tool builders start to use the Laser Transducer - Zocca (Como) for a 3-Axis Precision Grinder, Lucas (U. S.) for a milling center, Charmilles (Geneva) for an Automated Checking System of wire erosion machines, etc. Many other companies in Europe and the U. S. are now conducting extensive evaluation programs.

CONCLUSION

The modern Laser Interferometer has become a valuable tool in many metrology rooms of the world - either in standard organization like Istituto Colònetti, NPL or Machine Tool builders and Machine Tool users. Its contribution to the machine shop is also very important - calibration of a modern Machine Tool can be done with reported time saving of 6 to 10.

It is a key element built in machine where precision is a necessity.

The Laser Interferometer is helping daily to achieve efficient accuracy in all ranges of the technical world.

Coherant Optical Engineering, F.T. Arecchi and V. Degiorgio (eds.)

HIGH ACCURACY LONG DISTANCE METROLOGY

Bruno Querzola
CISE - P. O. B. 3986 - 20100 Milano, Italy

High accuracy telemetry consists of distance measurements where the fractional accuracy in ranging is better than 10 ÷ 100 ppm (part per million). The methods used are based on propagation time measurements of an e. m. wave over the examined path length.

The main utilization fields and requirements are listed in Table I.

MILITARY USE[1,2]

- "ALL WEATHER" OPERATION (NOT POSSIBLE AT ALL),
- VERY HIGH PROBABILITY OF DETECTION (NO LOSS OF MEASUREMENTS),[3]
- VERY LOW FALSE ALARM RATE,
- SET OF "ALL OBSTACLES" (INCLUDING TELEPHONE WIRE PAIRS),
- INFORMATION RATE ⩾ 1/sec, (IN CONNECTION WITH TARGET SPEED),
- MAXIMUM RANGE: AT LEAST 1÷10 Km,
- LOW ACCURACY IN RANGING (A FEW METERS),
- SMALL SIZES AND WEIGHTS,

TOPOGRAPHIC USE

- EASE OF USE AND HANDLING,
- LOW WEIGHT AND SIZES,
- $10^{-4} \div 10^{-5}$ FRACTIONAL ACCURACY ($\Delta L/L$);
- LOW RANGE (A FEW Km.)
- LOW PRICE,
- HIGH LIFE TIME OF THE SOURCE,
- POSSIBILITY OF DAY-OPERATION.

GEODETIC USE

- FRACTIONAL ACCURACY OF 10^{-6} OR BETTER,
- LONG RANGES (A FEW Km ÷ 100 Km),
- TRANSPORTABLE,
- INDEPENDENT POWER SUPPLY.

SATELLITE RANGING USE

- VERY LONG RANGE (FROM HUNDREDS TO MORE THAN TENS OF THOUSANDS OF Km),
- A FEW METER ACCURACY (LESS THAN TEN cm, FOR MORE SOPHISTICATED GEODESIC APPLICATIONS),
- RANGING EVALUATION WITH ONLY ONE SHOT OF THE PULSED LASER,
- LOW OUTPUT BEAM DIVERGENCE (FROM TEN TO 200 μRAD IN CONNECTION WITH ANGULAR ACCURACY OF THE TRACKING SYSTEM),

Table I : Utilizations of optical rangefinders and main requirements.

As the required accuracies are worse than a few meters on ranges of several chilometers the military rangefinders cannot be considered as high accuracy instruments. By laser telemetry it is possible to determine the modulus of the station-satellite vector with high accuracy. Accurate distance measurements can be performed with optical beams by using the different methods reported in Table II.

1) INTERFEROMETRIC TECNIQUES (UP TO 50 m. IN FREE AIR).
2) PHASE COMPARISON TECNIQUE WITH MODULATED C.W. BEAMS.
3) OPTICAL RADAR TECNIQUES WITH OPTICAL PULSES (FROM 100 m. TO OVER 100 Km).

ACCORDING TO THE ACCURACY IN RANGING, METHODS 2 AND 3 ARE BOTH EXPLOITABLE FOR:

o SYSTEMS WITH ONLY ONE OPTICAL CARRIER AND NO REFRACTIVITY CORRECTION.	FRACTIONAL ACCURACY WORSE THAN 1 ppm ($10^{-4} \div 10^{-6}$ ppm).
o SYSTEMS WITH TWO OR MORE OPTICAL CARRIERS AND WITH AIR REFRACTIVITY CORRECTION.	FRACTIONAL ACCURACY BETTER THAN 1 ppm.

Table II : Basic methods in optical beam ranging.

We neglect the interferometric method, which is described in another paper. The measurements concerning techniques 2 and 3(Table II) can be performed by evaluation of the transit time τ from the source to the target and back to the receiver of an amplitude or polarization modulated light. The pathlength L and the transit time are related through:

$$L = \frac{c}{2n_G} \tau \tag{1}$$

where c is the light speed in vacuum and n_G is the average over the whole path of the air group refractive index. As to the phase comparison telemeters with only one c.w. optical carrier a review of the methods and a description of some typical block diagrams are reported in ref. 4. The first phase comparison Bergstrand geodimeter, the phase comparison with frequency down conversion arrangement, the AGA Geometer mod. 8, the Geodelite mod. 3G and the Mekometer are described in the above mentioned reference. Some models of phase comparison rangefinders are compared in Table III.

Range accuracy values reported in the sixth column account for intrumental error only, accuracies better than 1 ppm being precluded to single wavelength telemetry, as the mean refractive index along the propagation path depends on local atmospheric conditions, above all on air pressure and temperature. This dependence makes the overall inaccuracy of one-wavelength telemeters higher than a few ppm. In spite of this accuracy limitation these telemeters are the only ones used nowadays as they are commercially available, whereas systems with air dependence correction are not on the market.

DEPENDENCE OF THE AIR REFRACTIVE INDEX ON ATMOSPHERIC CONDITIONS

The particular expression of the air refractive index can enable to draw the air refractive index and hence the geometric path-length, only as a function of transit times of optical pulses at two different wavelengths. The dependence of the refractive index both on wavelength of radiation and atmospheric parameters has suggested that the problem should be considered under two viewpoints:

1) search of analytical relationship showing dependence of the refractive index on wavelength under fixed standard atmospheric conditions;
2) derivation od expression for n_G accounting for variations of p, T, f (partial pressure of water vapour) and p_{CO_2} partial pressure of CO_2).

The former problem was tackled by several authors whose results allowed Edlen[5,6] to give a formula for the refractive index in standard conditions. If the atmospheric conditions differ from standard conditions, the correlation to be made on refractivity is given by the simplified formula:

$$(n_G-1)10^6=(n_{SG}-1)\frac{288°K}{760\,\text{torr}}\ \frac{p}{T}\ 10^6-(17.0-0.186\sigma^2)\frac{f}{T} \qquad (2)$$

where $(n_{SG}-1)$ is the group refractivity for standard condition from Edlen formula; the term in p/T stands for air pressure and temperature contributions and the term in f/T is the contribu-

	model	company	weight (Kg)	range (Km)	distance accuracies	Sources	modulation frequency number	modulation frequency field (MHz)	APPLICATION FIELD	APPROX PRICE
GEODIMETER	8	A.G.A. LIDINGO (Sweden)	23	60	± 5 mm+ +1 ppm	He-Ne LASER	4	30	GEODESY	No longer under construction. Replaced by MOD. 6BL.
//	710	//	2.5	5	//	//	3	15	TOPOGRAPHY	88.000 Swedish Crowns
//	12	//	2.8	1.7	± 5 mm+ +10 ppm	Ga As LASER	3	30	TOPOGRAPHY	20.000 Swedish Crowns
GEODOLITE	3G	SPECTRA PHYSICS (USA)	70	60	± 1 mm+ +1 ppm	He-Ne LASER	5	50	GEODESY	
MEKOMETER	ME 3000	KERN (SWISS)	18.7	3	± 0.1 mm+ + 1 ppm	XENON LAMP	5	500	TOPOGRAPHY GEODESY VIBRATIONS	100.000 Swiss Francs
TELLUROMETER	MA 1000	PLESSEY N.Y. (USA)		2	± 1.6 mm+ + 1 ppm	LAMP	5		TOPOGRAPHY	
RANGER	4 models	K.E.		4 6 12 60					TOPOGRAPHY GEODESY	
GEODIMETER	6BL	AGA LIDINGO (Sweden)	15	25 (40)	± 1 mm+ +1 ppm	He-Ne LASER	3	30	TOPOGRAPHY GEODESY	60.000 Swedish Crowns

Table III: Comparison among some models of c. w. telemeters.

tion of water vapour to the group refractivity out of standard conditions. The last term of eq. (2) in f/T must be negligible to allow the use of the " 2λ method". The last column of Table IV (where u is the relative humidity and $\lambda_r = 1/\sigma_r = 6943$ Å) shows that the presence of water vapour increases the fractional error $\Delta L/2$ in a not negligible way for air temperature higher than 10 ÷ 20°C.

T (°K)	u (%)	f (Torr)	$(17.0-0.186\,\sigma_r^2)$ f/T	$\frac{\Delta L}{L}$ (f/T)
273	50	2.3	0.14 ppm	0.21 ppm
	100	4.6	0.26	0.41
283	50	4.6	0.26	0.42
	100	9.2	0.54	0.85
293	50	8.7	0.49	0.79
	100	17.5	0.98	1.6
303	50	16	0.88	1.47
	100	32	1.75	2.93

$\frac{\Delta nG}{\Delta f} \sim 0.06$ ppm/Torr ; $\frac{\Delta nG}{\Delta f} \sim 1$ ppm/°K ; $\frac{\Delta nG}{\Delta p} \sim 0{,}35$ ppm/Torr

Table IV : Dependence of the fractional error in distance measurements on air water vapour content and air temperature, and typical variation (expression) of n_G with p, t and f.

In the same table the dependence of n_G on air pressure and temperature arising from the p/T term of (2) and the dependence of n_G on water vapour partial pressure are reported.

THE "TWO-WAVELENGTH" METHOD

In ref. 7, 8 it is shown the possibility to express geometric path-length L as a function only of two simultaneous measurements t_2 and t_1 at two different wavelengths without any dependence on the atmospheric quantities p and T. We consider two measuremnts being simultaneously if the mean index of refraction along the path n (that is the p/T ratio) keeps constant during both of them.

$$L = \frac{c}{2}\left(t_1 - \frac{t_2 - t_1}{\alpha}\right) \text{ where } \alpha = \frac{n^2_{SG} - n^1_{SG}}{n^1_{SG} - 1} \qquad (3)$$

The "2 λ method" consists in using this formula instead of eq. (1). α is a parameter depending only on index of refraction for standard conditions averaged the path at λ_1 and λ_2.

SOURCES OF ERRORS IN THE "2 λ METHOD"

The contribution to the inaccuracy in $\Delta L/L$ due to the evaluation of n_{SG} from eq. (2) is negligible ($\lesssim 0.05$ ppm). The errors Δt_1 and Δt_2 on the measures of t_1 and t_2 are caused both by instrumental effects and by atmospheric propagation processes.

By differentiating eq. (3) one gets the expression for the fractional error on distance L as a function of Δt_1 and Δt_2 :

$$\frac{\Delta L}{L} = \frac{c}{2L}\sqrt{\left(1 + \frac{1}{\alpha}\right)^2 \Delta t_1^2 + \frac{1}{\alpha^2}\Delta t_2^2} \simeq \frac{\sqrt{2}}{\alpha}\,\frac{\Delta t}{t} \qquad (4)$$

where the errors Δt_1 and Δt_2 are uncorrelated and we assume, in deriving the last form of eq. (4), that $\Delta t_1 \simeq \Delta t_2 = \Delta t$, $L \simeq ct/2$ and $1/\alpha \gg 1$. Expression (4) allows to perform transit time measurements with fractional errors $\Delta t/t$ less than $\alpha/\sqrt{2}$ times $\Delta L/L$, that is $\Delta t/t \lesssim \alpha/\sqrt{2}$ ppm to achieve a relative error in the distance measurement less than 1 ppm. In Table V a comparison is made among three prototypes based on the "2 λ method".

	KIND OF TRANSIT TIME MEAS.	λ_1(A)	λ_2(A)	SOURCES	α	$\frac{\Delta t}{t}$ (for $\frac{\Delta t}{t}$ = 1ppm)	MEASURABLE DISTANCES ($\Delta L/L \leq$ 1 ppm)
GEORAN [10]	PHASE COMPARISON	5145	4580	2 WAVEL. FROM ONE ARGON ION LASER	0.02	0.015 ppm	from 1 Km to 20 ÷ 30 Km
INSTRUMENT DEVELOPED AT BOULDER (U.S.A) [11]	PHASE COMPARISON	6328	4416	He - Ne and He - Cd LASER	0.047	0.03 ppm	
HIGH PEAK POWER PULSED RANGEFINDER [12]	DIRECT MEAS. BY PULSE PROPAGATION TIME MEASUREMENT	6943	3472	RUBY LASER AND SHG	0.12	0.09 ppm	from 10 Km to more than 100 Km

Table V : Some features of three different prototypes based on the "2 λ method".

ERRORS IN TRANSIT TIME MEASUREMENTS

In the analysis of the kind of errors contributing to transit time measurements a distinction can be made between:

- timing errors from receiving equipment (Table VI);
- timing errors from atmospheric effects (Table VII).

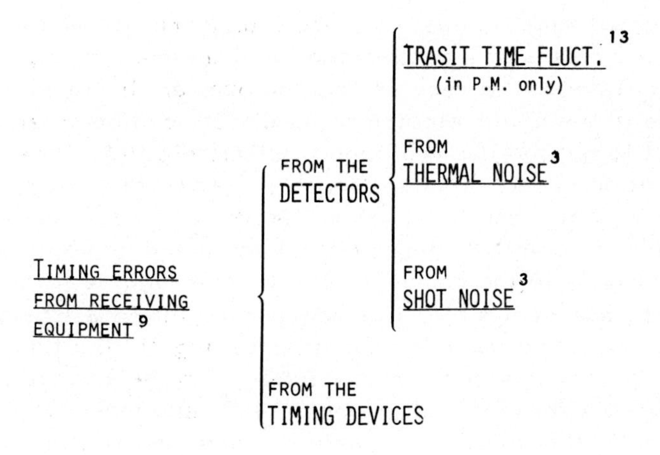

Tavle VI : Timing errors from receiving equipment.

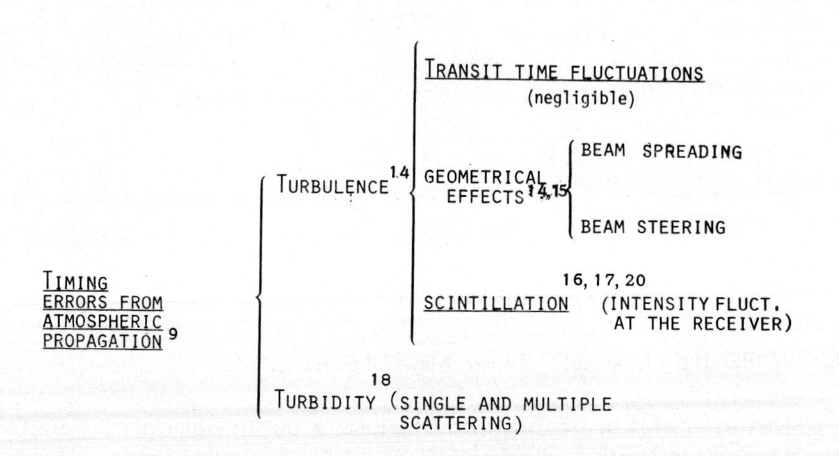

Table VII : The main causes of the transit time error are reported in ref. 9.

The transit time fluctuations in the photomultiplier can be decreased by means of a proper choice of the photomultiplier and the increase of the received photon number: in the same way a decrease in the photodetector noise effects can be obtained. This effect is not too dangerous for telemetric link since intensity fluctuations reach a saturation [17,19] after few kilometer path-lengths. All these turbulence effects can be decreased by properly dimentioning the transmitting and receiving optics of telemetric links. In the same way pulse deformations from turbidity can be made negligible. Further errors in transit time measurements can arise from "threshold crossing" in the timing device ("2 λ "pulsed systems) in connection with the emitted pulse duration when a few photons are received[9] and from reflector positioning. The whole error arising from stochastic error sources can be reduced by averaging over a few seconds for c.w. telemeters and over hundreds or thousands of shots for pulse telemeters. By averaging over a set of N measurements the error can be reduced by $N^{-1/2}$. In such a way the arrival time fluctuations associated with the discriminator can be decreased to a few psec (see the column of $\Delta t/t$ in Table V) as the fastest pulse discriminators have a time walk of few hundreds of psec.

LASER BEAM ATTENUATION ALONG THE PROPAGATION PATH

The main causes of laser beam attenuation are reported in Table VIII.

BEAM ATTENUATION[3,19]	TURBULENCE	(GEOMET. EFFECTS)	
	TURBIDITY	SINGLE AND MULT. SCATTERINGS	
		AIR ABSORPTION	

Table VIII : Laser beam attenuation sources.

The water vapour content (turbidity) is the main attenuation factor which increases the geometrical effects that can lead the laser beam energy out of the receiving optics. The "radar range equation"[21] can account for beam attenuation effects through L_M, the maximum range in absence of turbidity.

Fig. 1 shows a first comparison between continuous and pulsed " 2 λ systems" as to the achievable path-length. Up to now measurements over very large distances (exceedings 50-100 km or for ground to satellite links) are possible for pulsed systems only.

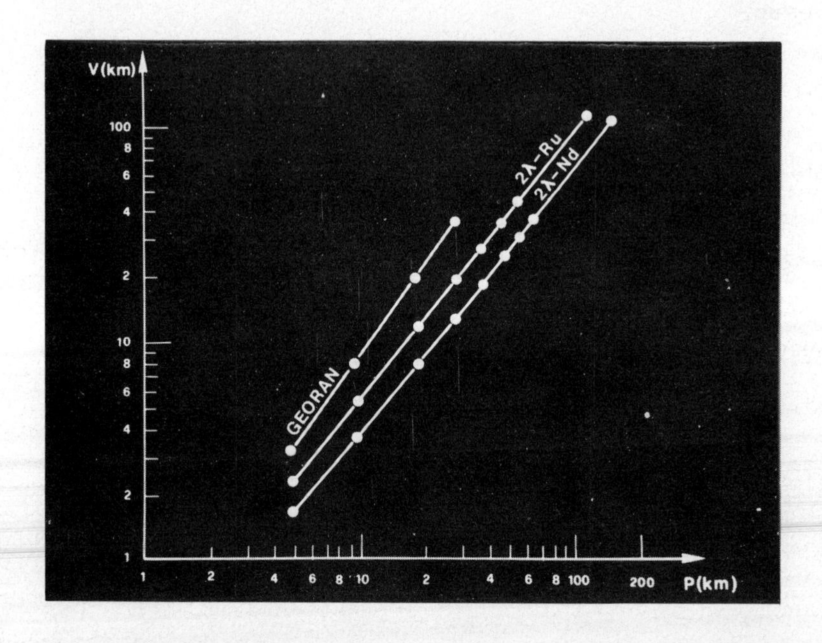

Fig. 1 - "Two-wavelength" phase comparison systems.

The main features of the two systems lately developed are reported in the first two lines of Table V. The former has been developed at the National Physical Laboratories (England), the latter at the Environmental Research Laboratories (Boulder, Colorado) where these systems have been experienced at first.

In the system shown in Fig. 2 the two laser beams are collimated and superimposed by means of a Wollaston prism; successively they go through a KDP electro-optic light modulator operating at a frequency of 2.8 GHz as a polarization modulator. Finally, the light is transmitted by an 8 inch Cassegrain Telescope. The light runs along the path to be measured and is sent back by a cat's eye retroreflector: it goes through the KDP modulator again and is detected by a pair of photomultipliers. The Wollaston prism can be used because the polarization of the light is modulated: this prism matches the beams for transmission and divides them upon reception. The evaluation of L consists of two transit time measurements at two different wavelengths through the "2 λ equation" (3).

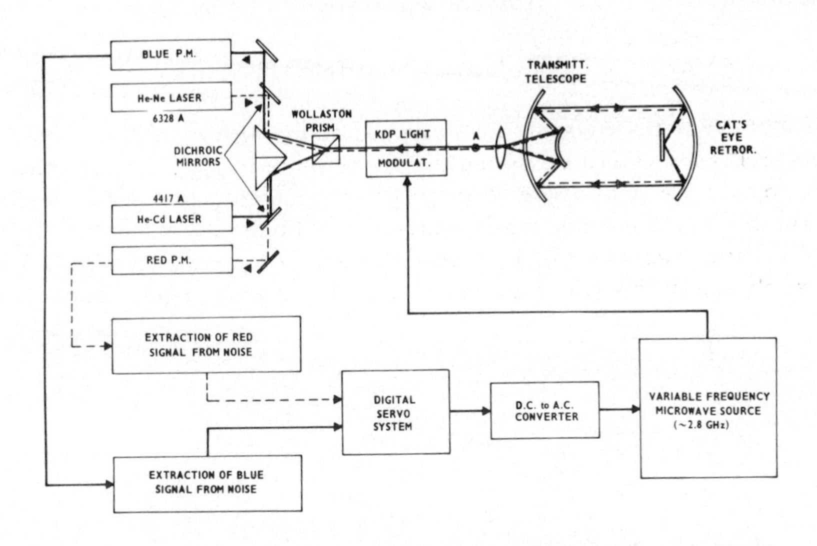

Fig. 2 - Block diagram of a two c. w. laser ranging instrument (developed by K. B. Earnshaw Hernandez-Boulder, Colorado).

The two transit time measurements are performed through the phase comparison between the returning polarization modulated beam and the microwave signal at the modulator by using 4 frequencies around 2. 8 GHz. The phase comparison is such that, if, for example, the signal at the modulator is 180° out of phase as to the returning polarized light, the light that has gone through the modulator twice will have no polarization change as to the emitted beam. This is the only condition giving a null signal at the output of the two photomultipliers. In this condition we have no error signal timing the two modulator frequencies by the servo system. Changing the modulation frequency, an exact odd number of half wavelengths of modulation can be made to occur over the optical path, thus giving rise to a null condition on the received light. A previous two c. w. system developed at Boulder,[22] which exploits the previously described technique of nulling operation, performs one transit time measurement only. The second measurement is carried out by adjusting the distance between two prism of an optical delay line. In Georan 1 [11] both the two null conditions are obtained through two separate mechanical adjustements authomatically controlled by the error signal at the output of the two different wavelength photomultipliers.

TWO-WAVELENGTH PULSED TELEMETER[9, 12]

Our first task was to analyze experimentally weather high pulsed lasers could be successfully used with the "2 λ method" and to setup a prorotype of such an instrument. In the block diagram of Fig. 3 a ruby laser emits a red beam λ_r = 6943 Å which partially converts (4%) into an ultraviolet beam (λ_u = 3472 Å). In such a way the two wavelengths arise from a laser source only.

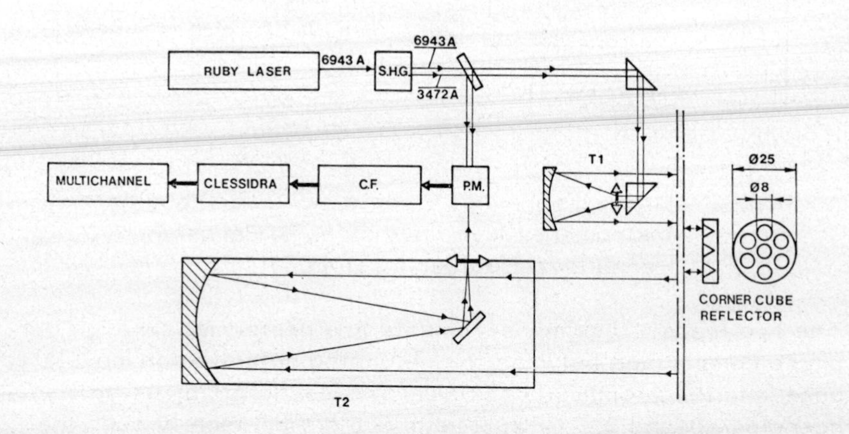

Fig. 3 - Block diagram of a "2 λ" high peak power experimental setup: T1 and T2 are the Newtoman transmitting and receiving telescope; P. M. is a 56 TVP photomultiplier; C. F. is a constant fraction discriminator.

The output pulses of our PTM[12] ruby laser have a 1.5 nsec rise-time and 10 MW peak power on 10 mm^2 cross section (corresponding to an intensity of 100 MW/cm^2). A beam splitter sends a part of the generated pulse (start) to the same 56 TVP photomultiplier which detects the stop pulse. The measurements were performed from a laboratory placed on "Monte San Primo" (1120 mt) close to the Como lake. The corner reflectors were put a two different distances at about 5 km and 13 km respectively. As a pulse discriminator we use an ORTEC "Constant fraction timing discriminator" mod. 453 which provides a time measurement almost independent from the pulses amplitude. The uncertainty in the switching time of the discriminator is the main source of instrumental error. As the electronic timer we use the "clessidra per picosecondi" an electronic timer built-up in CISE which can measure, with high resolution (50 psec/channel), the time interval between a start and a stop pulse. The timer outputs are recorded in the memory of a multichannel analyzer.

In Fig. 3 the statistical distributions of the transit times for the two wavelengths over a distance of about 5 km are reported.

The two sets of measurements, each consisting of about 150 counts (thus the measurement distance accuracy is improved a factor $\sqrt{150}$) at λ_r=6943 Å and λ_u=3472 Å were performed in cascade as only one timing device was available.

The transit time difference between the red and u. v. measurements is about 1 nsec, the refractive index difference being 33.10^{-6}.

The uncertainty of the mean value of these distribution was around 12 psec.

The above results lead to the conclusion that the major contribution to the uncertainty of the measurement comes from instrumental errors, the contribution due to atmospheric propagation being of less importance, at least over the measured distances.

As a conclusion, with our experimental condition (λ_r=6943 Å, λ_u=3472 Å, α = 0.121), fractional accuracies in rangings below one part per million can be obtained at distances larger than 20 km. By reducing the laser pulse intensity fluctuations and by increasing the number of counts of a factor two, the error should be lowered down to 3 psec, thus allowing the same accuracy over path-lengths of 5 km. The limits to the exploitation of a larger number of counts are set, at the present time, by the low pulse repetition rate of our source. Laser sources with a repetion rate higher (one pulse every 0.1 sec with a Nd-YAG laser source) than the present one can be exploited. An additional improvement can be obtained by gating the photomultiplier so that the noise in the detection channel decreases the probability of false measurement[9] consequently.

The state of art in the field of laser sources offers the possibility of ranging, with pulses as short as 10-100 psec.

There are some restrictions however in the availability of:

- high gain fast photomultipliers;

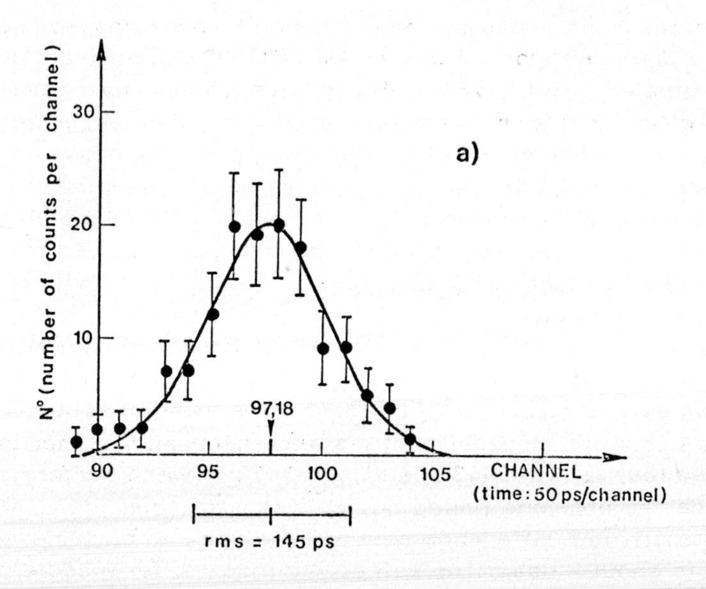

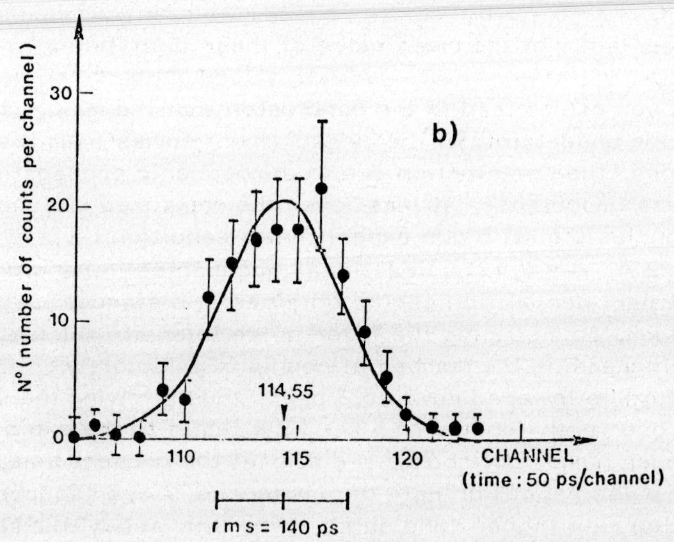

Fig. 4 - Statistical distribution of propagation times for the two laser beams at λ_r=6943 Å (a) and λ_u= 3472 Å(b). The measurements were performed on Oct. 25, 1973 over the Monte San Primo-Monte Garnasca path, close to lake Como. The distance was about 5 km.

- fast pulse electronic amplifiers,
- very fast discriminators and timing devices.

To overcome such restrictions a new alternative technique[23] has been studied measuring the pulse transit time along the round trip. The technique makes of a swept image converter tube (the streak camera ITL model STM 180, for example).

THREE-WAVELENGTH TELEMETERS WITH 0,1 ppm FRACTIONAL ACCURACY IN RANGING

In order to reach a fractional accuracy in ranging better than 1 ppm a two-wavelength technique must account for the p/T term of (2). The f/T term prevents from reaching accuracies of 0.1 ppm as the refractive index varies 0.1 ppm/mbar approximately with water vapour partial pressure and water vapour atmospheric variations[16] can exceed many mbar. Using three wavelengths G.D. Thayer[24] has demonstrated the possibility of correcting the distance measurement as to the water vapour refractive index dispersion: such a system[25] has been recently setup at the University of Washington through exploitation of:

- Two laser sources in the visible field from He-Ne (6328 Å) and He-Cd (4417 Å) lasers, respectively,
- a microwave beam at 9.9 GHz as a third wavelength.

This wavelength is chosen owing to the much stronger dependence of refractive index variations on water vapour content in the microwaves than in the visible.

REFERENCES

1) - P. Hermet,"Design of a rengefinder for military purposes", Appl. Opt., 11, 273 (1972).
2) - C.M. Kellington, "An optical RADAR system for obstacle avoidance and terrain following", VS. Army Avionics Laboratory USAECOM, Fort Moumanth, N.J. 07703 (1975).
3) - W.K. Pratt, "Laser communication systems", J. Wiley and Sons, inc. N.Y., London (1969).
4) - A. Sona, "Laser in metrology: distance measurements", Laser Handbook, North-Holland, Publ. Co., 1469-1476 (1972).
5) - B. Edlen, "The dispersion of standard air", J. Opt. Soc. Am., 43, 339 (1953).
6) - B. Edlen, "The refractive index of air Metrologia", 2, 12 (1966).

7) - P.L. Bender, J.C. Owens, "Correlation of optical distance measurements for the fluctuating atmospheric index of refraction", J. Geoph. Res., 70, 2461 (1966).

8) - M.C. Thompson, L.E. Wood, "The use of atmospheric dispersion for refraction correction of optical distance measurements", Proceed. Int. Association of Geodesy Symp. of Electromagnetic Distance Measurements - Oxford (G.B.), 165 (1965).

9) - B. Querzola, "High accuracy distance measurements by two-wavelength pulse laser source", to be published.

10) - K.B. Earnshaw and E.N. Hernandez, "Two laser optical distance-measuring instrument that correct for the atmospheric index of refraction", Appl. Opt., 11, 749 (1972).

11) - G. Shipley and R.H. Bradsell, "A new two colour laser ranger", International Symposium on Terrestrial Electromagnetic Distance Measurements, Stockholm (1974).

12) - B. Querzola, "High accuracy distance measurements by a two-wavelength pulsed laser", International Symposium on Terrestrial Electromagnetic Distance Measurements, Stockholm (1974).

13) - E. Gatti and V. Svelto, "Review of theories and experiments of resolving time with scitnillation counters", Nuclear Instruments and Methods, 43, 248 (1966).

14) - V.I. Tatarsky, "Wave propagation in a turbulent medium" McGraw-Hill; London, New York (1961).

15) - P. Beckmann, "Signal degeneration in laser beams propagated through a turbulent atmosphere", Radio Science, 69D, 629 (1965).

16) - D.L. Fried, "Aperture averaging of scintillation", J. Opt. Soc. Am., 57, 169 (1967).

17) - D.A. Wolf, "Saturation of irradiance fluctuations due to turbulent atmospherc", J. Opt. Soc. Am., 58, 461 (1968).

18) - J.R. Kerr, P.J. Titterton and C.M. Brown, "Atmospheric distortion of short laser pulses", Appl. Opt., 8, 2233 (1969)

19) - F. Albertin, B. Querzola, "Propagazione di fasci laser attraverso l'atmosfera", Alta Frequenza, 121 and 350 (1972).

20) - Y. Kinoshita, T. Asakura, M. Suzuki, "Fluctuation distribution of gaussian beam propagation through a random medium J. Opt. Soc. Am., 58, 768 (1968).

21) - M.I.T. Radar School Staff, "Principle of radar", Third edition - McGraw-Hill, London, N.Y. (1952).

22) - J.C. Owens, "The Use of atmospheric dispersion in optical distance measurements", Invited paper presented to the

"International Association of Geodesy", 25 Sept. 7 Oct. 1967, Lucerne Switzerland (1967).

23) - K.E. Golden, D.E. Kind, S.L. Leonard and R.C. Ward, "Laser ranging system with 1 cm resolution", Appl. Opt. 12, 1447 (1973).

24) - G.D. Thayer, "Atmospheric effects on multiple frequency range measurements", ESSA Tech. Rep. IER 65-ITS 53 U.S. Gov. Print. Off. Washington, D.C. (1967).

25) - G.R. Hugget, L.E. Slater, "Electromagnetic distance--measuring instrument accurate to 1×10^{-7} without metereological correction", Internazional Symposium on Terrestrial Electromagnetic Distance Measurements , Stockholm (1974).

Coherant Optical Engineering, F.T. Arecchi and V. Degiorgio (eds.)

APPLICATIONS OF HOLOGRAPHY TO THE RESTORATION OF WORKS OF ART

F. Gori

Istituto di Fisica Tecnica. Facoltà
di Ingegneria. Università di L'Aquila

People involved in conservation and restoration of works of art are currently faced with a number of problems. They range from defect detection to restoration operations. In such a difficult task, optical methods (in a broad sense, i.e., including ultraviolet and infrared techniques) can be of help (1,2). In particular, we think holography deserves consideration. In this lecture, we will give a few examples of problems holography can help to solve, giving particular attention to holographic flaw detection in paintings.

We first observe that some researches in conservation are of a rather general nature. They refer, for example, to the study of the behavior of typical materials used in artworks or in restoration, say marble, wood, canvas, under the action of chemical or biological agents as well as mechanical or thermal stresses. Very often, the modifications induced in these materials by the applied agent finally lead to morphological changes of the specimen under test. In these cases, holographic interferometry can be used as a tool for detecting such changes. For example, the canvas of old paintings suffers from creeps and a classical restoration procedure is to line the old canvas with a new lining canvas. Lining canvas is a complex material and to describe its behavior suitable mathematical models are sometimes used. In order to test these models, holographic interferometry can be used (3). In fig. 1, an example of double exposure interferogram is reported that refers to a lining canvas under the action of a central load normal to the canvas plane. The unique capability holographic interferometry has of giving a two-dimensional map of object deformations can be fully exploited.

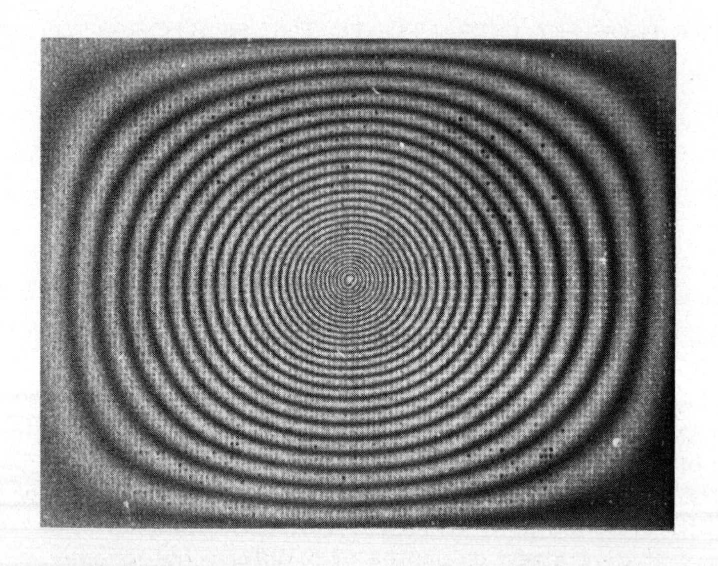

Fig. 1

A second class of problems encountered in restoration deals with single works of art and encompasses defect detection, testing of restoration processes and so on. For these problems, holography can also be useful. As an example, let us consider again the lining canvas. Once the lining canvas has been glued to the old painting, one has to be sure that tensive stresses are uniformly distributed along the painting frame. A simple holographic test can be made to verify whether the correct distribution has been obtained. An example of interferogram obtained with a canvas with unequal stretching along the frame diagonals is shown in fig. 2. (see ref. 3)

Another application of holography is the detection of flaws in old paintings. We will refer to paintings on wooden panels, but paintings on walls or canvas could also be considered. In a panel painting, the wood support is coated with a number of superposed priming layers made from mixtures of gesso and glue, the painted surface being on the top of the primers. A typical damage suffered from this kind of painting is the formation of detached regions generally localized between the wood and the first primer.

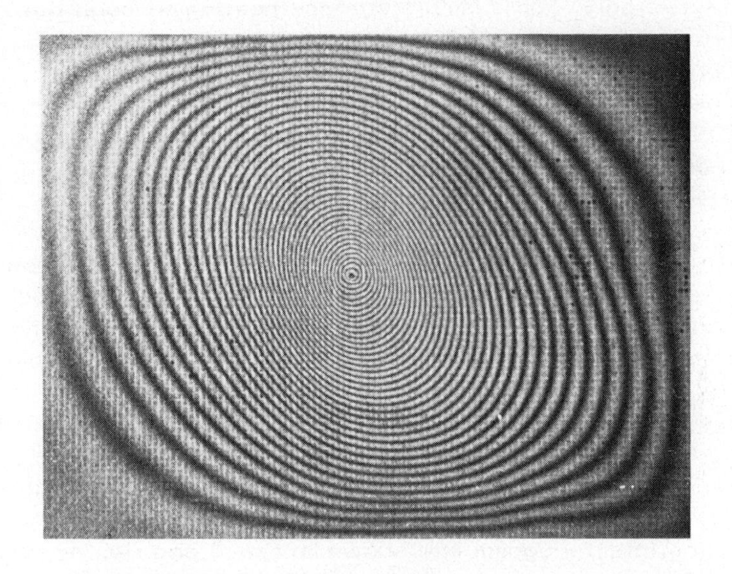

Fig. 2

If a panel painting is subjected to humidity or temperature variations, it responds to them with a very slow process involving complex deformations. The adjustment process can last some days. An important consequence of this large response time is that if the room parameters are not controled the painting suffers a perpetual adjustment process due to periodical daily temperature and humidity variations. We will call such a process an ambient drift. Simple interferometric measurements (4) reveal that local displacements of a wooden panel can occur with a velocity of a few microns per minute.

Let us now discuss the possibility of detecting detachments through holographic interferometry (3, 5). Suppose the painting is suffering some kind of deformation. Regardless of the detailed features of the process, it is to expected that the detached regions behave differently from the nondetached ones. As a consequence, detachments should be revealed as local changes of the holographic inteference pattern. The existence of ambient drift suggests that double-exposure holography can be used without further stress application. However, the sensitivity of this procedure is rather poor. In fact, as room parameter variations act on both the wood support and the layered structure, the method must be thought of as differential. The sensitivity can be increased by producing some kind of stress that acts directly on the surface layers only, for example surface heating through the application of warm air or infrared radiation. Due to the lack of adhesion between the priming layers and the wood support, the detached regions disperse heat at a lower rate than the surrounding regions. Therefore, both the local temperature rise and the thermal expansion effects are higher for the detached regions. Double-exposure holograms can be made during the heating process or during the subsequent cooling process. We will briefly refer to such processes as thermal drifts. We now describe some experiments made to test the holographic detection method. Preliminary experiments were conducted in order to obtain some information about the behavior of unprimed wooden panels. A wood support of 40cmx50cmx4cm was glued to a heavy base at the ends of the longer median line and heated with warm air. Double-exposure holograms were made during the thermal drift, with a time interval of some minutes between the exposures. The photograph of the panel and an example of double-exposure interferogram are shown in fig. 3 and fig. 4, respectively. Comparison between the two figures clearly shows that the interference fringes are affected by the wood grain. The same support was subsequently coated with the usual priming layers and the holographic test was repeated. The kind of interferogram given by the primed support is shown in fig. 5. It is seen that the fringes become quite regular, the wood grain effect being smoothed out by the primers. Nevertheless, a systematic deformation of fringes in the form of a hook is still present on the right of the vertical median line. This is because the wood supported was actually made of two vertical slabs of unequal width glued together. The line traced out by the fringe hooks corresponds to the junction line between the two slabs. This two-slab structure was chosen in order to reproduce the actual support of an ancient panel painting on which subsequent experiments were carried out. The painting, namely Santa Caterina by Pier Francesco Fiorentino (fifteenth century), was glued to a heavy base in the same manner as the test support. Some double-exposure holograms were made exploiting ambient drift. An example of the resulting interferogram is shown in fig. 6. Some irregularities of the fringes reveal the presence of detached regions. Nevertheless, the detection is poorly defined and presumably limited to the largest detachments. A relevant improvement of the sensitivity was obtained, as expected, with the use of thermal drifts.An example is shown in fig. 7.

Fig. 3

Fig. 4

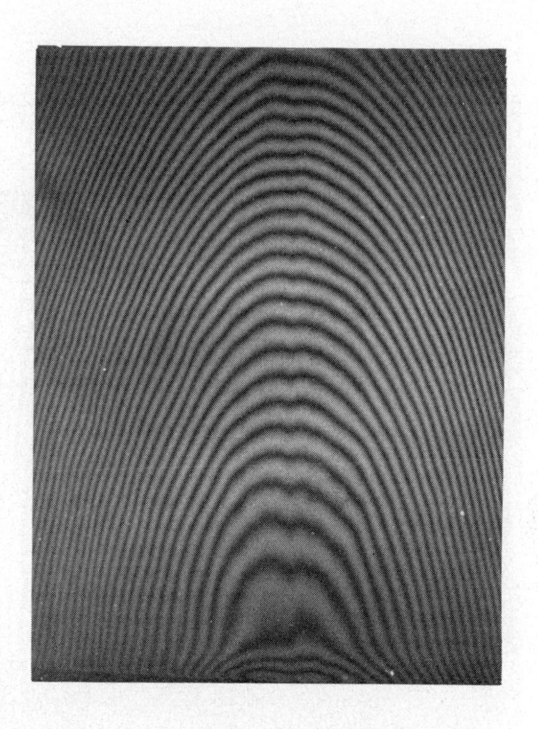

Fig. 5

Fig. 6

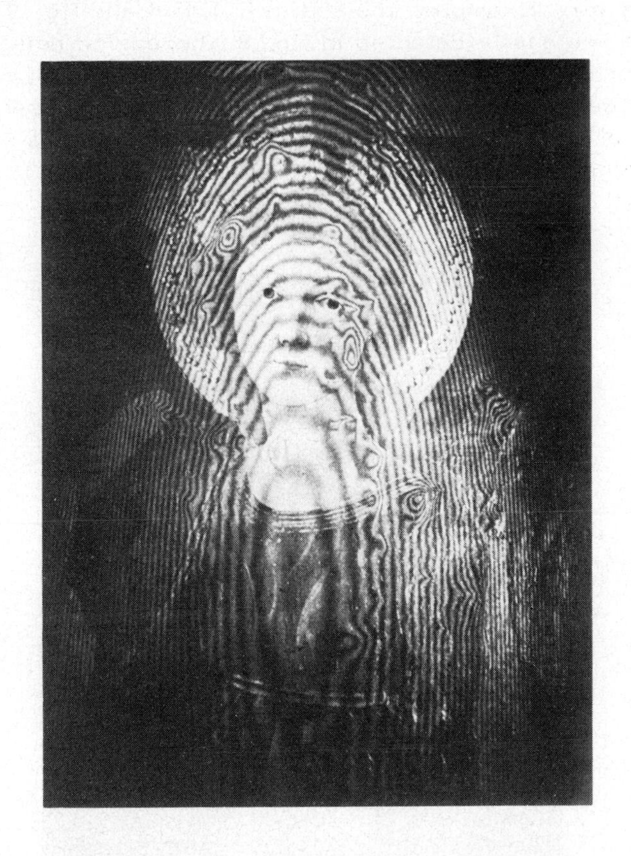

Fig. 7

As it can be seen, many more detached regions are revealed and they are more defined. It will be observed that a number of detachments lie along a vertical line. This is the line corresponding to the junction line between the slabs of the support.

Subsequently, a number of artworks including paintings as well as wooden statues were tested through holographic interferometry (6). Examples are shown in fig. 8 and fig. 9. Detached regions are easily detected in fig. 8 whereas no detachment is apparent in fig. 9. However, careful examination of fig. 9 shows that sudden changes of fringe position exist corresponding to cracks of the priming layers (at the bottom of the painting and at the right upper quadrant).

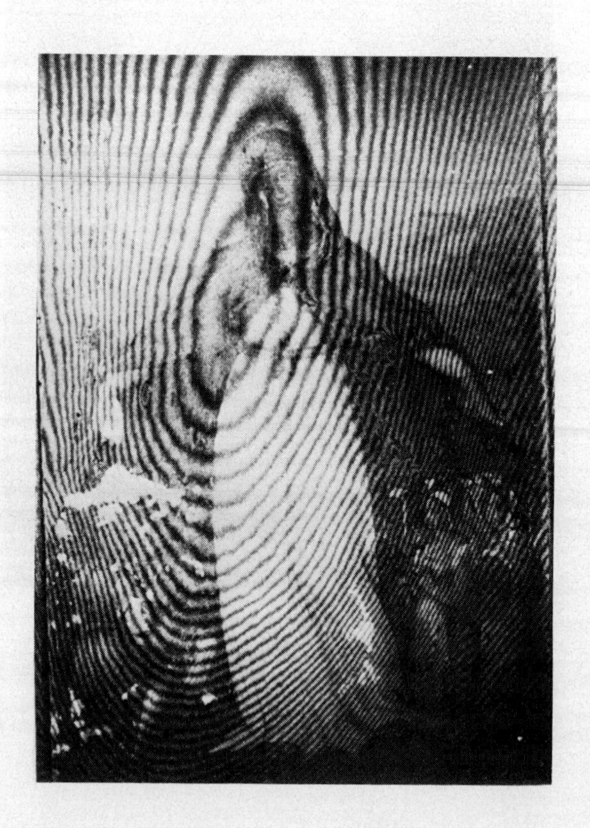

Fig. 8

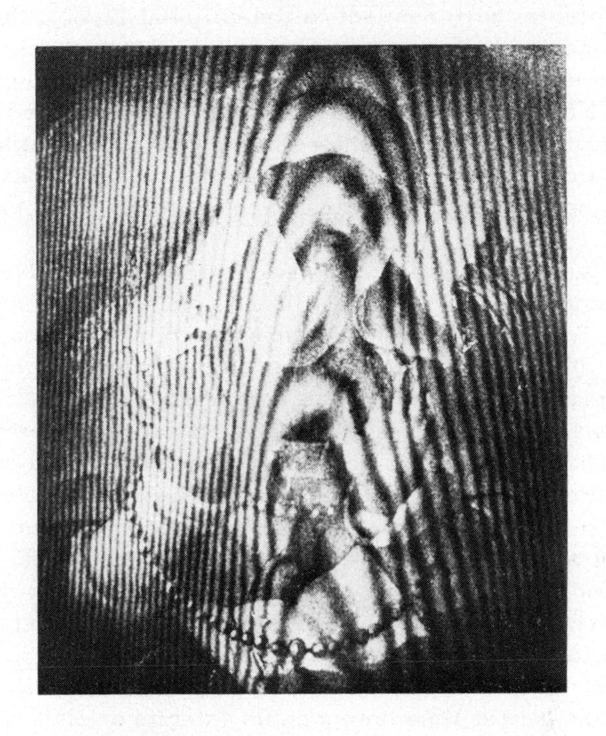

Fig. 9

An obvious question to ask is wheter the fringe anomalies really correspond to detached regions. In order to answer this question, the works of art were restored through injection of suitable adhesives. After that, the holographic tests were repeated. In all cases, fringe anomalies were greatly reduced or disappeared altogether (6). This was especially true for those works of art where the priming layers had reduced thickness. On the other hand, artworks with heavy priming layers still exhibited slight fringe anomalies after the restoration. This can be explained as follows. For paintings with thin primers the effect of even small detachments can be seen on the fringes whereas for thick primers larger detachments are needed in order to give detectable fringe deformations. As a consequence, the amount of adhesive injected during the restoration process was greater for thick primer paintings than for thin primer paintings. Now, the injected adhesive is an island of inhomogeneous material with respect to the original layers as far as thermal and mechanical properties are concerned. Such different behavior is therefore revealed by holographic interferometry if relevantamounts of adhesive are to be injected. We note, by the way, that holographic interferometry could be used for an exhaustive study of the kind of adhesive that best matches the thermal and mechanical properties of the original mixtures of gesso and glue.

Now a few words about experimental details. Nearly all the experimental work was done with an argon laser operating at 5145 Å. To avoid careful adjustment of optical paths, an intracavity etalon was mounted so that single frequency operation and hence long coherence time was achieved. Typically, the output power was between 400 and 700 mW. It can be noted that flat objects like paintings do not really require a large coherence time so that they could be handled without using etalon. This would increase the output power for a factor of about two. Exposure times were in the order of 1/10 sec with 10E56 Agfa plates for the largest paintings we tested (surfaces of about 1 m^2). For larger paintings or for work outside the lab, pulsed ruby lasers could be used. The temperature rise we used was between 10° and 15°C. Recent experiments (7) show that these figures can be reduced if real time holographic interferometry is adopted.

All the work described herein was done in cooperation among the Istituto di Fisica Tecnica di L'Aquila, the Istituto di Fisica di Ingegneria di Roma and the Istituto Centrale del Restauro.

REFERENCES

1) - J. R. J. van Asperen de Boer, Appl. Opt. 7, 1711 (1968).
2) - S. Keck, Appl. Opt. 8, 41 (1969).
3) - S. Amadesi, F. Gori, R. Grella, G. Guattari and P. Pasquini, in Problemi di conservazione, G. Urbani, Ed. (Ufficio del Ministro per il coordinamento della Ricerca Scientifica e Tecnologica, Roma, 1973).
4) - S. Amadesi, F. Gori, R. Grella, and G. Guattari, Twenty-Sixth National Congress of ATI, L'Aquila (1971).
5) - S. Adamesi, F. Gori, R. Grella, and G. Guattari, Appl. Opt. 9, 2009 (1974).
6) - F. Gori and G. Urbani, Le Scienze 13, No. 74 (1974).
7) - S. Aurisicchio, A. Finizio and G. Pierattini, Consiglio Nazionale delle Ricerche, S. S. A. 76. 1; Roma (1976).

Coherant Optical Engineering, F.T. Arecchi and V. Degiorgio (eds.)

SCANNING TECHNIQUE FOR INTERPRETING DOUBLE EXPOSURE HOLOGRAMS

Vittorio Fossati Bellani

C.I.S.E. - P.O. Box 3986 - 20100 MILANO (Italy)

INTRODUCTION

In this seminar we review the principles of the scanning technique for extracting quantitative information from double-exposure holograms. Since its introduction (1), this technique turned out to be a practical tool for very fine measurements of displacements, deformations, vibrations, etc.
The scanning technique is also suitable for beeing used in an automatic read-out system for interpreting holograms which is right now under development and will be outlined in a later section.

THEORY

The interference fringes in a double exposure hologram are due to coherent superposition of light beams coming from the same object point which is put in different positions during each exposure. The fringe order N is related to the displacement $\vec{d}$ of the considered point by the well-known formula

$$\vec{d} \cdot (\hat{\varrho}_o + \hat{\varrho}_{ill}) = N \lambda \tag{1}$$

where $\hat{\varrho}_o$ and $\hat{\varrho}_{ill}$ are, respectively, the unit vectors in the direction of observation and of illumination. Eq.(1) means that for a fixed viewing direction the fringe order depends only on the component of $\vec{d}$ along the bisector of the angle between $\hat{\varrho}_o$ and $\hat{\varrho}_{ill}$ (this component is usually called the sensitivity vector). The fringe order N can be determined only if there is a fixed reference point by counting the number of fringes between this point and the one which is beeing examined. Otherwise N is not known and eq. (1) is useless. When the direction of the displacement is not known "a priori" at least three different points of view are needed for determining $\vec{d}$ and this might require the recording of two or more holograms. Alternatively one can follow the principle first introduced by Aleksandrov and Bonch-Bruevich (2). Generally the fringes are not localized on the surface of the object under test because the sensitivity vector changes by moving the line of sight so that the fringe order changes and the fringes appear to move across the object point. Therefore, for a viewing direction defined by the unit vector $\hat{\varrho}_i$, eq. (1) becomes

$$\vec{d} \cdot (\hat{\varrho}_i + \hat{\varrho}_{ill}) = N^{(i)} \lambda \qquad (1')$$

where $N^{(i)}$ is the new fringe order.
Subtracting eq. (1') from eq. (1), we obtain

$$\vec{d} \cdot (\hat{\varrho}_o - \hat{\varrho}_i) = \pm N_i \lambda \qquad (2)$$

where $N_i = N - N^{(i)}$ is the number of fringes which have crossed the object point while changing the viewing direction from $\hat{\varrho}_o$ to $\hat{\varrho}_i$. The uncertainty in the sign of eq. (2) is due to the impossibility of distinguishing which was the first exposure, but can be easily removed with physical arguments. By chosing three proper observation directions $\hat{\varrho}_i$, we obtain three independent equations (2), whose resolution yields the value of the three components d_x, d_y, d_z of $\vec{d}$.
No fixed reference point is needed with this technique, since it is not necessary to know the absolute fringe order.
Anyway the accuracy is usually better for the components of $\vec{d}$ parallel to the plate rather than for the normal one, due to the small size of the hologram plate. For this component better accuracies could be obtained either by putting into the scene a fixed reference point or by recording two holograms in two different positions with respect to the object (3).

SCANNING TECHNIQUE

The possibility of practical use of this principle has for a long time been limited by the difficulty of counting fringes easily and accurately.
In attempting to overcome this problem the scanning technique exploits the real image projected by a hologram illuminated with a conjugate reference beam. If this illuminates a small region of the plate, the real image exhibits a fringe pattern corresponding to the viewing direction defined by the observed point and

by the center of the illuminated spot on the hologram.
If the spot is moved across the plate, the fringe pattern on the real image changes as a consequence of the variation of the line of sight, so that by moving the observation direction from $\hat{\varrho}_o$ to $\hat{\varrho}_i$ the fringe order changes from N to $N^{(i)}$ and the number N_i of eq. (2) can be determined by counting the number of fringes crossing the observed point.
The scanning of the plate is accomplished by means of an oscillating mirror whose axis can be oriented in any wanted direction. The laser beam, after reflection by the scanning mirror is focused so to be always conjugate to the reference beam. This can be obtained with a spherical mirror or a well corrected lens. The setup with the mirror is illustrated in fig. 1. The conjugate reference beam, by crossing the hologram, reconstructs the real image (fig. 2). The fringes are counted by putting a photodetector at the point under test and analyzing at the oscilloscope the output signal. This exhibits a sinusoidal shape so that the number N_i of fringes can be counted for each scanning just by looking at the cycles of the waveform.

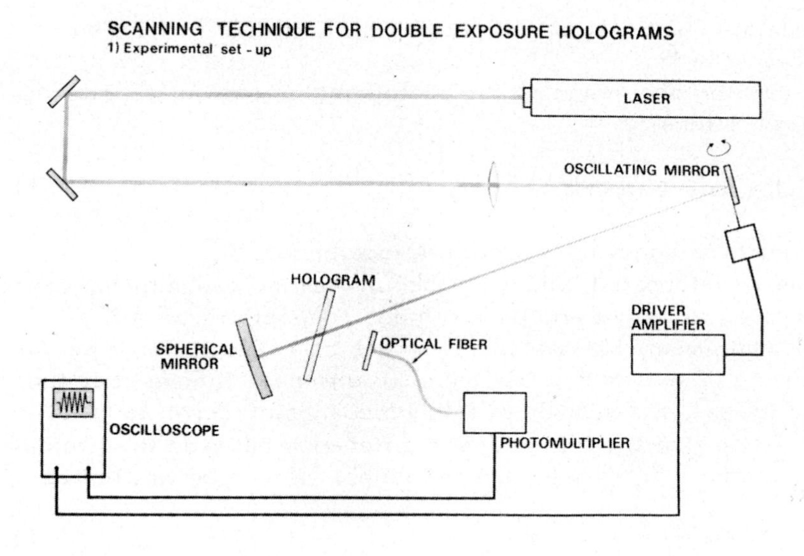

Fig. 1 - Setup for scanning technique

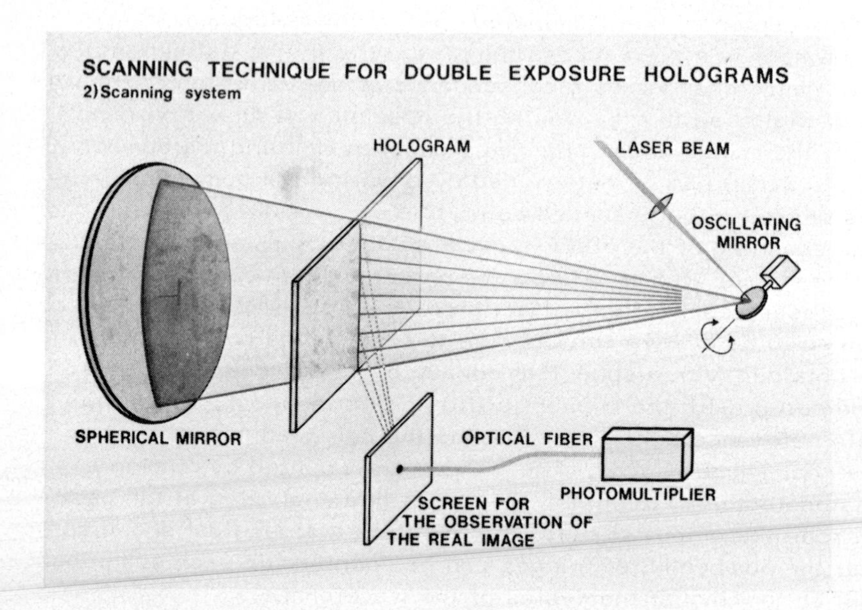

Fig. 2 - Real image reconstruction system

AUTOMATIC EVALUATION OF THE FRINGE PATTERN

During one scanning the photodetector actually records a light intensity

$$I = \cos 2\pi N \qquad (3)$$

where N is again the absolute fringe order.
This waveform exhibits a number of maxima and minima, corresponding to bright and dark fringes respectively.
Each minimum, for instance, occurs at a time t_i for a certain viewing direction $\hat{\varrho}_i$. Between two different minima, say the ith and the jth, the number of fringes changes from N_i to $N_j = N_i + j - i$ so that the fringe order difference between the two minima is $N_j - N_i = j - i$. Therefore eq. (2) can be written as

$$j - i = \frac{\vec{d}}{\lambda} \cdot (\hat{\varrho}_j - \hat{\varrho}_i) \qquad (i,\ j = 1,\ n) \qquad (4)$$

for each scanning. In eq. (4) n is the total number of minima in one scanning. In practice the viewing direction can be easily

related to the angular position of the scanning mirror as shown in ref. (4).
Eq. (4) can be estabilished for each scanning, building up an over-determined linear system which can be solved by the least square method.
The experimental setup is shown in fig. 3. The output of the photodetector is sent to an analog to digital converter which samples the waveform I (t) of eq. (3) and relates it to the sampling time t and consequently to the corresponding viewing directions.
A minicomputer processes these data choosing the minima of I (t), measuring the geometrical parameters of the experiment and solving the linear system of equations (4).
A further step consists in providing the read out system of automatic positioners and scanners so that manual operation can be reduced as much as possible. A block diagram is shown in fig. 4.

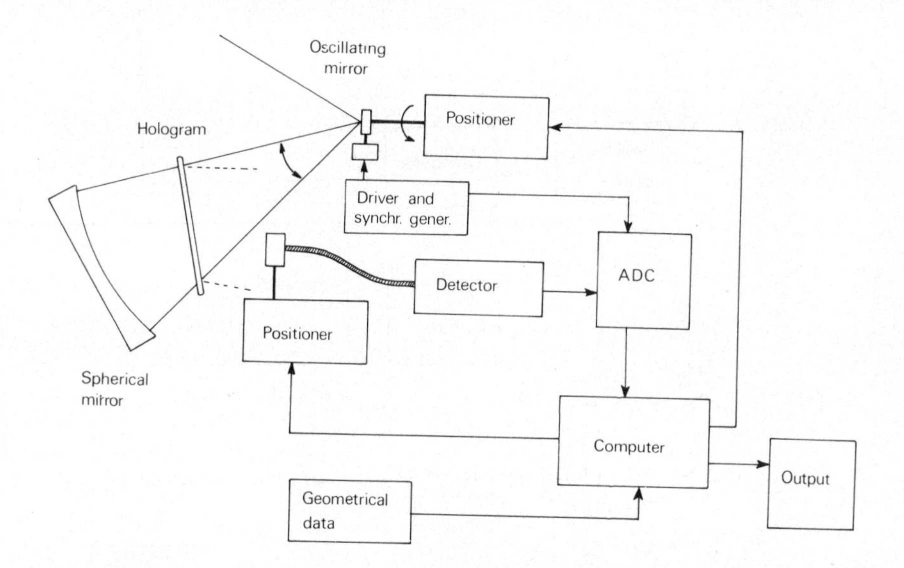

Fig. 3 - Setup for automatic read-out system

Referring to this figure it can be seen that the final system requires the following steps:

a) detection of signal and analog to digital conversion
b) collection of digital data and processing by a computer
c) positioning of detector and scanning mirror.

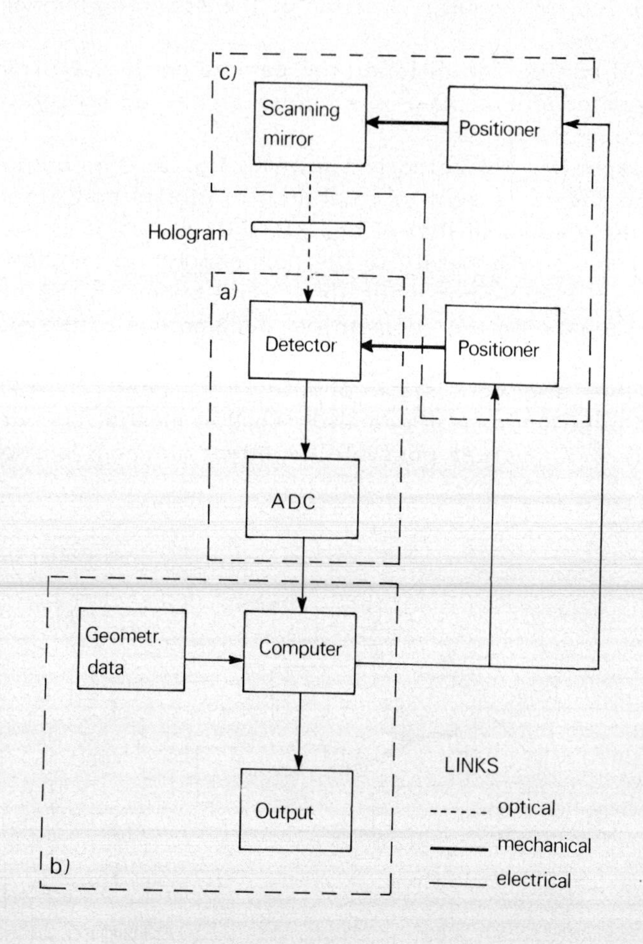

Fig. 4 - Block diagram of the automatic read-out system

For step a) we use an analog to digital converter which is allowed to work by a pulse generator only during the proper part of the signal, to be choosed by the operator. The digitized data are then processed by a minicomputer working on line (step b). The only operator's task here is to supply once for all the computer with all the geometrical parameters of the experiment. After processing the data of one scanning the computer drives a stepping motor for repositioning the oscillating mirror till the total number of required scannings has been performed for one point. Then the detector is moved by another computer driven stepping motor to a new point and the whole procedure starts again.

REFERENCES

1) - V. Fossati Bellani and A. Sona, Appl. Opt. 13, 1337 (1974).
2) - E.B. Aleksandrov and A.M. Bonch-Bruevich, Sov. Phys. Tech. Phys. 12, 258 (1967).
3) - K. Biedermann and L. Ek, Proc. of the Intern. Conf. on Holography and Optical Data Processing, Jerusalem 22-26 August 1976.
4) - V. Fossati Bellani, Proc. of the Intern. Conf. on Holography and Optical Data Processing, Jerusalem 22-26 August 1976.

Coherant Optical Engineering, F.T. Arecchi and V. Degiorgio (eds.)

VELOCIMETRY AND DIGITAL CORRELATIONS

M. Corti

CISE - P.O. Box 3986 - 20100 Milano, Italy

INTRODUCTION

The measurement of object velocities is an important application of laser light scattering techniques, as discussed previously in this School. The light scattered from the moving object is slightly shifted in frequency, due to the well known Doppler effect. Thus the knowledge of the frequency shift and the geometry of the experiment allows us to compute the velocity directly. The main advantage of this technique is essentially that it does not perturb the system being studied. Very high and very low velocities can be measured; the lower limit being around 10^{-3} cm/sec. The size of the testing volume may be quite small, of the order of 10^{-3} cm in diameter.

Since the first experiment by Yeh and Cummins [1] in 1964, important applications have been developed and a large number of technical papers have appeared on various journals. Review articles [2], books [3, 4, 5] and bibliographic searches [6] may guide the reader through the available material. Let us limit ourselves to give a general outline of the method and to discuss with some details the so called photon correlation velocimetry which is a rather recent development [7]. This techniques is particularly useful for systems from which very little scattered light is available like for istance supersonic gas flows in wind tunnels or blood flow in retina vessels in human eyes.

BASIC OPTICS

The light intensity scattered from a pure fluid or a gas is much too weak to allow velocity measurements. Some seeding particles are therefore necessary, either naturally occurring or specially introduced into the flow. The particles should be chosen in such a way that they faithfully follow the motion of the medium [8]. For example particle diameters up to 10 µm can be appropriate for liquid flows, while for supersonic gas flows the diameter should not exceed 0.5 μm. Since the scattered light intensity decreases strongly with the particle diameter [8], the type of seeding is an important factor which determines the sensitivity of the equipment for velocity flow measurements.

The Doppler frequency shift of the light scattered from a particle moving at a velocity $\vec{V}$ is given by the scalar product

$$\Delta\omega = (\vec{K}_s - \vec{K}_o) \cdot \vec{V}$$

where $\vec{K}_o$ and $\vec{K}_s$ represent the incident and scattered wave vectors. The shifts involved in most practical cases fall in the range of 10^{-3} Hz to 10^7 Hz which is unaccessible to optical interpherometers while it is ideally covered by self beating spectroscopy techniques. In this latter case the scattered field is mixed with the incident field on a square law detector, like a photosensitive surface; the resulting photocurrent is modulated at the difference frequency of the two fields. Electronic means are then used to measure the frequency. Many optical schemes have been proposed to perform the mixing but the most commonly used scheme nowaday is the so-called Doppler difference arrangement. The laser output is split in two beams which are directed at the test volume from slightly different angles. The scattered light from a particle moving in the crossing region of the two beams contains two Doppler shifted frequencies which beat together on the photodetector. This arrangement has many advantages: it is less sensitive to the mechanical vibrations of optical mounts; light paths are automatically matched and the measured shift is independent of observation angle. It is therefore possible to use large collection optics without introducing an instrumental spread in the measured Doppler frequency.

A vector diagram is drawn below, where $\vec{K}_1$ and $\vec{K}_2$ are the wave vectors of the two incident beams, ϑ is the angle between them, $\vec{V}$ is the particle velocity and $\vec{K}_s$ is the scattering wave vector

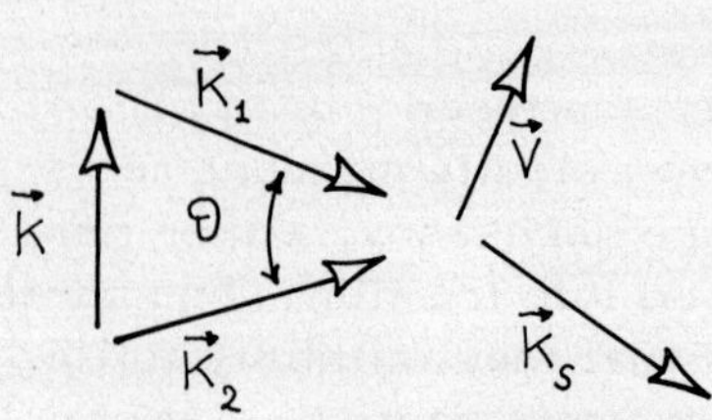

The two Doppler shifted frequencies

$$\Delta\omega_1 = (\vec{K}_s - \vec{K}_1) \cdot \vec{V}$$

$$\Delta\omega_2 = (\vec{K}_s - K_2) \cdot \vec{V}$$

combine on the photodetector to give:

$$\Delta\omega = \Delta\omega_1 - \Delta\omega_2 = (\vec{K}_2 - \vec{K}_1) \cdot \vec{V} = \vec{K} \cdot \vec{V}$$

with $\vec{K} = \vec{K}_2 - \vec{K}_1$ and $|\vec{K}| = (4\pi/\lambda)$ sen $\theta/2$

A typical setup is shown in Fig. 1

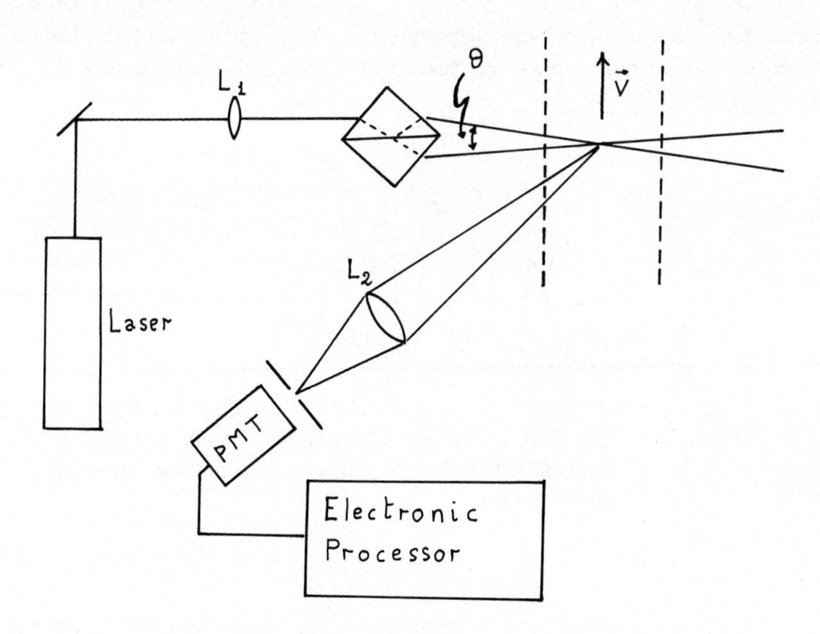

Fig. 1 - Basic Doppler difference setup

We can change the angle ϑ and the position of the intersecting region, by moving the beam splitter with respect to the incident laser beam. The lens 1 ensure that each beam is focused in the crossing region. The slit in front of the photomultiplier stops the light originated outside the crossing region.

If we place a screen across the intersection region we see

a set of bright and dark fringes spaced by $s = \lambda/(2 \text{ sen } \vartheta/2)$. A particle going through this region gives rise to a light intensity signal modulated at the Doppler difference frequency. For example at the green Argon-laser line and an angle of 5° between the incident beams, the Doppler-difference frequency is about 200 KHz for a particle moving at one meter per second in the direction of the vector $\vec{K}$.

DOPPLER SIGNAL PROCESSING

A typical signal is drawn in Fig. 2; its duration is due to the finite transit time of a single particle through the intersection region. The modulation depth is maximum for particles diameters smaller than fringe separation. The optics should be carefully aligned in order to focus the two beams into the intersection region.

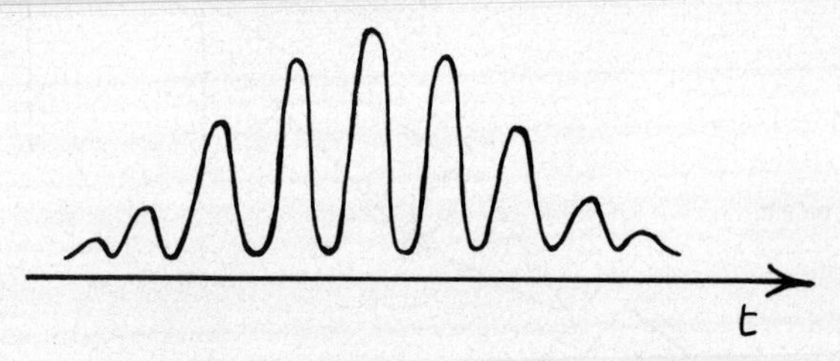

Fig. 2 - Typical Doppler signal; its finite duration reflects the finite dimension of the intersection region.

Various electronic instrumentation can be used to measure the Doppler frequency. The spectrum analyser has been widely used. It is made up by a narrow band filter which is swept across the frequency spectrum; its output is squared and recorded as a function of time. The frequency tracker is another good instrument available on the market. It is in principle a spectrum analyser with the addition of a feed back loop which keeps the center of the sweeping narrow filter at the peak of the power spectrum. It is capable of following average velocity variations in time.

Doppler frequency measurements become a serious problem when the intensity of the light signal is very small. Le us compare for example the signal values for two particle diameters [8]. For a 1 μm particle illuminated by 1 mW of laser light of wavelength $\lambda = 0.6$ μm, focused down to 100 μm, a photodetector 1 meter away from the scattering volume with a collection aperture of 10 cm in diameter records around 10^8 photoevents per sec. If the particle stays in the beam 10^{-4} sec (that is with fluid velocity of 1 meter/sec) the available photoevents are 10^4, for each particle crossing. In the case of 0.2 μm diameter the same calculation gives on the average 1 photoevent per particle crossing. For the larger particle the electric signal looks quite similar to the one drawn in Fig. 2. The detector is a photomultiplier, the electrical integration time is chosen to be as long as possible to reduce the shot-noise, but anyway shorter than the Doppler period. For this type of high signal any frequency analysis technique will be adequate. The situation is completely different for the smaller particle.
The frequency tracker cannot operate in this case and the spectrum analyser would require an extremely long measuring time to sort out the Doppler frequency spectrum from the large shot noise background. The experiment would be feasible with an N parallel channel spectrum analyser because the measuring time would be reduced by a factor N. Such an instrument with N narrow band filters centered at different frequencies would be quite expensive, unflexible and difficult to calibrate [9]. The problem is better solved by measuring the Fourier transform of the frequency spectrum, that is the correlation function of the electrical signal.

PHOTON CORRELATION SPECTROSCOPY

In the last few years quite efficient fast correlators became available [10, 11, 12] . They are equivalent to the parallel channel spectrum analyser, because they work in real time. The processing time is extremely short so that no information is practically lost. Digital correlators are ideally suited for weak optical signals; let us look at the problem with some detail.

Weak light beams are generally detected by high-gain photomultiplier tubes which yield a current pulse for each photon absorbed by the photosensitive surface. If the optical signal to be analysed is so intense that many photons are absorbed within the integration time of the photodetector, the output electric

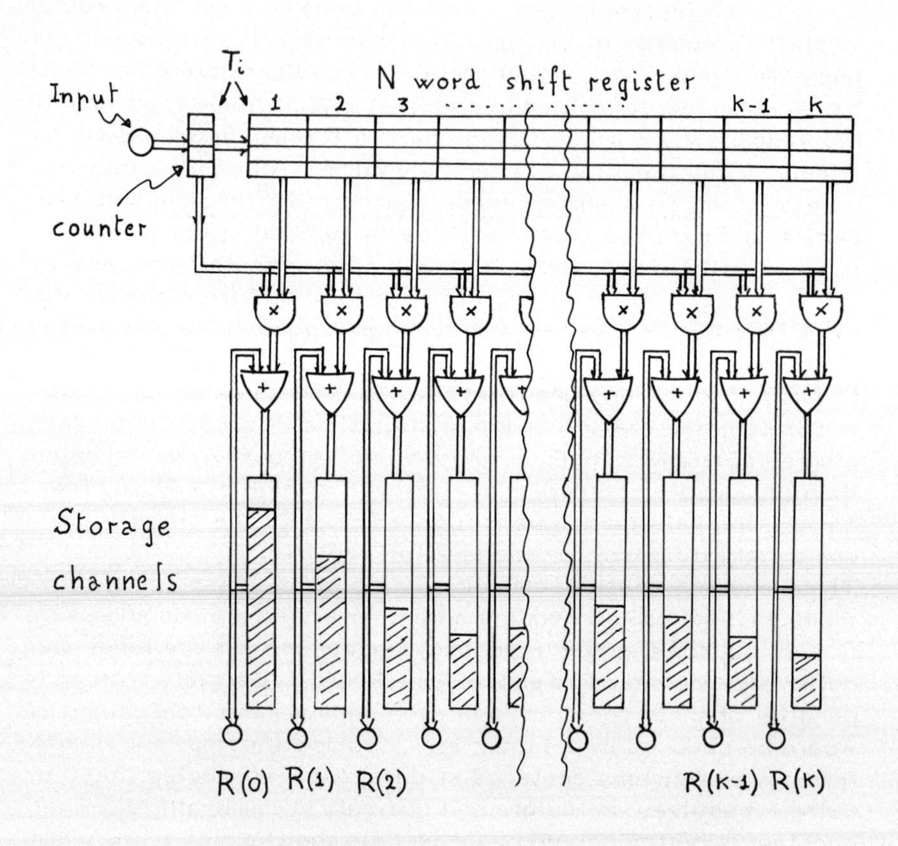

Fig. 3 - Schematic block diagram of an ideal N-channel digital correlator.

current i (t) can be considered as an analog signal proportional to the intensity of the light beam. If we assume that i (t) is a stationary random variable, its autocorrelation function $R(\tau) = \langle i(t)\, i(t+\tau)\rangle$ depends only on the delay τ and can be evaluated by a time average. In pactice, $R(\tau)$ is measured by sampling i (t) periodically and performing the appropriate multiplications among the samples. Since it is easier and faster from an electronic point of view to perform multiplications among digital signals, the result of each sampling operation is quantized through an analog-to-digital converter.

If, however, the optical signal is so weak that it is very unlikely to detect more than one photon within the integration time of the photodetector, the output electric current consist of a random train of nonoverlapping pulses. In this case the efficiency of correlators utilizing analog-to-digital conversion is very low, because the gating time is uncorrelated with the photoelectron-pulse arrival time. It is therefore more convenient to exploit the fact that the signal is already in digital form.

The photoelectron pulses are passed through a discriminator which generates a train of uniform logic pulses with the same relative times as the original photoelectrons. Time is divided into uniform sample time intervals of duration T and the number of pulses n_T in each interval can be counted [13].

A schematic block diagram of an ideal N-channel digital correlator is shown in Fig. 3. A counter, a N-word shift register, N multiplication elements and N storage channels are the main parts of it. Auxiliary electronics like clock, delay generators and input output facilities are not shown. Data are collected in the counter during the sample time T. At the end of each sampling interval the counter content is multiplied by the delayed version of the signal stored in the delayed channels of the shift register. These products are added to the storage channels. Then the information contained in the counter is shifted into and within the shift register. The counter is now ready for a new information coming in. If all the transfer and multiplication operations are made in a time interval small compared with the sampling time T, the correlator is said to work in real time. In order that the measured correlation function be coincident with the effective correlation function of the optical signal, the electronic apparatus should be able to register any possible number of photoelectron arrivals in the sampling time. In practice the maximum number of recorded photoelectron pulse is limited by the dead time of the apparatus. Furthermore, if real time operation is desired at high sampling frequencies, another limit is set by the fact that large numbers require long multiplication and registration times. Fast processing can be achieved if n_T is restricted to assume only the values 0 or 1. This is the so called double clipping mode of operation. Of course the practical simplification is paid for by a distortion of the measured correlation function. The distortion, however, depends strongly on the average value $\langle n_T \rangle$ and can be made very small by working with $\langle n_T \rangle \ll 1$ [12]. This is the actual case for many experimental conditions.

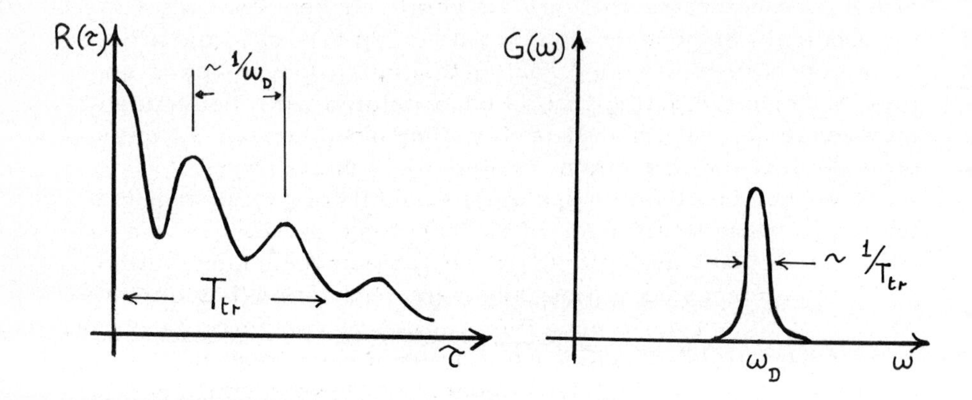

Fig. 4 - An example of a correlation function and of a frequency spectrum.

For a laminar flow Fig. 4 shows the typical output $R(\tau)$ and $G(\omega)$ of a correlator and a spectrum analyser respectively. In the case of stationary flows the two outputs are Fourier transform pairs, as stated by the Wiener-Khinchine theorem.

$$R(\tau) = \int_0^\infty G(\omega)\cos(\omega\tau)d\omega \quad ; \quad G(\omega) = \frac{2}{\pi}\int_0^\infty R(\tau)\cos(\omega\tau)d\tau .$$

The oscillatory part of the correlation function is related to the Doppler frequency shift ω_D. For a uniform velocity flow the damping is mostly due to the finite duration of the signal (see Fig. 2), that is the transit time T_{tr} of the particle through the scattering region. It is good practice to set the experimental conditions such that T_{tr} is much longer than the oscillation period T_D of the Doppler signal, say $T_{tr} \sim 100\ T_D$ or, equivalently, $r \sim 100\ s$ where r is the diameter of the scattering region and s is the fringe spacing defined previously. The latter considerations are of primary importance if we are interested to measure particle velocity spreads like in turbulent flows. In this case the measurement accuracy increases if the transit time is much longer than the damping time due to the particle velocity spread. For optimum accuracy, of the order of a few percent, care should also be taken in minimizing some instrumental damping factors due to the optical setup [2] and to the particle dimensions [14].

Expressions of the correlation function can be found, for instance, in Ref. 2. It is assumed a gaussian profile for the laser beams. As an example, the correlation function of the light scattered from particles moving at constant velocity is given by:

$$R(\tau) \div \exp\left(-\frac{\omega^2 \tau^2}{\eta^2}\right) \cdot \left[1 + 1/2 \cos\left(\frac{2\pi u}{s}\right.\right]$$

This formula is valid for $r > 2s$; u is the velocity component in the direction of the vector $\vec{K}$. While geometrical parameters, like beam dimension and intersection angle, are measured separately, velocity data are obtained by least square fitting of the measured correlation function with the above formula.

A less straightforward procedure is followed for the turbulent case. A model for the velocity distribution should be assumed and the corresponding expression for the correlation function is calculated. Fitting to the measured correlation function allows then to calculate the distribution parameters. The curve fitting procedure is suitable for steady flow measurement because the computation time is usually rather long, of the order of several seconds. The attainable accuracy is good even for extremely low light signals. For example, a 1 % accuracy in the velocity determination should be feasible for a total of 10^5 photoelectrons collected in the measurement time, that is about 30.000 Doppler cycles for a light signal which generates an average of 3 photoelectron per Doppler cycle [8]. Much effort is put nowadays into the search of computation schemes capable of giving the average velocity parameters in a short time. This is important for on-line processing of non steady flows. Fourier transform of the measured correlation functions with some simplifying algorithms is claimed to give velocity determinations in a few milliseconds [8]. For this purpose fast digital correlators with associated computer are under development.

We conclude by observing that the sensitivity of photon correlation velocimetry, undoubtly useful in aerodynamics, can also be extremely interesting in other fields of research. For instance, by illuminating a retina vessel through a human eye, one can measure the blood velocity in vivo. In this case blood cells scatter light quite efficiently, but the illuminating light should be kept very small, less than 10 μWatt, in order to avoid eye damage [15].

REFERENCES

1) - Y. Yeh, H. Z. Cummins
Appl. Phys. Lett. 4, 176 (1964).
2) - J. B. Abbis, T. W. Chubb and E. R. Pike
Opt. and Laser Techn. 6, 249 (1974).
3) - F. Durst, A. Melling and J. H. Whitelaw
Principles and Practice of Laser-Doppler Anemometry, Academic Press, London (1976).
4) - B. M. Watrasiewicz and M. J. Rudd
Laser Doppler Measurements, Butterworths (1976).
5) - T. S. Durrani and C. A. Greated
Laser Systems in Flow Measurements, Plenum Press 1977.
6) - Bibliography of Laser Doppler Anemometry Literature, F. Durst and M. Zaré eds. Published by DISA Elektronik A/S. DK-2740 Skovlunde, Denmark, October (1974).
7) - E. R. Pike
J. Phys. D. 5, L23 (1972).
8) - E. R. Pike in Photon Correlation Spectroscopy and Velocimetry, H. Z. Cummins and E. R. Pike eds. (Plenum Press, to appear in 1977).
9) - R. J. Baker and G. Wigley
Proceedings of the LDA-Symposium, Copenhagen 1975, p. 350; P. O. B. 70, DK-2740 Skovlunde, Denmark.
10) - C. J. Oliver in Coherence and Quantum Optics
L. Mandel and E. Wolf eds. (Plenum Press, New York, 1973), p. 395.
11) - R. Asch and N. C. Ford
Rev. Sci. Instrum. 44, 506 (1973).
12) - M. Corti, A. De Agostini and V. Degiorgio
Rev. Sci. Instrum. 45, 888 (1974).
13) - Dead time considerations are here omitted; see Ref. 12 for details.
14) - R. V. Edwards, J. C. Angus, M. J. French and J. W. Dunning Jr. J. Appl. Phys. 42, 837 (1971).
15) - T. Tanaka, C. Riva, I. Ben-Sira
Science 186, 830 (1974).

Coherant Optical Engineering, F.T. Arecchi and V. Degiorgio (eds.)

INCOHERENT SPATIAL FILTERING
DIGITAL HOLOGRAPHY

A. W. Lohmann

Physikalisches Institut der Universität Erlangen
8520 Erlangen, Germany

INTRODUCTION

These notes cover the two lectures on "incoherent spatial filtering" and on "digital holography". These two subjects are pursued for the same reasons, which will be given below under the heading of "trends in optical image processing". Thereafter, we will present the essential concepts of our two themes. The details can be found in the following references :
"Incoherent matched filtering with Fourier holograms"; (1)
"Real time incoherent optical-electronic image subtraction";(2)
"How to make computer holograms"; (3)
"Hybrid Image Processing". (4)

TRENDS IN OPTICAL IMAGE PROCESSING

Optical image processing with laser light and holograms has excited many researchers and on-lookers. Five and ten years ago, many great things were promised by the optical community. Today, some of the influential outsiders and some insiders as well are somewhat sceptical about the future of optical image processing. Why is that so?

First of all, it would be wrong to resign, because there are some definite contributions of optics to data processing, as for example image synthesis for side-looking radar. But still, we have to admit, that optics was somewhat oversold, which could happen because it is so fascinating to look at a holographic image.

The arguments may differ, but most experts of optical processing would probably agree, that progress has slowed down considerably during the last five years. In my opinion, the main reason is that the field of coherent optical processing has been harvested by now, at least within the realm of "classical" optical components. Some people feel, there will be more progress again when modern transducers become widely available. These transducers have to convert electronic signals into coherent optical signals, or incoherent optical signals (such as TV output) into coherent optical signals. It looks as if these modern transducers, sometimes called "light valves", may soon become generally available at a reasonable price and with satisfying specifications.

Two other trends are promising as well, in my opinion. These trends can be implemented with today's hardware at a reasonable price. These trends are :

I. From coherent to incoherent signals;
II. From all-optical to hybrid systems.

Trend (I) is promising, because the "coherent noise" is avoided. Furthermore, we do not need those expensive "light valves", that are not perfect in every aspect as yet. An incoherent optical processor is fed with an input image in the form of a TV image, for example. Practically every spatial filtering operation, which has been performed so far in coherent light, can be performed also incoherently. An exception is the processing of phase objects as inputs. These claims will be substantiated in the attached reprints.

Trend (II) is necessary because the range of operations that can be performed by all-optical processors is limited. A digital computer is certainly more flexible. Also TV electronics offers some nonlinear capabilities such as hard-clipping and rectification, which are not easy to perform optically, to say the least. On the other hand optical processing has some well known advantages such as speed and low price. Because all three technologies (optics, TV, digital computer) have their specific advantages we should try to merge these technologies into "hybrid processors".

SURVEY OF THE TWO LECTURES

1. Incoherent spatial filtering
1.1 The filter function OTF for incoherent image formation; (see also W. Lukosz on "Fourier Optics") [5]
1.2 Matched filtering in incoherent light [1]
1.3 Subtraction of two incoherent images [2]

2. Digital holography
2.1 Fundamentals, practical realisation [3]
2.2 Survey of applications [3]
2.2.1 Display
2.2.2 Interferometric test objects
2.2.3 Holographic memory
2.2.4 Scanner
2.2.5 Distorter
2.2.6 Spatial filter
2.3 Application for character recognition [4]

REFERENCES

1) Incoherent matched filtering with Fourier holograms; A. W. Lohmann, H. W. Werlich, Appl. Opt. 10, 670 (1971).
2) Real time incoherent optical-electronic subtraction; S. R. Dashiell, A. W. Lohmann, J. D. Michaelson, Opt. Com. 8, 105 (1973).
3) How to make computer holograms; A. W. Lohmann, Proc. SPIE Seminar, Vol. 25, p. 41 (1971)
4) Hybrid image processing; B. Braucker, R. Hauck, A. W. Lohmann, PSE Journal, to appear in 1977.
5) A. W. Lohmann, Appl. Opt. 16, 261 (1977).

Coherant Optical Engineering, F.T. Arecchi and V. Degiorgio (eds.)

TWO-DIMENSIONAL DIGITAL FILTERING AND DATA COMPRESSION WITH APPLICATION TO DIGITAL IMAGE PROCESSING

V. Cappellini

Istituto di Elettronica - Facoltà di Ingegneria and Istituto di Ricerca sulle Onde Elettromagnetiche del C.N.R.
Firenze, Italy

1. INTRODUCTION

The methods and techniques of digital image processing are of increasing interest due to several reasons: high efficiency which is obtainable; great expansion of digital computers of standard or special type; decreasing cost of hardware implementation for the development of high integration digital circuits and particularly of microprocessors.

Among the different methods and techniques of digital image processing, two-dimensional digital filtering and data compression are becoming of great importance.

In this paper some methods and techniques are described for designing two-dimensional digital filters and data compression systems for image processing. Results obtained in processing real images (in particular biomedical images and images received from earth resource satellites) are shown.

2. TWO-DIMENSIONAL DIGITAL FILTERING

Many types of two-dimensional (2-D) digital filters with finite impulse response (FIR) and infinite impulse response (IIR) have been developed [1] .

Some important design methods of 2-D FIR digital filters are: methods using "windows"; frequency sampling methods; approximation methods. Filters of this type can present adequate efficiency (in particular linear phase characteristics) and relatively simple hardware and software implementation: they are useful for many actual digital image processing applications.

Some approaches to design 2-D IIR digital filters are: differential correction methods; factorization methods; rotation methods (from 1-D to 2-D). These approaches and other ones are up to now not so well defined as FIR methods, due to the difficulties of factorizing 2-variable transfer functions and to the resulting problems of designing filters with sufficient accuracy and stability. Therefore, while from a theoretical view-point IIR filters can have higher efficiency than FIR filters, they are not so easily available for digital image processing.

In the following we will describe in some details a special class of FIR filters using a new type window (Cappellini window) and a new class of IIR filters using a trasformation from 1-D to 2-D.

2.1 A Class of 2-D FIR Digital Filters

A 2-D FIR digital filter can be defined by the following relation [1]

$$y(n_1,n_2) = \sum_{k_1=0}^{N_1-1} \sum_{k_2=0}^{N_2-1} a(k_1,k_2)x(n_1-k_1,n_2-k_2) \tag{1}$$

where $x(n_1,n_2)$ are the input data, $y(n_1,n_2)$ are the output data and $a(k_1,k_2)$ are the coefficients defining the 2-D frequency response of the digital filter.

In the window method, we are considering, we start from an impulse response, which is to be truncated introducing the minimum error in the frequency response. To this purpose the coefficients $a(k_1,k_2)$ are defined as $h(k_1,k_2)w(k_1,k_2)$, where $h(k_1,k_2)$ are the impulse response samples and $w(k_1,k_2)$ are the samples of the window function, having zero value in the region out of the truncation.

Many window functions are known. Recently an high-efficiency window was introduced (Cappellini window). The 1-D form of this window (2-D form can be easily obtained through rotation) has the property of giving a minimum value of the uncertainty product of modified type in which, differing from the classical definition of Heisemberg, only positive values of frequency are used [2] [3] . Resulting 2-D FIR digital filters have higher efficiency than other known filters using windows in a wide range of required characteristics.

2.2 A Class of 2-D IIR Digital Filters

A 2-D recursive IIR digital filter can be defined by the relation [4] [5]

$$y(n_1,n_2) = \sum_{k_1=0}^{N_1} \sum_{k_2=0}^{N_2} a(k_1,k_2)x(n_1-k_1,n_2-k_2) -$$

$$- \sum_{l_1=0}^{M_1} \sum_{l_2=0}^{M_2} b(l_1,l_2)y(n_1-l_1,n_2-l_2) \qquad (2)$$

where: $x(n_1,n_2)$ and $y(n_1,n_2)$ are, respectively, the input and output data; $a(k_1,k_2)$ and $b(l_1,l_2)$ are the coefficients defining the digital filter (l_1 and l_2 being not both equal to zero).

The method we proposed [6] [7] is based on transformations of the squared magnitude function of 1-D digital filters to 2-D domain (through the McClellan transformation) and uses a suitables decomposition in four stable filters (Pistor method).

The method has the interesting property of high speed in the design procedure, approximating any circular symmetry 2-D transfer function.

3. DATA COMPRESSION

Data compression is a transformation performed to reduce the amount of not useful or redundant data.

Several data compression methods and techniques have been introduced of reversible and not reversible type. Some important methods are: adaptive sampling, predictors, interpolators, encoding, multiple filtering, use of transformations (Fourier, Walsh...).

Let us describe in more detail some methods using the Fast-Walsh-Transform (FWT). The main characteristics of the defined methods [8] consists in dividing the transformed image data into several squares of small size (4x4, 8x8, 16x16) and in employing for each of them a minimum word-lengh (a bit number sufficient to represent the maximum absolute value in the square plus 1 bit for the sign). An additional fixed length word is inserted before the square amplitude values in order to specify the number of bits used to represent these square values. In a modified method we used, no value is maintained for those sub-areas or squares in which the addition of the absolute values of the transformed data results below a given threshold.

4. APPLICATIONS TO DIGITAL IMAGE PROCESSING

The above considered methods of digital filtering and data compression can be applied to digital image processing in many interesting areas.

We applied extensively these methods to biomedical image processing, in particular in nuclear medicine, performing low-pass filtering, inverse filtering and data compression. The methods (especially of data compression) were also applied to images received from earth resource satellites (LANDSAT). Good results were obtained in comparison with other more standard methods of digital filtering and data compression.

REFERENCES

[1] Special issue on 'Digial Signal Processing', Proc. IEEE, April 1975.

[2] W. Hilberg, P.G. Rothe, "The General Uncertainty Relation for Real Signals in Communication Theory", Information and Control, vol. 18, pp. 103-125, 1971.

[3] E. Borchi, V. Cappellini, P.L. Emiliani, "A New Class of FIR Digital Filters Using a Weber-Type Weighting Function", Alta Frequenza, vol. 44, n. 8, 1975.

[4] G.A. Maria, M.M. Fahmy, "An L_p Design Technique for Two-Dimensional Digital Recursive Filters", IEEE Trans. ASSP, vol. 22, pp. 15-21, 1975.

[5] D.E. Dudgeon, "Two-Dimensional Recursive Filters Design Using Differential Correction", IEEE Trans. ASSP, vol. 23, pp. 264-267, 1975.

[6] M. Bernabò, V. Cappellini, P.L. Emiliani, "A Method for Designing 2-D Recursive Digital Filters Having a Circular Symmetry", Proc. of Florence Conf. on Digital Signal Processing, September 1975.

[7] M. Bernabò, P.L. Emiliani, V. Cappellini, "Design of 2-Dimensional Recursive Digital Filters", Electronics Letters, vol. 12, n. 11, May 1976.

[8] V. Cappellini, F. Lotti, C. Stricchi, "Some Image-Compression Methods Using the Fast-Walsh-Transform", Proc. of 1975 Mid-west Symp. on Circ., Montreal, August 1976.

Coherant Optical Engineering, F.T. Arecchi and V. Degiorgio (eds.)

COHERENT OPTICAL IMAGE DEBLURRING

J.W. Goodman
Department of Electrical Engineering
Stanford University
Stanford, California 94305
U.S.A.

I. BACKGROUND ON THE IMAGE DEBLURRING PROBLEM

Suppose that an "object" intensity distribution is blurred by some degradation associated with the system that forms its image. We assume that the degraded optical system is linear and space-invariant, and therefore that the "image" intensity distribution $i(x,y)$ is related to the object through a convolution equation,

$$i(x,y) = \iint_{-\infty}^{\infty} o(\xi,\eta)s(x-\xi,y-\eta)\, d\xi d\eta \tag{1}$$

where $s(x,y)$ is the spread function of the imaging system. We would like to obtain an estimate $\hat{o}(x,y)$ of the true object $o(x,y)$ from the measured image $i(x,y)$, or equivalently to perform a deconvolution operation.

One general approach to this problem is the use of inverse filtering. Let $I(\nu_X,\nu_Y)$ and $\theta(\nu_X,\nu_Y)$ represent the "contrast spectra" of the image and object intensity distributions, i.e.,

$$I(\nu_X,\nu_Y) = \frac{\iint_{-\infty}^{\infty} i(x,y)e^{-j2\pi(\nu_X x+\nu_Y y)}\,dxdy}{\iint_{-\infty}^{\infty} i(x,y)dxdy}$$

$$\theta(\nu_X,\nu_Y) = \frac{\iint_{-\infty}^{\infty} o(x,y)e^{-j2\pi(\nu_X x+\nu_Y y)}\,dxdy}{\iint_{-\infty}^{\infty} o(x,y)dxdy} \tag{2}$$

Then if $S(\nu_X, \nu_Y)$ represents the optical transfer function of the degrading system, i.e.,

$$S(\nu_X, \nu_Y) = \frac{\iint_{-\infty}^{\infty} s(x,y) e^{-j2\pi(\nu_X x + \nu_Y y)} dxdy}{\iint_{-\infty}^{\infty} s(x,y) dxdy} \quad (3)$$

these various spectral functions are related by the simple equation

$$I(\nu_X, \nu_Y) = S(\nu_X, \nu_Y)\theta(\nu_X, \nu_Y) \quad (4)$$

which is entirely equivalent to Eq.(1). As a first approach to the deconvolution operation, we might use as our estimate of the object spectrum

$$\hat{\theta}(\nu_X, \nu_Y) = \frac{I(\nu_X, \nu_Y)}{S(\nu_X, \nu_Y)} \quad . \quad (5)$$

Stated in different words, we might obtain our estimate of the true object by passing the image $i(x,y)$ through a linear space-invariant filter with transfer function

$$H(\nu_X, \nu_Y) = \frac{1}{S(\nu_X, \nu_Y)} \quad . \quad (6)$$

For obvious reasons, this type of filter is known as an inverse filter.

While simple in its conception, the inverse filter suffers from several well known defects. First, for many degradations of interest (e.g., defocusing, motion blur), the optical transfer function has isolated zeros, thereby requiring the inverse filter to have poles. Thus the dynamic range of the deblurring transfer function must be infinite. Secondly, and even more serious, the inverse filter takes no account of the noise inherent in the measurement of the image $i(x,y)$. Thus the inverse filter is found to boost frequency components that may consist mainly of noise, and the resulting deblurred image may be so noisy as to be entirely useless.

The solution to this dilemma is to use a least-mean-square error filter, or Wiener filter, which produces an estimate $\hat{o}(x,y)$ that minimizes $E[|\hat{o}-o|^2]$, where $E[\cdot]$ is a statistical expectation operator. In deriving the general form of this kind of filter, it is generally assumed that the image $i(x,y)$ is related to the true object $o(x,y)$ by

$$i(x,y) = \iint_{-\infty}^{\infty} o(\xi,\eta)s(x-\xi,y-\eta)d\xi d\eta + n(x,y) \qquad (7)$$

where $n(x,y)$ is a statistically stationary noise process with power spectral density $\Phi_N(\nu_X,\nu_Y)$, and independent of $o(x,y)$. (Some modified least-mean-square-error filters derived for object-dependent noise have been discussed by Walkup [Ref. 1], but are not considered here). In addition, the object $o(x,y)$ is regarded as a stationary random process with power spectral density $\Phi_o(\nu_X,\nu_Y)$. The transfer function for the least-mean-square error filter is then found to be

$$H(\nu_X,\nu_Y) = \frac{S^*(\nu_X,\nu_Y)}{|S(\nu_X,\nu_Y)|^2 + \dfrac{\Phi_N(\nu_X,\nu_Y)}{\Phi_o(\nu_X,\nu_Y)}} . \qquad (8)$$

Several properties of the least-mean-square-error restoration filter are worth noting. First, at all frequencies for which the signal-to-noise ratio Φ_o/Φ_N is large compared with $|S(\nu_X,\nu_Y)|^{-2}$, we have

$$H(\nu_X,\nu_Y) = \frac{S^*}{|S|^2} = \frac{1}{S(\nu_X,\nu_Y)} . \qquad (9)$$

Thus the transfer function coincides with that of an inverse filter at such frequencies. However, at frequencies where the signal-to-noise ratio Φ_o/Φ_N is very small, the transmission of the filter is also very small, thus reducing the amount of noise present in the restored image.

As a typical example, we show in Figure 1 the transfer

functions corresponding to (a) a geometrical optics approximation to a focusing error, (b) an inverse filter to remove such error, and (c) a least-mean-square-error restoration, assuming "white" spectra for the object and noise. Note that only the moduli of the transfer functions are plotted; a phase shift of π radians is associated with alternate lobes of these patterns.

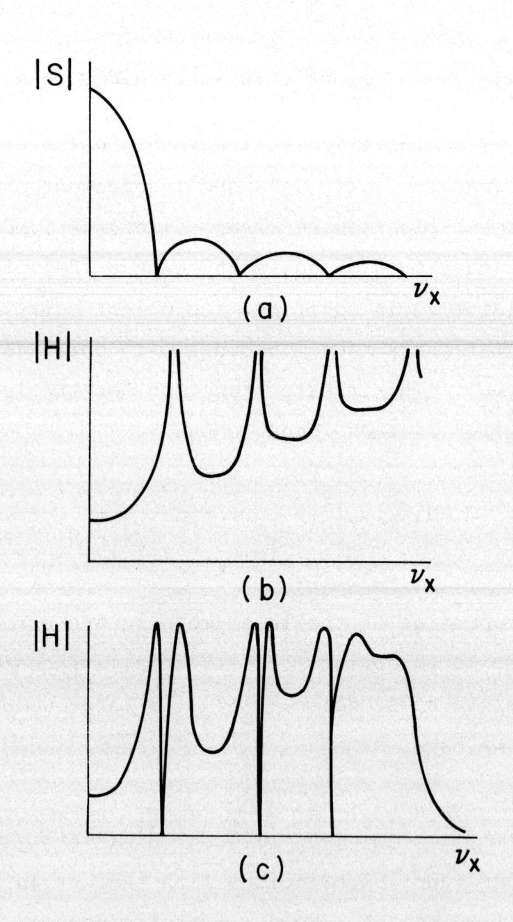

Figure 1: Transfer Function Examples

II. FILTER REALIZATION FOR COHERENT OPTICAL SYSTEMS

A. Optical Filtering System: Coherent optical approaches to image deblurring are based on the optical system shown in Figure 2. The blurred image $i(x,y)$ is entered as a field distribution across plane P_1. This field is Fourier transformed by lens L_2, producing a field incident on plane P_2 that is the Fourier spectrum of the blurred image. This spectrum is modified by an absorbing and phase-shifting structure placed in plane P_2 and having a complex amplitude transmittance proportional to $H(\nu_X,\nu_Y)$. Lens L_3 then transforms the transmitted spectral components, producing an output field distribution across P_3 that is proportional to the deblurred object distribution (provided the effects of the zeros of $S(\nu_X,\nu_Y)$ are not too severe.)

B. Properties of Photographic Film

In order to understand the various approaches to the realization of image deblurring filters, it is necessary first to understand some basic properties of photographic emulsions. Since coherent optical filtering systems operate linearly on fields, rather than intensities, it is the amplitude transmittance of a transparency that is of chief interest. We discuss here the relationship between amplitude transmittance and the exposure to which the emulsion was originally subjected.

If the film or plate is exposed within the linear region of the conventional H&D curve (i.e., where density is proportional to the logarithm of exposure), the amplitude transmittance of the resulting transparency can be related to the original exposing intensity I by [Ref. 2]

$$t_A = kI^{-\gamma/2} \tag{10}$$

where k is a constant and γ is the so-called "gamma" of the process. On the other hand, if a positive transparency is made, the relation becomes

$$t_A = k'I^{-\gamma/2} \tag{11}$$

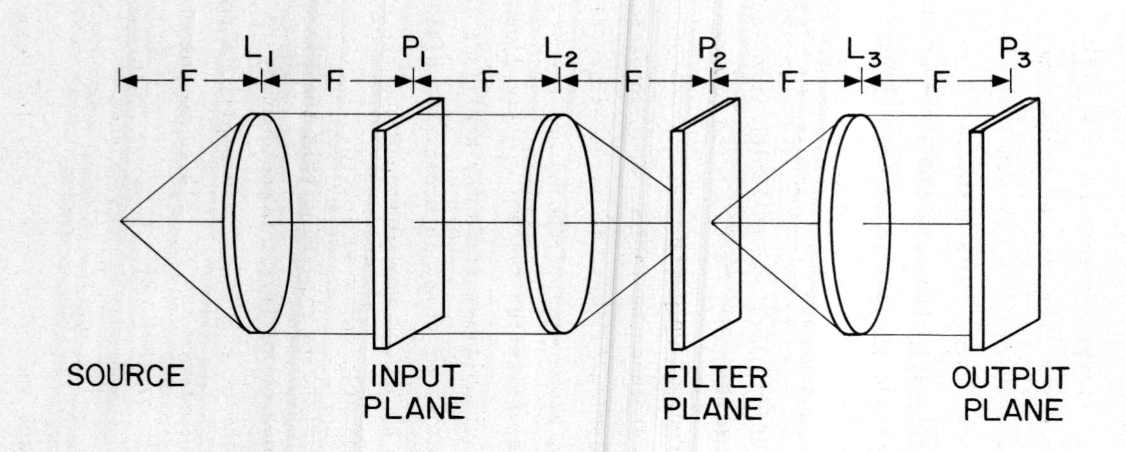

Figure 2: General Coherent Optical System for Image Deblurring

where k' is again a proportionality constant and γ is the product of the gammas of each of the two photographic steps involved, and is in this case a negative number.

When holograms are recorded, it is preferable to describe the film in terms of the relation between changes of amplitude transmittance Δt_A and changes of exposing intensity ΔI. Over a limited dynamic range we find that

$$\Delta t_A = \beta \Delta I \,. \tag{12}$$

This is the region in which holograms are usually recorded.

It is worth emphasis again that coherent optical processing systems operate linearly on complex amplitudes, not on intensities. Thus the transparency that serves as an input to the optical system must have the property that the transmitted complex amplitude is proportional to the original image intensity $i(x,y)$. Referring to Eq.(11), this requirement can be satisfied over a reasonably large dynamic range by using as the input a positive transparency with a photographic gamma of -2. In a similar fashion, the estimate $\hat{o}(x,y)$ of the original object appears at the output as a complex amplitude distribution, but all recording processes respond to intensities rather than amplitudes. An output photograph with intensity transmittance proportional to the modulus $|\hat{o}(x,y)|$ of the desired output can be obtained by producing a positive transparency with a gamma of -1/2 from the output of the processor. Note that any negative regions of $\hat{o}(x,y)$ will be folded up to appear as positive responses.

C. Filters Which Attenuate by Means of Absorption

Passive optical transparencies must have an amplitude transmittance which is less than or equal to unity. Hence, rather than boosting the high-frequency components in the restoration process, coherent optical systems attenuate the low-frequency components. The required attenuation can be accomplished either by absorption or by diffraction. We consider the former type of filter first, the latter second.

The earliest coherent optical deblurring filters were those of Marechal and Croce [Ref. 3] and Tsujiuchi [Ref. 4]. The filter structure placed in the frequency plane was a sandwich of an absorbing photographic transparency with amplitude transmittance proportional to $|S(\nu_X,\nu_Y)|^{-1}$, and a transparent phase shifting plate to change the transmitted phase by π radians in alternate ring lobes of the inverse filter H. The phase-shifting plate was made by vacuum deposition of a dielectric on a glass substrate. The dynamic range in the frequency plane over which the desired transfer function H was realized was rather small, the transfer function of defocusing error being restored over 2 or at the most 3 lobes, corresponding to a dynamic range of perhaps 10:1 in amplitude transmittance through the focal plane.

The use of holographic filters to achieve the required π radian phase shifts was almost simultaneously thought of by Stoke and Zech [Ref.5] and Lohmann and Werlich (Ref. 6] in 1967. The initial approach then was the following. As shown in Figure 3, a photographic record of the degrading spread function (a positive transparency with a gamma of -2) is entered in plane P_1 of the optical system. In the first recording step, the pinhole in plane P_1 is blocked. Thus a field proportional to $S(\nu_X,\nu_Y)$ falls on the photographic plate, which is processed to yield a negative transparency with a gamma of 2. This particular transparency has an amplitude transmittance

$$t_1(\nu_X,\nu_Y) \cong \frac{k}{|S(\nu_X,\nu_Y)|^2} \quad . \tag{13}$$

Next a second photographic plate is exposed, but this time with the pinhole in plane P_1 open. The light from the pinhole serves as a reference beam, and a hologram is thus recorded. If the exposure variations lie within a linear region of the t_A vs. E curve, the variations of amplitude transmittance are of

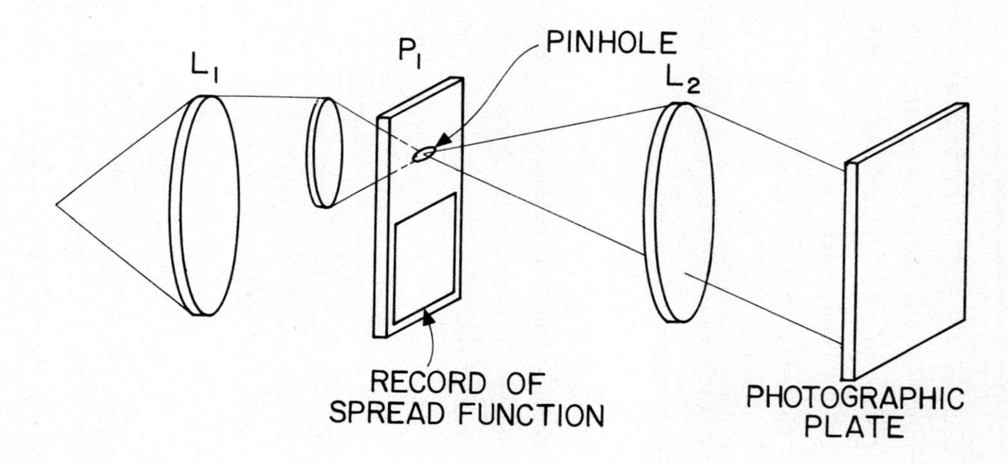

Figure 3: Holographic System for Recording Image Deblurring Filters

the form

$$\Delta t_A = \beta a^2 |S(\nu_X,\nu_Y)|^2 + \beta a A\, S(\nu_X,\nu_Y)\, e^{-j2\pi x_o \nu_x}$$

$$+ \beta a A\, S^*(\nu_X,\nu_Y) e^{+j2\pi x_o \nu_x} \tag{14}$$

where A is the amplitude of the reference beam at the recording plane, a is the amplitude of the object bean at $\nu_X = \nu_Y = 0$, and x_o is the x-coordinate of the reference point source. Restricting attention to the "conjugate" term of transmittance, i.e., the last term of Eq.(14), which can be isolated by examining only one first-order component of the diffracted light in the output plane, we find an effective amplitude transmittance for the holographic filter given by

$$t_2(\nu_X,\nu_Y) = \beta a A\, S^*(\nu_X,\nu_Y) \ , \tag{15}$$

where the exponential term may be suppressed if we examine only the one diffraction order of interest.

If the two filters are now placed in contact, the total amplitude transmittance through the focal plane becomes

$$t_1 t_2 = k\beta a A \frac{S^*(\nu_X,\nu_Y)}{|S(\nu_X,\nu_Y)|^2} = \frac{k\beta A a}{S(\nu_X,\nu_Y)} . \tag{16}$$

Thus the sandwich of these two filters yields an inverse filter.

The chief limitation of this technique lies in the fact that the desired forms (13) and (15) for t_1 and t_2 can be maintained over only a limited dynamic range of amplitude transmittance. Most restrictive is the absorbing filter described by Eq.(13). To understand the difficulty, we first note that the relation between amplitude transmittance t_A and photographic density D is

$$t_A = 10^{-D/2} . \tag{17}$$

Now in practice, D can be easily controlled only over the range

0 to 3, but with great difficulty can be taken as high as 4. The accuracy of control in the range 3 to 4 may be poor. In any case, optimistically assuming a maximum density of 4, t_A can be seen to vary from 1 to 10^{-2}. With reference to Eq. (13), a dynamic range of 100:1 in t_A implies a dynamic range of only 10:1 in $|S(\nu_X, \nu_Y)|$. Thus the inverse filter will have its correct transmission only over a 10:1 dynamic range in the frequency domain. With the more realistic assumption of a maximum density of 3, the dynamic range drops to 5.6:1. Such a small dynamic range is not even sufficient to improve the resolution in a defocused photograph by 2 to 1.

An increased dynamic range in the frequency domain is achievable by a method introduced by Stoke and Halioua [Ref. 7]. Again an absorbing filter is used, but of a slightly different kind than described above. The first step of this process is to record a negative transparency using exposure $|S(\nu_X, \nu_Y)|^2$ but this time with a photographic gamma of unity. Thus the amplitude transmittance of the absorbing transparency is

$$t_1(\nu_X, \nu_Y) \cong \frac{k}{|S(\nu_X, \nu_Y)|} . \qquad (18)$$

Now this transparency is used in the process of recording the holographic transparency, as illustrated in Figure 4. Again the spread function $s(x,y)$ appears in plane P_1 and is Fourier transformed by lens L_1. The resulting field $S(\nu_X, \nu_Y)$ in the frequency plane P_2 falls on the absorbing filter t_1 described by Eq. (18). The field transmitted by this filter is

$$\begin{aligned} U_t(\nu_X, \nu_Y) &= a\, S(\nu_X, \nu_Y) \cdot \frac{k}{|S(\nu_X, \nu_Y)|} \\ &= kae^{j\phi(\nu_X, \nu_Y)} \end{aligned} \qquad (19)$$

where $\phi(\nu_X, \nu_Y)$ is the phase distribution associated with the spectrum $S(\nu_X, \nu_Y)$. The remainder of the optical system in

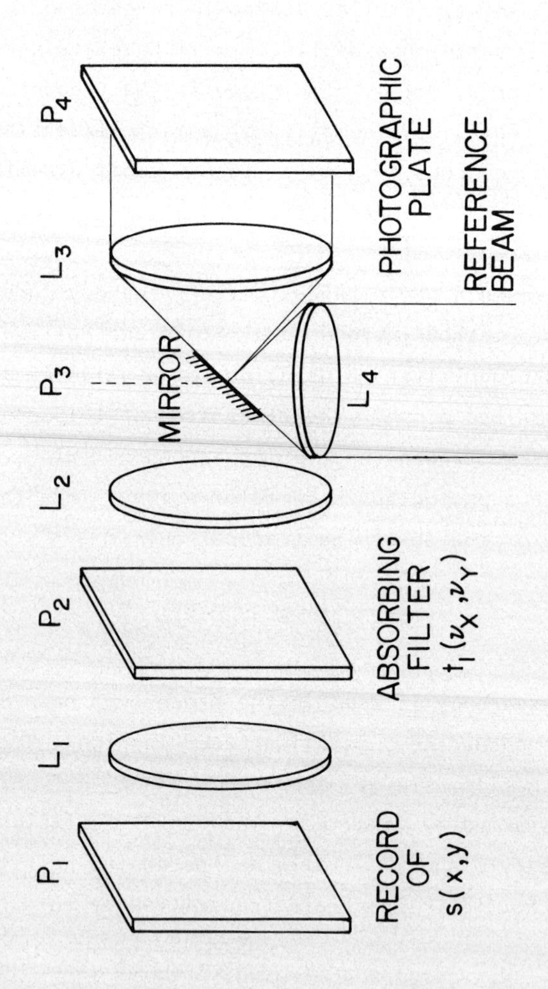

Fugure 4: Stroke and Halioua Deblurring System

Figure 4 serves only to record a hologram of the field $U_t(\nu_X,\nu_Y)$, the intensity incident on the final photographic plate being

$$I(\nu_X,\nu_Y) = A^2 + k^2a^2 + 2Aak\cos[2\pi x_o\nu_X + \phi(\nu_X,\nu_Y)]. \quad (20)$$

Assuming that exposure of the hologram is again carried out to make Δt_A proportional to ΔI, the one component of transmittance of interest is

$$t_2(\nu_X,\nu_Y) = k\beta Aa e^{-j\phi(\nu_X,\nu_Y)} . \quad (21)$$

Now a sandwich of the absorbing filter and holographic filter yields a total amplitude transmittance

$$t_1t_2 = \frac{k\beta Aa e^{-j\phi(\nu_X,\nu_Y)}}{|S(\nu_X,\nu_Y)|} = \frac{k\beta Aa}{S(\nu_X,\nu_Y)} . \quad (22)$$

Again, an inverse filter has been realized. However, this time the dynamic range over which the filter works properly can be much larger. If a maximum density of 4 can be achieved, the achievable dynamic range of the inverse filter is theoretically 100:1, or sufficient for about a 10:1 improvement in the resolution of a defocused picture. A maximum density of 3 implies a maximum dynamic range of about 32:1. Examination of the experimental results presented in the literature suggests that about a 40:1 dynamic range has been achieved with this method.

D. Filters Which Attenuate by Means of Diffraction

In 1970, Ragnarsson [Ref. 8] introduced an entirely new method for realizing image deblurring filters. His method requires only a single holographic filter, rather than a sandwich of two filters. Furthermore, attenuation is achieved by means of diffraction rather than by means of absorption.

To record this type of filter, the simple holographic arrangement of Figure 3 is used. However, unlike the usual holographic recording conditions, which use a reference beam intensity somewhat greater than the object beam intensity at the recording plane, the exact opposite condition is used. If a dynamic range of 100:1 in amplitude transmittance is desired, the reference beam should be 10,000 times weaker than the object beam at the center of the frequency plane.

The Ragnarsson filter is a bleached filter. To understand how it works, we must adopt the following postulates, all of which can be verified in practice:

(1) for a bleached photographic filter,

$$t_A = e^{j\phi} \cong 1 + j\phi \quad , \quad \phi << 1 \tag{23}$$

(2) The phase shift ϕ of a bleached photographic transparency is proportional to the density D of the transparency before bleaching,

$$D = k\phi \tag{24}$$

(3) If the transparency is exposed in the linear region of H&D curve,

$$D = \gamma \log E + D_o \tag{25}$$

where D_o is a constant.

With these three postulates we find that variations Δt_A of amplitude transmittance are proportional to variations $\Delta(\log E)$ of log exposure. But in addition, for small excursions ΔE about a mean exposure $\bar{E}$, we have

$$\Delta \log E \cong \frac{\Delta E}{\bar{E}} \cdot \log_{10} e \tag{26}$$

and thus

$$\Delta t_A = k' \frac{\Delta E}{\bar{E}} \quad . \tag{27}$$

Now concentrating only on the component of transmittance that diffracts light into the desired first-order component, and dropping the associated exponential, we have

$$\Delta t_A(\nu_X,\nu_Y) \cong \frac{k' A a S^*(\nu_X,\nu_Y)}{A^2 + a^2|S(\nu_X,\nu_Y)|^2} \tag{28}$$

where A is the amplitude of the reference beam and a the amplitude of the object beam at $\nu_X = \nu_Y = 0$. If we let $K = A^2/a^2$, i.e., K is the "beam ratio" of the holographic recording at $\nu_X = \nu_Y = 0$, then (28) becomes

$$\Delta t_A(\nu_X,\nu_Y) \;\tilde{=}\; k' \frac{A}{a} \frac{S^*(\nu_X,\nu_Y)}{K + |S(\nu_X,\nu_Y)|^2} . \qquad (29)$$

Thus the holographic filter so recorded has the form of a least-mean-square-error restoration filter. This filter is appropriate for the case of signal and noise power spectra of identical shape and for a ratio of noise to signal power spectra equal to K.

Examination of the experimental results of Ragnarsson indicates that he achieved a dynamic range of about 40 to 1 in amplitude transmittance in the frequency plane. More recent work of Tichenor [Ref. 9] has extended the dynamic range of these filters to 100:1 in amplitude by means of a relatively complicated two step filter recording process. A 100:1 dynamic range is sufficient to improve a defocused image by a factor of 8 to 1 in linear resolution.

III. NOISE LIMITATIONS OF COHERENT OPTICAL IMAGE-DEBLURRING SYSTEMS

In this section we summarize some of the sources of noise that are encountered in coherent optical image-deblurring experiments. In general, these sources of noise become more and more troublesome as the dynamic range of the deblurring filter increases, i.e., as the degree of attempted restoration increases. More detailed discussions of this subject, with experimental results shown, are found in Ref. [10].

A. Noise Inherent in the Photograph to be Deblurred

The most troublesome type of noise encountered is that present in the original photograph that is to be restored. In practice, such noise typically appears as film-grain noise. However, a more fundamental point-of-view regards this noise as arising from the finite amount of light flux utilized in recording the original photograph and the quantum fluctuations associated with that light flux. These quantum fluctuations are in a sense "amplified" by the imperfect detective quantum efficiency of the detector.

Noise inherent in the original photograph is extremely difficult to combat. As a general rule of thumb, when frequency components of the image have their contrast so severely depressed that they fall below the noise level, then no linear restoration technique can restore image information associated with these components. In experiments, image restoration techniques have been most successfully applied to photographs which were intentionally recorded in such a way as to minimize noise (e.g., by use of fine-grain, slow emulsions to record the original pictures).

B. Noise Defects Associated with the Input Transparency

In experimental coherent optical image deblurring, a troublesome source of noise can be optical defects associated with the input transparency. These are not defects associated with the recorded image *per se*, but rather defects arising from our inability to perfectly enter the blurred image into the processing system.

A common type of defect arises from dust specks that settle on the input transparency, or fall into the processing baths and eventually stick to the transparency. In addition, small bubbles that may exist in the emulsion or base of the input transparency are a similar defect. Each such defect acts as a small impulsive noise source, generating an impulse response of the deblurring filter. Since the desired component of light output is highly attenuated by the deblurring filter, the strength of the noise impulse responses can easily be so great as to obscure the wanted output. Tichenor [Ref. 9] has observed this type of noise with a deblurring filter having a dynamic range of 100:1. Evidence of this type of noise is also present in the results of Ragnarsson [Ref. 8]. In general, the severity of this type of noise increases as the dynamic range of the deblurring filter increases.

C. Noise Arising from Fourier Plane Scattering

The image deblurring filter is usually recorded on a photographic plate. As a consequence, the spectrum of the blurred image incident on the filter will suffer a small degree of random scattering by the filter itself. Even though the filter may be recorded on high-resolution photographic plate, the scattered light can be a significant source of degradation of the deblurred image. As shown by Tichenor [Ref. 9] the severity of this type of noise in the image depends on both the dynamic range of the deblurring filter and the size of the image field to be deblurred. The latter dependence is particularly important. A small image field can be deblurred with relatively little observed Fourier plane scattering noise, but a large field deblurred by the same filter may be totally obliterated by noise. These predictions have been verified experimentally by Tichenor; see [Ref.10].

The dependence of Fourier plane scattering on image field suggests that any coherent optical deblurring system should have a variable iris for controlling the size of the input.

REFERENCES

1. J. Walkup and R.C. Choens,"Image Processing in Signal-Dependent Noise", Optical Engineering, Vol. 13,pp.258-266, May-June 1974.

2. J.W. Goodman, Introduction to Fourier Optics, McGraw-Hill, (1968).

3. A. Maréchal and P. Croce, "Un filtre de frequences spatiales pour l'amelioration due contraste des images optiques", C.R. Acad.Sci.Paris, 127, 607 (1953).
4. J. Tsujiuchi, "Correction of optical images by compensation of aberrations and by spatial frequency filtering", in Progress in Optics, (Ed. E. Wolf), Vol. II, 131, North Holland Publishing Co. (1963).
5. G.W. Stroke and R.G. Zech, "A posteriori image-correcting 'deconvolution' by holographic Fourier-transform division", Phys.Lett., 25A, 89 (1967).
6. A. Lohmann and H.W. Werlich, "Holographic production of spatial filters for code translation and image restoration", Phys.Lett. 25A, 570 (1967).
7. G.W. Stroke and M. Halioua, "A new method for rapid realization of the high-resolution extended-range holographic image-deblurring filter", Phys.Lett. 39A 269 (1972)
8. S.I. Ragnarsson, "A new holographic method of generating a high efficiency, extended range spatial filter with application to restoration of defocussed images", Physica Scripta, 2, 145 (1970).
9. D. Tichenor, Extended Range Image Deblurring Filters, Ph.D. Thesis, Department of Electrical Engineering, Stanford University (1974).
10. J.W. Goodman, "Noise in coherent optical information processing", Optical Information Processing, (Y.E. Nesterikhin, G.W. Stroke and W.E. Kock, Eds.), Plenum Publishing Co., 85 (1976).
11. J.L. Horner, "Optical spatial filtering with the least mean-square-error filter", J.O.S.A., 59, 553 (1969).

Coherant Optical Engineering, F.T. Arecchi and V. Degiorgio (eds.)

OPTICAL DATA PROCESSING

F.B. ROTZ

Harris Electronics Systems Division, Electro-Optics Department - Melbourne, Florida

In a typical data collection system, we find that we have many users wanting to extract specific information from large data bases. The total problem is one of data collection, data storage, data processing, data dissemination, and data utilization. The principal topic of this paper is to discuss parallel data processing of imagery and widebandwidth electronic signals. Parallel processing is clearly a requirement when one considers that a single image formed by a sensor in just one part of the spectral region may contain from 10^6 to 10^8 pixels (picture resolution elements). Furthermore, if we consider that many sensors may be acquiring data simultaneously, we find that the total amount of data collected per unit time is very large indeed. Similarly, the extraction of information from signals having wide bandwidths or large time bandwidth products requires the use of new technologies.

The key question to ask is how we can process these data to extract the information of interest to a particular user. The solution to this problem presupposes that the user knows what information he is trying to extract and that he has some knowledge of the sensor characteristics. Our efforts here ought to focus on determining what kinds of parallel processing systems are available for potential application to image processing. I shall review the fundamental features of coherent optical data processing systems and give some representative illustrations of how they can be used.

Optical processing systems are inherently parallel processors. They are capable of performing operations on images containing millions of pixels in extremely short time intervals. The throughput is, in fact, generally limited by peripheral devices and not by the processing speed of the optical system itself. If, in addition, the optical system is coherently illuminated, we have the advantage of being able to operate on the two-dimensional Fourier transform of the image and can therefore perform two-

-dimensional spatial filtering operations in parallel.

The basic theory of optical spatial filtering is well developed. The data to be processed modulate a coherent beam of light. The light diffracted by the data is collected by a lens that displays the two-dimensional Fourier transform of the data at the image plane (frequency plane) of the primary light source. A transparency or mask placed in the frequency plane is the spatial filter that modified the Fourier transform of the data. The filtered data, displayed at the image plane, can then be detected for further processing.

Since the desired filtering operations frequently require the use of filters that have complex values (both phase and amplitude information), means for recording such filters are required. Spatial carrier frequency filters, in which a complex-valued function is encoded as a non-negative function, can be constructed by using a technique closely related to one used in holography for recording holograms. The filter construction system, therefore, consists of a signal beam in which the Fourier transform of the desired impulse response is displayed, and an off-axis reference beam. If the sum of the reference beam and the Fourier transform is detected by a square-law detector, the recorded function can be used directly as the desired filter. Alternatively, the filter may be generated by a computer and recorded by one of several means.

One application of coherent optical systems is immediately suggested if we note that the Fourier transform exists as a physical, measurable light distribution. From a direct measurement of the intensity of the Fourier transform we can learn something about the structure or texture of the information contained in an image. This information could be used to classify images or subregions of images according to their distinctive spatial frequency content. This could be the first step in many data reduction processes.

More powerful data processing operations can be implemented, however, if we use spatial filters in the frequency plane to modify the Fourier transform of the image. In this paper we review two related applications which illustrate some features of parallel processing. Both are concerned with image correlation, that is, to solve the problem of locating one image (or a part of an image) relative to another. The two images do not necessarily have to be produced by the same sensor nor do they have to be taken at the same time or from the same vantage point.

The first application could be termed image mapping. Suppose that we have a reference library of images obtained from a given sensor and that we construct spatial filters for these images. It may be noted that if we consider the optical processor as a computer, the the filters are equivalent to the memory of the computer. Each filter contains as much information as an entire image, e.g., 10^6 to 10^8 data points. Several hundred filters can be recorded on one role of film to give a mass storage capability. Once the filter library has been recorded, we are in a position to process data.

The steps required for image mapping include constructing spatial carrier frequency filters of many frames of imagery to form a reference library. A frame of imagery whose coordinates and orientation are to be determined is then cross-correlated with the reference images. By measuring the position of the cross-correlation peak, the position of the frame of imagery relative to the stored reference function is determined. We do not need to use all of the imagery to locate its position; rather, only a subregion can be used. The reduced area still provides a cross-correlation peak with sufficient signal-to-noise ratio to measure accurately the location of the entire frame. It is not important which part of the imagery is used; every part of the imagery contained in the corresponding image in the reference library should correlate equally well, and each part produces a cross-correlation peak having the same location so that the position of the mission imagery is unambiguous. The intensity of the cross-correlation peak will not vary as a function of which subportion was illuminated if the terrain has very small changes in elevation and in illumination; different camera viewing angles therefore cause only small changes in the geometry.

Since it may be an expensive proposition to obtain reference imagery covering large areas of the earth, we have investigated an alternative means for mapping. If we can find the center coordinates and orientation for atleast one frame of imagery, and if there is some overlap of the frames, wa can carry out the mapping by finding the position of each frame relative to the previous frame. The key to this technique is the use of an erasable recording material in the frequency plane which will be described later.

A related application is that of cloud motion analysis. Again, we generally have a sequence of frames of imagery and want to determine the motion of the cloud patterns. From these movements we can make a first order estimate of the wind velocities.

The precedure is similar to that described before except that now the correlation plane contains many subcorrelation peaks, one for each cloud pattern, that must be stored.

The motion of individual patterns can be determined by allowing only the pattern of interest to appear in an aperture placed in the input plane of the system. By moving the aperture in the input plane and recording the position of the cross-correlation peak in the output for each position of the aperture, a synoptic picture of the cloud motion over the entire field of view can be developed. Cloud rotation can be measured simply by rotating one frame until the cross-correlation in the output plane is maximized.

The magnitude of the cross-correlation is a measure of the similarity of the patterns in each frame. If the shape of the patterns changes significantly between frames, the correlation peak will be correspondingly reduced. Thus, this technique also offers an objective measure of the similarity between patterns which is useful in determining the rate at which clouds change shape. The same comments apply to measuring changes in, for example, the seasonal texture of terrain.

We have used photoplastic devices to construct spatial carrier-frequency filters in situ in real time. Spatial filters for use in applications such as pattern recognition, target detection, character recognition and signature detection have usually been recorded on a nonreusable material such as photographic film. The required filters were, therefore, recorded before the data was processed to form a reference library.

In some applications it is desirable to construct a filter in-situ for a pattern not contained in the library. Although this can be done by using self-developing materials such as photopolymers, these materials are not reusable. Reusable materials such as photochromics are either insensitive and are not practical for use in spatial filters. The photoplastic real-time device we developed is reusable. It can be recycled many times; moreover, its sensitivity is comparable to that of 649F photografic emulsion on which filters are typically recorded. This device considerably extends the flexibility and usefulness of coherent data processing systems because it gives us the opportunity to update the memory of the computer.

We have also made efforts to improve the flexibility of optical processing by developing a hybrid system in which the optical system performs the highspeed, fixed routine operations while a digital computer performs low-speed flexible routine operations.

Furthermore, such a system has considerable potential for use in a human interactive system. The readout system consist of the image orthicon, a threshold device, a cell generator, a PDP-8/1 computer, a graphics display unit, and a TV monitor.

The TV camera detects the light distribution at the output plane of the optical system; the resulting video signal is thresholded and fed into a cell generator which is a high-speed, double-buffered memory composed of shift registers and various other logic circuits. The cell generator organizes the thresholded video signal into a number of cells; each cell represents a small rectangular area at the output plane of the optical system. A cell is active if there is at least one video peak exceeding threshold in the area represented by the cell. The state of a cell is indicated by the state (0 or 1) of the corresponding flip-flop in the shift registers. The cell generator can be operated in an almost arbitrary format, but it normally creates a 32x32 array of cells.

The function of the electronic readout system is to collect, collate, and display the processed data. In a typical processing application the output of the optical system contains correlation peaks which must be detected and processed further. A basic concept in cross-correlation operations is that the output data rate can be reduced significantly relative to the input data rate. This concept is particularly important in optical processing because a frame of input data may contain as many as 10^8 pixels. Since correlation is a linear operation, the output detector must have the same number of resolution elements as the input and, unless data reduction takes place, extremely high data rates may be required to transfer the processed data to the user in a useful form. For example, if a frame of imagery is processed in 1 sec, data rates in excess of 100 Mb/s are required at the output. Clearly, such data rates cannot be handled by conventional data display systems.

A third area in which further progress is needed is that of input devices. Although photographic film is an ideal input media, we often would like to process electrical analog or digital data. Data composing devices such as the light valve, ferroelectric materials, thermoplastic materials, and the like are useful for inputting two-dimensional data whereas acousto-optic cells have potential for inputting one-dimensional time signals. As further improvements are made in input and output interface devices, we can expect to see even wider use made of coherent processing systems.

The examples given here by no means exhause the list of applications. But spectral analysis and correlation offer two powerful operations that are of great general utility. For example, another important application is image deconvolution (or image enchancement or image deblurring). This method belongs, however, to the correlation class of operations. Other applications are differientation, subtraction processes for change detection, and cross spectral estimation. Furthermore, the optical processor can perform nonlinear as well as linear operations.

In the future, then, we can expect to see optical processors used in conjunction with digital computers to carry out the required processing operation. Each system will be used to perform those operations for which it is best suited; the digital computer for numerical calculations and for operations not easily implemented optically, and the optical computer for performing integral operations and other parallel operations on the entire image simultaneously.

Coherant Optical Engineering, F.T. Arecchi and V. Degiorgio (eds.)

HOLOGRAPHIC MEMORIES

F. B. ROTZ

Harris Electronic Systems Division, Electro-Optics Department - Melbourne, Florida

INTRODUCTION

Since Leith and Upatnieks[1] showed how holographically stored information can be reconstructed separately from the direct and conjugate image beams, holography has provided potential solutions to many problems. Through continued development of electrooptic devices needed to implement practical systems, considerable progress has been made in realizing useful hardware for data storage and retrieval systems.

Holographic techniques offer several advantage over conventional techniques for archivally storing and retrieving both analog and digital information. The primary advantage stems from the increase in packing density that can be achieved holographically, particularly when the information is recorded as a Fourier transform hologram; the packing density is then maximized (except for object fields containing fine detail) for both analog and digital information[2]. Furthermore, if the information is recorded as a Fourier transform hologram, other advantage automatically accrue: (1) each picture element or bit contributes to every part of the hologram, providing a natural encoding process which causes the recorded information to be retrieved with greater freedom from the effects of imperfections such as dust, scratches or blemishes; (2) the shift-invariant theorem for Fourier transforms means that the readout beam does not have to track perfectly the recorded information so that alignment tolerances are substantially independent of packing density; (3) both the readin and readout can be arranged for low packing density, while retaining high packing density on the storage medium, which again relieves tolerances on alignment for input/output devices; (4) very high recording rates can be achieved for storing digital information by using the parallel, high-speed nature of optical systems.

In this paper we discuss several holographic storage and retrieval systems. Although each system is intended to satisfy a particular need, considerable variations can be envisioned to meet other requirements. These systems are (1) an analog storage and retrieval system for graphical information such as maps, charts, and engineering drawings, (2) a digital storage and retrieval system for rapid, random access to blocks of data without moving parts, (3) a mass memory system having the capability for storing both analog and digital information, and (4) a high-speed digital system in which the data are recorded on a roll of film.

HOLOGRAPHIC STORAGE AND RETRIEVAL OF ANALOG INFORMATION

Large graphics that contain fine detail (such as maps, charts and engineering drawings) are particularly difficult to store at high densities using conventional micrographic techniques. For example, an engineering drawing that is 24"x36" may contain detail as small as 0.004 inches; if this graphic is reduced 60X using conventional techniques so that the micro-image is 10 mm x 15 mm, the resolution required at the storage medium is 1.7 microns. Not only is this resolution difficult to preserve, it also leads to severe tolerences on depth of focus both in the recording and the retrieving processes and on lateral motion in retrieving.

Holographic techniques offer an attractive and practical solution to these problems. Suppose we first reduce the graphic approximately 20X onto 35 mm film, using conventional micrographic techniques (see Figure 1a). The resolution for this intermediate image is then approximately 5 μm. Other initial reduction ratios are possible; the ratio does not impact the ultimate packing density that can be achieved holographically. One advantage of this initial photoreduction is that smaller diameter optics can be used in the subsequent holographic systems; a second advantage is that most graphics originate as black lines on a white background so that a negative image provides an optimum redundancy removal process, i.e., the information (black lines) now appears as transparent regions within an opaque surrounding: a third advantage is that the negative transparency allows for recording of specular insted of diffuse information, thus eliminating the speckle phenomenon in the resultant display.

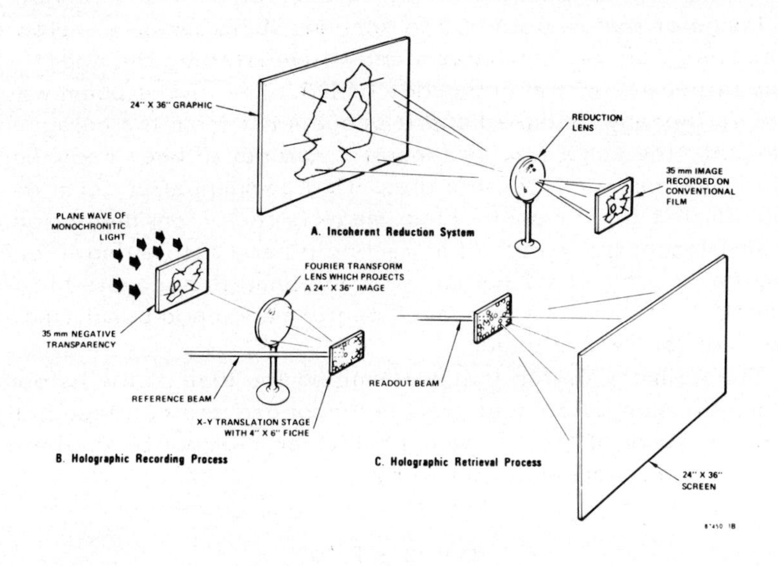

Fig. 1 - Holographic Storage and Retrieval of Analog Information

The negative transparency is then placed in a holographic recording system, as shown in Figure 1B, and illuminated by a plane wave of monochromatic light. Light diffracted by the transparency is collected by a Fourier transform lens which also projects the transparency onto a screen at the original size. If this Fourier transform lens is the same as the one used in the initial photoreduction step, and if the overall magnification is l:l, the resultant image is completely free of distortion. This results from the basic theory of optical design from which it can be shown that telecentric system is automatically free from distortion.

Suppose that we now place a photografic film at the back focal plane of the lens as shown in Figure 1B, and capture the light intensity at that plane, having added an off-axis reference beam. The reference beam interferes with the signal beam so that both the amplitude and phase of the signal beam wavefront can be recorded as a non-negative (intensity) function on photographic film. After a set of exposures are made on one 4" x 6" film chip (one hologram per graphic), the fiche can be developed in the usual way and placed in a storage and retrieval unit.

Figure 1C shows the basic retrieval process. This system is particularly simple since no lenses are required to reconstruct the image of the graphic on the screen. A hologram is selected by means of an x-y mechanism and illuminated by the readout beam (a replica of the reference beam). The signal beam wavefront is thereby released and it propagates from the hologram in exactly the same way as though it had never been recorded. This wavefront includes the distortion compensation term that ensures that the reconstructed image is free from distortion.

An alternative method for recording and retrieving information, using a two-lens holographic system, is suggested in Reference 2. It provides improved performance and simplifies the design of the lenses used.

The primary factor that determines the size of the hologram is the minimum detail that must be recorded and retrieved. If we let the size of the minimum resolution element be d, the diameter of the hologram is given by

$$\bar{d} = 2\lambda F/d$$

where $\bar{d}$ is the diameter of the hologram, λ is the wavelength of light, and F is the focal length of the Fourier transform lens. If $d = 5\ \mu m$ (corresponding to $100\ \mu m$ detail in the input graphic) and if $\lambda = 514.5$ nm, the diameter of the hologram is $\bar{d} = 0.2F$. Since we want the focal length of the lens to exceed the diagonal of the input transparency, we select F = 50 mm which gives a hologram diameter of $\bar{d} = 10$ mm. We see, therefore, that a hologram diameter of 10 mm should be more than adequate from a resolution viewpoint. We also see that the initial photoreduction ratio does not influence the hologram size; if a 10X system is used, both F and d increase by a factor of two so that $\bar{d}$ remains constant. A hologram diameter of 10 mm gives an equivalent reduction ratio of 60X for 24" x 36" graphic.

Compared to conventional micrographic techniques, the holographic storage and retrieval system has several features:

1. Each resolution element (pixel) of the graphic contributes the light to every part of the hologram. This natural encoding phenomenon gives immunity to dust, scratches or blemishes that could obscure significant areas of micrographic images.

2. The axial position tolerances of the hologram, both recording and retrieval, are much greater. Instead of maintaining a depth of focus adequate to record a $1.7\ \mu m$ spot (equivalent to a

f/1.7 lens system) the hologram need be positioned to an axial tolerance equivalent to an f/100 lens. Since the depth of focus is proportional to the square of the f/#, the holographic system has a tolerance that is 3,500 times less stringent (for an equivalent reduction ratio) than a conventional micrographic system.

3. An even larger reduction on the lateral positional tolerance of the hologram is possible. In conventional systems operating at a magnification of 60X, 0.05 inch movement of the film results in a 3 inch image movement. Such systems are, therefore, sensitive to vibrations and thermal variations, and require accurate x-y selection mechanism. A hologram recorded in the Fourier transform plane can be moved any distance (as long as it is fully illuminated) without introducing any image motion. Although it is usually better to record in a near Fourier transfrom plane to reduce the dynamic range required of the recording medium, even then the sensitivity to motion is significantly less. The distance that the image can move is, at worst, the same distance that the hologram moves. But in any practical system using near Fourier transform holograms, the image motion will be approximately 1 percent of the hologram motion. Thus, if the hologram moves 0.05 inch, the image will not move more than 0.0005 inch -- a factor of 6,000 times better compared to a conventional system at the same magnification.

Fig. 2 shows a reconstruction of an A-size drawing that had been stored in a hologram having a 4 mm diameter. As can be seen, resolution has been preserved throughout the storage and retrieval process. This technique can also be applied to graphics or documents having grey scale; hence, it is a holographic technique for storing analog information.

In majority of cases, installations of the first type consist of a hydraulic pump (of chain, piston, or paddle type), the choking gap (a pressure valve or other element), a reservoir, and measuring and control facilities. A system of this type can be termed a "circulatory pump-gap system". Figure 3 shows a pump-gap installation in which the investigated fluid in a closed system is pumped repeatedly under pressure through a choking gap. Installations of this type work at 1400 - 2800 rpm, in the pressure range 40 - 120 kg/cm^2, and they require from 0.5 to several tens of liters of the investigated fluid, one experiment taking from several hours to several days to complete.

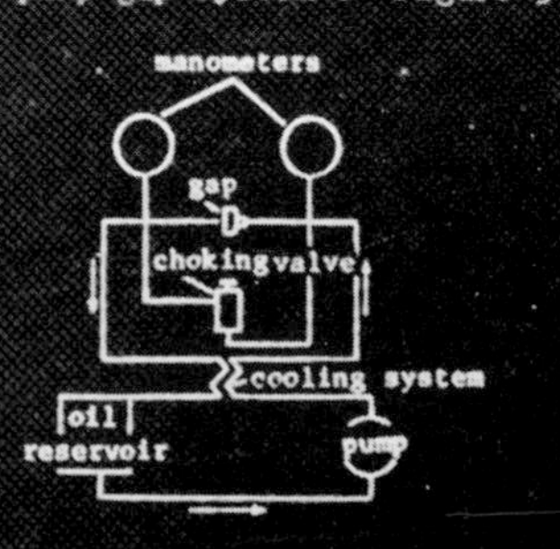

Figure 3. Scheme of a large installation of the pump-gap type for thinning of oils, from the data of the Aviation Institute.

An injector apparatus is another version of the first type. In the injector apparatus, in Figure 4, the effects of pressure are achieved by forcing the investigated fluid with the aid of a piston pump through an injector-sprayer, which serves here as a diaphragm and choking valve, and thus the fluid is thinned. This type of apparatus involves a pressure of 50 - 175 kg/cm^2, volume of the investigated sample of up to 500 cm^3, and time required to achieve a considerable degree of thinning — about four hours.

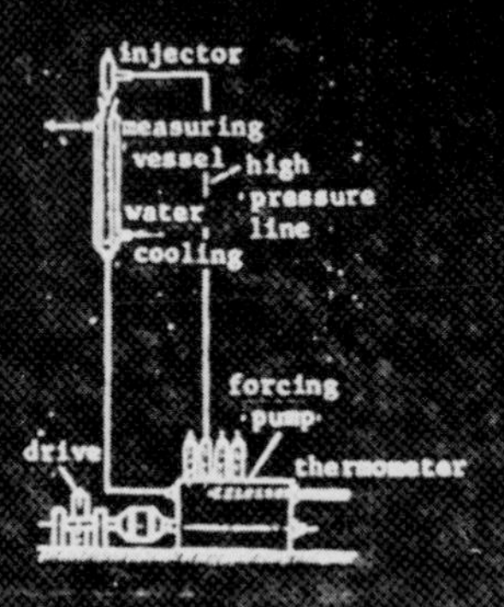

Figure 4. Scheme of the injection apparatus for thinning of oils, from the data of the Aviation Institute.

Fig. 2 - Reconstruction of an "A" Size Drawing

HOLOGRAPHIC STORAGE AND RETRIEVAL OF DIGITAL INFORMATION

Many of the advantages of holography cited in the previous section can also be used in systems used to store digital information. Because of its unique properties, it is not surprising that attempts have been made to apply holography to a broad range of memory and storage hierarchy. Activity has ranged from developing small-capacity, high-speed memories to large-capacity, read-only storage in the multi-terabit range. Additionally, significant activity has been directed toward solving the highly specialized problems associated with ultrahigh data rate recorders and reproducers. Memory systems are now, and will continue to be, the highest single cost item in the computer hardware structure. This, at least in part, accounts for the intensive research activity in optical alternatives to computer memory and storage.

Although research continues across the broad spectrum of memory hierarchy, some strong indicators point to very specific areas where the technology has a reasonable chance of success. Before we consider the basic characteristics of holographic memories, we identify the targets at which holographic memories have been aimed. Perhaps the two most widely used performance measures for memories are capacity and access time. Clearly there are many other factors such as transfer rate, size, power consumption, interface ease, reliability and reproducibility which may play equally important roles in characterizing memory performance. Similarly, memory cost or (more commonly) cost per bit, forms one of the important criteria for memory selection For purposes of this discussion, capacity and access time will be sufficient factors if we remember that even the ultimate in memory performance is unacceptable if the eventual costs are not consistent with what the marketplace can afford.

Figure 3 shows the present state-of-the-art in so-called conventional memory technology in terms of capacity and access time. The technology ranges from the relatively small but fast semiconductor memory through moving head disc memory to the larger and slower bulk storage devices such as magnetic tape. Clearly most memory and storage technology is confined to magnetic phenomena. The exceptions to the magnetic dominance have been at the low-capacity slow access end with the IBM 1360 and Precision Instrument Model 190 bit-by-bit optical technology.

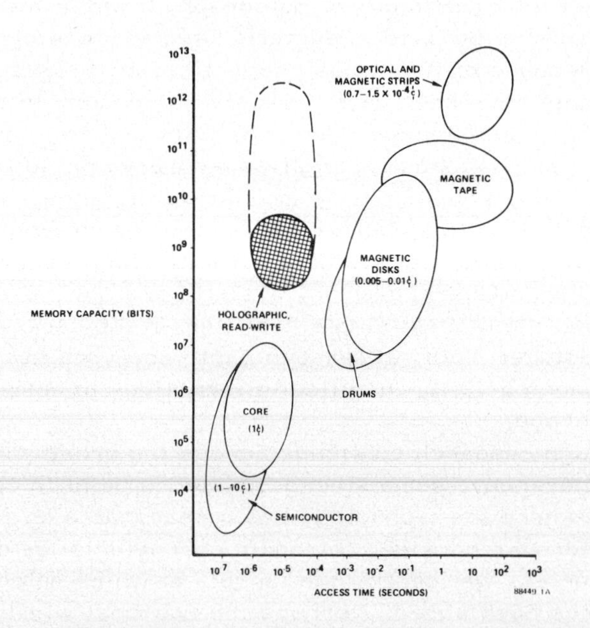

Fig. 3 - Current Memory Technology

The trends in memory and storage technology indicate a gradual (although sometimes rapid) trend up and to the left, i.e., toward larger and faster devices. Consequently, the aim of early holographic memory researchers was toward those areas where the payoff would be largest, i.e., $10^8 - 10^{10}$ bit capacities with 1 - 10 μsec access time for disc replacement. Similarly, the promise of extremely high data packing density afforded by holographic encoding encouraged research activity at the upper end of the spectrum to achieve terabit capacity.

A logical distinction between various holographoc memories can be made by first considering those which are truly read/write memories and are aimed at existing mainframe and peripheral technology and those which are directed at mass storage applications. While the basic technology in both applications is somewhat similar, the approaches to solutions require emphasis on different components.

Read/Write Holographic Memories

The major effort to date on read/write holographic memories has addressed capacities between 10^8 and 10^{10} bits with access times measured in microseconds. Several domestic companies, including Harris Corporation, Bell Labs, IBM, and RCA, as well as several foreign laboratories at Siemens, Nippon, Hitachi and Thomson-CSF, have either developed breadboard memories or are actively pursuing development of major components which are required in holographic memories. It is beyond the scope of this paper to discuss in detail the work of each laboratory and to dwell on the progress made in developing individual components. It is sufficient to say that, although progress on each holographic memory component has been significant over the past decade, a truly viable cost competitive memory has not emerged from the breadboard stage into the marketplace. Because of the interactive nature of all holographic memory components, a major advance in one area may produce only minor improvement in system performance.

Figure 4 shows some of the basic elements of a block oriented read/write memory. In many respects it is similar to the system shown in Figure 1B. The differences are that a page composer which converts an electrical bit stream into a two-dimensional spatial signal is used instead of a transparency to modulate a laser beam; the hologram array is stationary and the information from successive page composer inputs is deflected to any of several recording positions by a fast, random access acousto-optic device; the output is a photodetector array instead of a viewing screen; the recording medium has a read/write/erase instead of an archival read only capability.

The recording operation is basically the same as described before except that both the signal and reference beams are directed toward a storage position in the hologram array by an acousto-optic beam deflector which is not shown. Several trade-offs can be established to determine the optimum number of bits to be stored in each hologram and, based on fundamental optical considerations, the maximum capacity can be found.[3] For a 10^8 bit capacity system, a convenient number of bits per hologram requires a 128 x 128 element page composer and a 78 x 78 element hologram array. The memory capacity cannot be increased much beyond 10^9 bits without encountering rather severe lens design problems. To achieve a larger capacity memory, therefore, requires either a reversible volume storage medium or a moving recording medium.

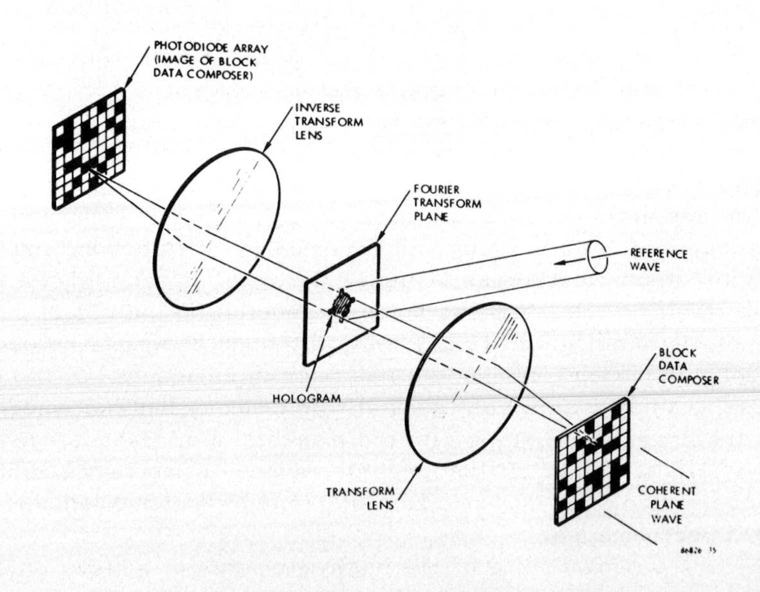

Fig. 4 - Basic Elements of a Block Oriented Memory

Perhaps the two components which have received the greatest attention are the input page composer and the holographic recording media. In both cases, the performance limitation has been dictated by the availability of suitable materials. The page composer should contain between 4000 and one million elements. Various materials have been or are being considered, including PLZT, liquid crystals, thin deformable membrane mirror arrays and cadmium sulfide. All currently suffer from one or more shortcomings including uniformity, speed, contrast ratio and stability. The holographic recording materials have also received considerable attention and all candidates fall short of the desired properties of high efficiency, high sensitivity and long lifetime. Significant progress has been made in developing beam deflectors and photodetector arrays, stimulated in part by the requirements imposed on these devices by holographic memory researchers. The performance of beam deflectors has nearly doubled in the past ten years and sophisticated multielement two dimensional photodetector arrays were essentially unavailable ten years ago. Research in all component areas is continuing. We can expect significant breakthroughs only by application of new materials or through better understanding and perfection of existing materials.

Until recently, heavy emphasis was placed on preserving high speed access with no moving parts while increasing capacity. Clearly the target was disc replacement. Analyses of the constraints imposed on common optical components such as lenses as well as a better appreciation of physical limitations imposed on the electro-optical components has led most investigators to revise their predictions of ultimate read/write holographic memory performance, regardless of cost considerations. The capacity of nonmechanical, practical read/write holographic memory with microsecond access time is likely to fall between 10^8 and 10^9 bits, not up to 10^{12} bits as was believed earlier. In any case, because of costly electro-optical components, this type of memory will be characterized by high cost per bit. Only in highly specialized applications where technical performance plays an overriding part will we see this memory used. Even so, it is not likely to emerge from the research laboratory before the end of this decade.

Read-Only Memories

Read-only holographic memories typically use film as the recording media. Once exposed, the film record is removed from the recorder, developed by normal techniques, and placed in a holding area until data retrieval is required. If any portion of the recorded data must be changed or updated, the entire record must be re-recorded and replaced within the memory. Read-only memories, therefore, are best suited for archival, non-dynamic memory applications or applications where updating is relatively infrequent.

Other than the recording media, the holographic exposure and data readout processes are similar to those used in a read/write memory. Similar devices are required to implement a read-only memory as are required to implement a read/write memory; hence, read-only memories are therefore constrained by similar device limitations. Possibly the most critical device limitation in read-only memory implementation has been the page composer. Since a one dimensional (instead of a two dimensional page composer) is typically required, device capability in this area has recently been improved enough to allow system applications.

One way to overcome the page composer limitation in holographic recording is to use a synthetic hologram approach. In this approach, a film intensity function is calculated by a special purpose digital processor and scanned onto the film by a scanning device. The resulting film exposure has nearly the same reconstruction properties as does an interferometrically generated hologram.

The recent advances in materials and components have allowed production of a few prototype holographic memories. For example, let us discuss some specific hardware systems being developed by Harris Corporation, Electronic Systems Division. Synthetic holography has been successfully applied to the recording and storage of digital data in the Human Read/Machine Read (HRMR) System developed for Rome Air Development Center. A research prototype, shown in Figure 5, was delivered in May 1973 and an engineering prototype is currently under development.

Fig. 5 - Human Read/Machine Read Research Prototype

The HRMR System addresses to document storage, retrieval and dissemination problem which is impacting both government and industrial complexes having large document data bases. The HRMR concept is based upon annotating a standard microfiche with the digital equivalent of the associated images. Optical readout of the digital directly from the microfiche facilitates storage, retrieval and dissemination of data to both local and remote locations.

A direct extension of the concept is the full utilization of the microfiche film chip for digital data recording. Thirty megabits of user data per film chip is presently being realized at a packing density exceeding one megabit per square inch. Since this packing density is significantly below theoretical limitations, considerable improvement can be anticipated as components and techniques are further refined.

Utilization of holography as the digital data recording technique in the HRMR System provides an inherent immunity to dust, scratches, and film imperfections associated with practical hardware which is capable of functioning in an operational environment. Only normal microfilm storage environmental conditions are required. The recorded data is archival and optical readout of data is nondesctructructive. This results in a virtually permanent record and contrasts with magnetic media which suffers from signal loss and deterioration due to readout and long term storage. Of further benefit, the positional invariance property of holography facilitates readout and allows relatively simple and economical hardware configurations.

Microfiche generation in the HRMR System is accomplished by means of a laser recorder which scans onto a film chip both the human readable images and the synthetically generated machine readable holograms containing digital data. Digital data is recorded sequentially onto the fiche in 500 kilobit blocks and at data rates compatible with magnetic tape drives. Fiche are automatically developed and all digital data is verified by means of parity bits which are appended during the recording process.

Since the HRMR System storage media is oriented around the standard microfiche film chip, there is a maximum compatibility with commercially available microfiche handling equipment. It has been a straighforward development to configure a medium scale microfiche storage and retrieval device capable of handling approximately 7000 microfiche. Total digital store of this module is $2(10^{11})$ bits.

In the present HRMR System configuration, the mass memory is on-line to a PDP-11/45 computer. Random fetch of any 500 kilobit data block stored within the memory is provided with a minimum access time of 3 seconds and with a maximum access time of less than 15 seconds. Transfer rates are compatible with DEC Unibus cycle times, but can be tailored to any host computer's channel characteristics and data absorption rates.

Because the holographic techniques used in the HRMR mass memory is read-only, utilization of the system is projected to be oriented primarily toward archival data store applications in which the data placed in the memory is non-dynamic. Such data bases are quite common in both governmental and industrial organizations and are typically characterized by large magnetic tape libraries. A magnetic tape, having typical block sizes and utilizations, can be holographically recorded on one to two fiche. The 7000 fiche storage capacity of the holographic memory provides on-line access to approximately 3500 magnetic tapes with access times a fraction of manual retrieval, mount and read times. Based upon user requirements, additional holographic memory modules can be added to increase this capacity at least an order of magnitude.

Possibly the most significant characteristic associated with the HRMR System's holographic memory is simplicity of operation. The HRMR configuration has integrated into a mini-computer system, a $2(10^{11})$ bit memory and has made this data store available at a storage cost of approximated y 2.5×10^{-6} cents per bit. In contrast to many other conventional mass memory approaches, which record sequential blocks and must retrieve blocks sequentially, the HRMR approach provides random access to data blocks without a sacrifice in overall access time.

While the use of the synthetic holography for storage and retrieval of digital data on microfiche provides a solution to document-oriented mass memory requirements, different recording techniques and physical record formats are more suitable to other types of applications. For example, the storage and retrieval of digital data in very large data records at extremely fast recording and readout data rates can be handled using roll film and interferometric holography.

High-Speed Digital Recorder

Two existing techniques for recording digital data at high speeds are multi-track longitudinal magnetic recorders and laser beam recorders which record binary spots directly onto film. In each case, as the data rate increases the packing density must increases to maintain a reasonable tape or film velocity. But as the packing density increases the tolerance on the lateral position, axial position and skew of the recording medium decreases significantly, often to the point where the data can be recovered only by the machine on which it was recorded.

Again, holographic techniques offer fundamental features for solving these problems. As we showed before, a high packing density can be achieved with relaxed tolerances on the position and angle of the recording medium. High speed recording can be obtained by using the parallel nature of the optical system. These features are noted in Figure 6 which compares the recorded data format of a spot recorder and a holographic recorder.

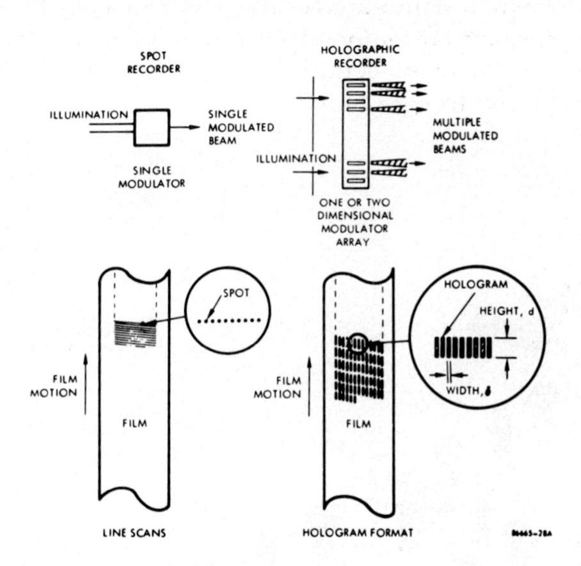

Fig. 6 - Data Formats for a Spot Recorder and a Holographic Recorder

In a spot recorder a single beam of light is temporally modulated and scanned onto film, generally by means of a rapidly rotating multifacet mirror, to produce scan lines of binary data as the film moves. In a holographic recorder a single beam of light is both temporally and spatially modulated by a page composer to produce multiple beams. These beams are then recorded holographically on film by means of a multi-facet mirror. Because each hologram may contain up to 128 bits, the angular velocity of this mirror is lower, by a factor of 128, than that of a spot recorder for same data rate.

Tolerances on readout are reduced because a typical hologram height of the order of 800 μm whereas a scan line is typically 5-10 μm wide. Therefore, it is considerably easier to accurately address a row of holograms than a scan line. Tolerance to skew is similarly decreased because a small angular change in the film upon readout represents a much smaller change in overlap in the readout beam. Lateral and focal tolerances are also eased as described before.

Figure 7 shows the basic elements of a high speed digital recorder. The recording process begins with the demultiplexing of the incoming data stream into 128 channels of time ordered data. Sufficient buffering (32 bits per channel) is provided to bridge over the small deadtime of the scanner, and to absorb small fluctuations in data rate.

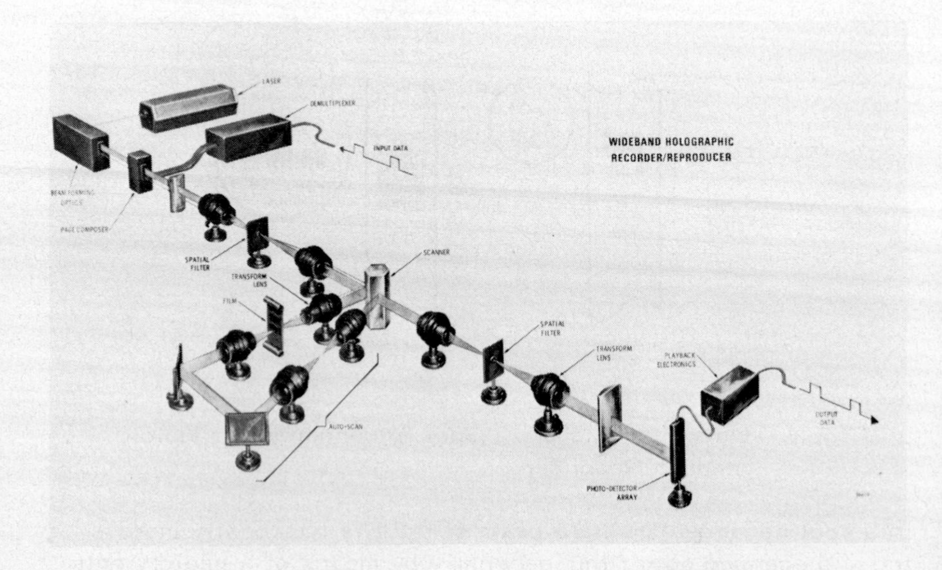

Fig. 7 - Basic Elements of a High Speed Digital Recorder

The parallel electronic information is converted to parallel optical signals by the acousto-optic page composer as shown in Figure 6. The presence or absence of an acoustic wave packet indicates a "logic one" or a "logic zero" respectively. After the acoustic waves have propagated into the optical aperture, the laser beam is gated on by a single acoustic modulator located immediately following the laser. This beam propagates through the beam forming optics where both reference and signal beam illuminations are produced. The signal beam is appropriately shaped to uniformly illuminate the 128 parallel acoustic beams and the reference beam. After passing through the reference and signal beam acoustic modulators, the two beams are brought to focus at a filtering slit. The slit is imaged onto the film after being reflected from a spinner facet. The rotation of the spinner generates a master scan on the moving film as shown in Figure 6.

The first hologram is recorded immediately after the data handling logic receives a data synchronization pulse from the control electronics. Data records are continually recorded in holograms across the film until a counter in the control logic reaches a fixed count. Data records are not clocked out of the data generator until next synchronization signal is received. This process is then repeated and the next hologram row is recorded.

After the film is developed, this system can be used as a reader by illuminating only the reference beam. The same scanner is used to cause the recovered data pattern from each hologram to fall on a photodetector array. Note that, since the system is symmetrical about the axis of the scanner, the photodetector array is in the image plane of the acousto-optic page composer. Thus, 128 bit are detected in parallel and multiplexed into a replica of the input data stream.

If the readout rate is much lower than the recording rate, the recording and reading functions may be separated. The generous tolerance allowances afforded by the holographic recording technique implies significant reader simplicity relative to spot or magnetic recorders operating at high packing densities. Typical recording rates are 400 (10^6) bits/sec and readout rates are 40 (10^6) bits/sec.

In general, one-dimensional holograms lead to simpler system implementations because the film provides motion of the recording medium relative to the hologram recording plane. Short pulse lasers are, therefore, not required and the data can be read out with linear photodetector arrays. Roberts has described an alternative technique, using two-dimensional holograms, for recording data at high rates. This technique does not require the use of a spinning mirror but does require very high-speed, two-dimensional photodetector arrays.

CONCLUSION

During the past several years we have witnessed a considerable effort which was and still is being undertaken by many research laboratories to apply the principles of holography to a broad spectrum of memory and storage applications. The research is directed toward both the intermediate-capacity, fast access time read/write memory market as well as the large-capacity, longer access time read-only storage devices. Effort is also continuing in the specialized area of ultra-high speed transfer of data into and out of large intermediary bulk stores.

Read/write memories have not yet emerged from the research laboratory into the commercial marketplace. Efforts have been hampered primarily by the unavailability of suitable materials which are needed to configure several key memory components. Even if we assume that material and other technology problems are overcome, the prospects that holographic memories will seriously challenge other existing and emerging technologies before the end of this decade, indeed if ever, is unlikely. The inertia of magnetic technology coupled with remarkable yearly improvements in packing density, access time and transfer rates presents a formidable challenge to those who desire to penetrate that particular segment of the market.

The prospects for read-only holographic memories which have multi-microsecond access time and 10^8 bit capacity appear to be better because problems associated with high-speed page composers and reuseable storage media are obviated. Unfortunately, the read-only property will limit its usefulness to special applications where data volatility, extreme environments and data security overshadow cost considerations.

The prospects for application of holographic techniques to read-only bulk storage to be much better. Systems with capacity between 10^{11} and 10^{13} bits and with multi-second access time are currently being built as engineering developmental units. At these capacities, costly electro-optical components can be justified. On a more modest scale, 10^7 bit capacity 1.5 second access time read-only storage units are already commercially available for application to the point-of-sale credit card verification problem.

REFERENCES

1) - E.N. Leith and J. Upatnieks, JOSA 53, 1377 (1963); JOSA 54, 1295 (1964).

2) - A. Vander Lugt "Packing Density for Holographic Systems" Appl. Opt. 14, 1081 (1975).

3) - A. Vander Lugt, Appl. Opt. 12, 1675 (1973).

4) - A.M. Bardos, Appl. Opt. 13, 841 (1974).

Coherant Optical Engineering, F.T. Arecchi and V. Degiorgio (eds.)

DIGITAL IMAGE PROCESSING

Hans Georg Musmann

Lehrstuhl für Theoretische Nachrichtentechnik
und Informationsverarbeitung
Technische Universität Hannover, Germany

Abstract - A survey of the concepts of digital image processing is presented with a concentration on picture digitization and picture coding. In connection with picture digitization the problems of sampling, quantization and image representation in orthogonal bases are discussed. In connection with picture coding some information theory background is introduced to explain the coding with variable and fixed codeword lengths.

I INTRODUCTION

Digital image processing has become attractive for applications in various fields like medicine, earth observation, printing and communications. There are mainly two reasons for this, the greater flexibility of digital image processing in comparison to analog processing and the increasing speed and power of the digital processors.

Corresponding to the great variety of applications there are also many different concepts and mathematical methods of digital image processing. To get a survey we can subdivide these concepts into four major areas as shown in Fig. 1.

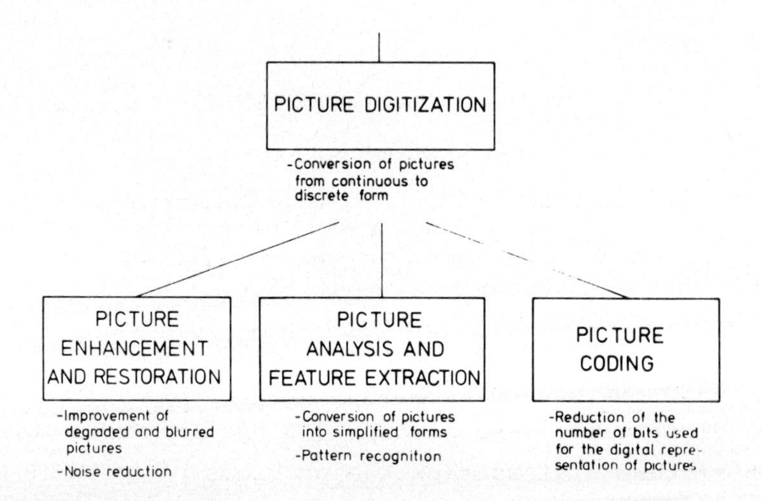

Fig. 1 Digital Picture Processing Techniques

Digital image processing generally starts with the digitization of the picture to be processed. The analog picture is scanned line by line. In sampling the signal of a line we get a sequence of samples where each sample represents one picture element. For a digital representation of the picture each sample amplitude must be quantized and coded into a binary number. The resulting binary data can be regarded as a matrix of numbers representing the picture. This matrix can be stored and processed in a digital computer.

Picture enhancement and restoration is one of the main areas of digital image processing. By picture restoration degradations introduced into the picture by the imaging process are compensated while by picture enhancement the picture is put into a form more suitable for human viewing or further machine processing. A typical example is edge sharpening.

Picture analysis and feature extraction is another main area of digital image processing. By picture analysis and feature extraction the interesting information of a picture is extracted and a simplified description of the picture is derived. Describing a picture only by the contours of the image objects is one example. Often feature extraction is a preprocessing stage for pattern recognition.

Picture Coding is closely related to image digitization and digital image representation. The aim of picture coding is to represent a picture by as few binary data as possible without degrading the picture quality in order to save storage capacity, transmission channel capacity or transmission time.

In this paper the concepts of picture enhancement and restoration as well as of picture analysis and feature extraction will only be outlined. Whereas some problems of more general interest in the area of picture digitization and picture coding will be treated in more detail. In connection with picture digitization the mathematical background of image representation in orthogonal bases will be presented. For an easier understanding of the problem of picture coding the fundamentals of information theory will be introduced. Aspects and problems concerning the picture processing equipment and the hardware implementation of processors will not be treated .

For a more general introduction in picture processing the reader is referred to the books of Rosenfeld /1/ and Huang /2/. In a special volume of the IEEE Press

Book Series edited by Jayant /3/ the reader finds selected papers on signal digitization and coding.

II PICTURE DIGITIZATION

The conversion of an analog picture into a digital form mainly includes two processing steps, a quantizing and a coding procedure. The quantization includes a quantization of the image plane into lines and picture elements (pels) and a quantization of the luminance range of the single picture elements. The first quantizing procedure is called scanning and sampling and determines the spatial resolution while the second one determines the luminance or amplitude resolution. To achieve a digital representation of the picture a binary number or, more generally, a binary codeword corresponding to the quantized luminance level is assigned to each picture element of the picture. It is not necessary also to assign address codewords to the picture elements for marking the spatial location, if the picture elements are always stored or transmitted in a special arrangement, e. g. in matrix form.

The digitization of a picture starts with the scanning and sampling procedure. The picture is scanned line by line and the output signal of the scanner is sampled with a fixed sampling frequency. Each sample of a picture represents one picture element. If the scanner produces a picture signal u(t) with bandwidth W then this signal can be written in form of the following expansion of orthogonal functions /4/

$$u(t) = \sum_{n=-\infty}^{+\infty} u(t_n) \frac{\sin 2\pi W(t-t_n)}{2\pi W(t-t_n)} \quad (1)$$

$$\text{where} \quad t_n = n/2W \quad (2)$$

According to this so called sampling theorem (1) the original signal u(t) can be uniquely reconstructed from the sample values $u(t_n)$ taken from u(t) in equally spaced time intervals of length 1/2W. If the sampling frequency f_s is greater or equal to 2W then the sampling procedure generates no quality degradations of the original picture. So the sampling theorem sets an upper bound to the required spatial resolution. The sampled signal has a periodic power spectrum with period W. For reconstruction of u(t) the amplitude samples $u(t_n)$ must be filtered by a low pass filter of bandwidth W. A reduction of the sampling frequency below the rate of $f_s = 2W$ will cause an aliasing error and picture quality degradations in areas of high detail in the reconstructed picture. The aliasing error arises from high frequency components which are transposed into the baseband W by the sampling process as illustrated in Fig. 2a.

A picture signal generally has a periodic structure in its power spectrum with peaks at multiples of the line frequency originating from the periodic line scanning. This structure can be exploited to reduce the effects of aliasing. If the sampling frequency f_s is chosen in such a way that the peaks of the overlapping spectra do not cover each other then the interfering components can partially be eliminated with help of a comb filter as indicated in Fig. 2b.

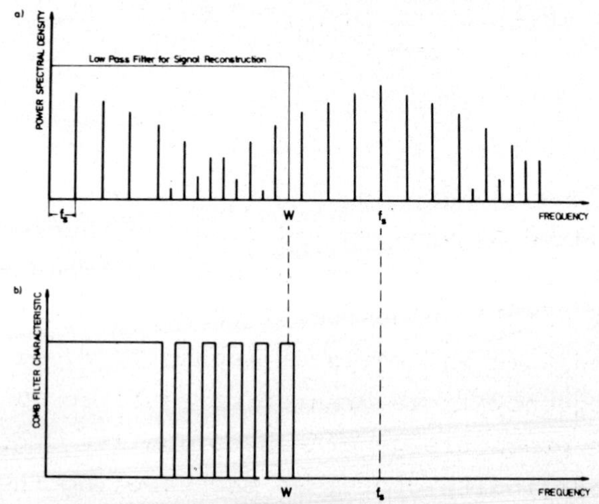

Fig. 2 a) Simplified Spectrum of a Sampled Picture Signal With Overlapping High Frequency Components Due to $f_s < 2W$

b) Idealized Comb Filter for Reducing the Aliasing Components

Sometimes a certain amount of picture quality degradation is acceptable for the receiver of the picture which can finally be a human viewer or a digital processor. In these cases the spatial resolution and also the amplitude resolution of a picture can be reduced to an extent which is limited by the allowable distortion given by the receiver. With the distortion criterion the receiver sets a lower bound to the spatial resolution.

For human viewing there is a natural bound given by the resolution of the eye. The spatial resolution of the human eye is $\delta = 1{,}5'$. Looking on a document which is in a distance of about 3o cm from the eye this resolution corresponds to a minimum distance of d = o,131 mm for two picture elements to be separated, see Fig. 3.

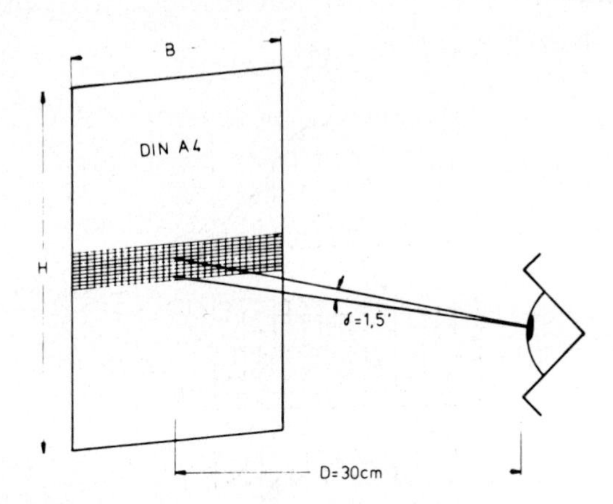

Fig. 3 Spatial Resolution of the Eye, $d = D\,tg\delta$ = o,131 mm = minimum distance for two picture elements to be separated

This resolution results in about 8 x 8 = 64 pels per mm^2. The eye can distinguish between about 256 luminance levels as can be demonstrated in subjective tests using critical test pictures with fine grey tone shadings. To represent these amplitude levels by binary numbers or codewords we need an 8 bit codeword for each picture element. For quantizing the luminance amplitude a uniform quantizer with equally spaced quantizing intervals is used. Fig. 4 illustrates the sampling, quantizing and coding procedure. Also the inevitable quantizing error is shown.

A reduction of the amplitude resolution below 256 quantizing levels generates visible quantizing errors.

The effects of aliasing errors and quantizing errors are shown in the examples of Fig. 5. The aliasing error produces distortion in high detail areas and along diagonal edges. The quantizing error produces contours in low detail areas.

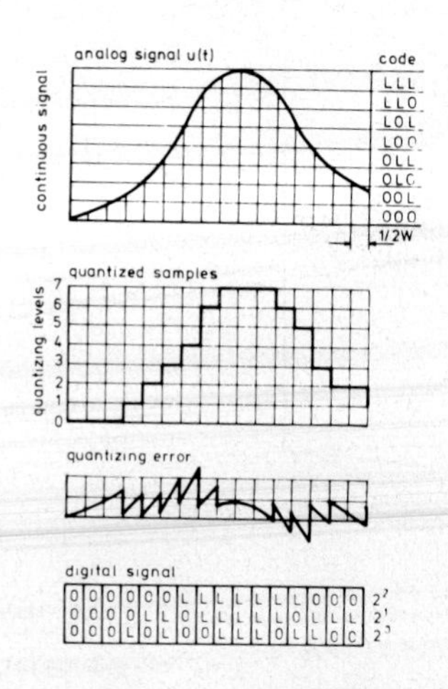

Fig. 4 Representation of the Sampling, Quantizing and Coding Procedure

The described digitization procedure including scanning, sampling, quantizing and coding results in a special digital representation of the picture called Pulsecodemodulation (PCM). The outcoming digital picture can be regarded as a matrix of N x N terms each matrix term representing the luminance of one picture element.

A problem of the recent research is the digitization of colour pictures. From subjective tests it is known that an arbitrary colour can be reproduced by an additive or subtractive mixture of three primary colours. While in printing techniques a subtractive mixture of four primary colours is used in colour television an additive mixture of three primary colours is applied. Here the digitization of colour pictures shall be explained using the examples of colour television.

Fig. 5 Examples of a Processed Picture with Different Spatial and Amplitude Resolutions
Row 1, 2, 3: 24o x 16o, 8o x 6o, 4o x 3o pels per picture
Column 1, 2, 3: 8 bit, 5 bit, 3 bit amplitude resolution

In colour television the three primary colours are electrically represented by the luminance signal U_Y and the two chrominance difference signals U_{R-Y} and U_{B-Y}. The luminance signal U_Y for itself produces a grey tone picture on the monitor. Fig. 6 shows the three-dimensional colour space U_Y, U_{R-Y}, U_{B-Y} as it can be reproduced by a television system.

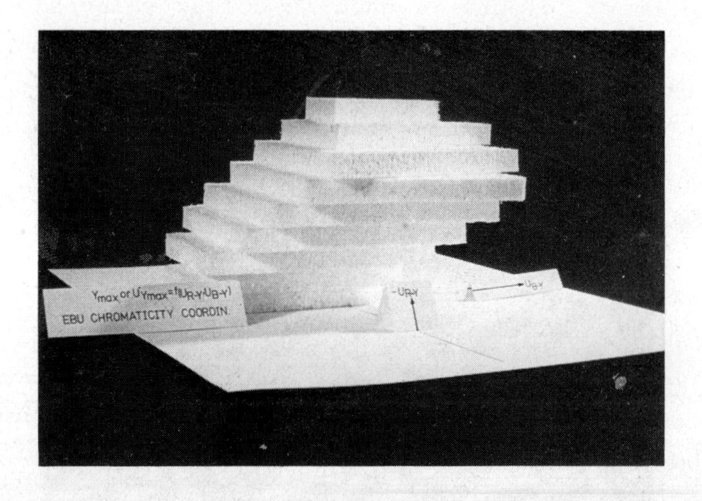

Fig. 6 Reproducible Colour Space of the TV Signals U_Y, U_{R-Y}, U_{B-Y} (from Stenger /5/)

Corresponding to 256 quantizing levels for U_Y there are 256 layers with constant luminance signal U_Y. For simplification of the model only 9 of these layers have been realized. Fig. 7 shows one of these layers reproduced in colour. The luminance of this example is relatively great, it is a layer from top of the model. The center of the coordinates U_{R-Y}, U_{B-Y} is the white point. The bright surrounding background of the layer demonstrates the range covered by the two chrominance signals but which cannot be reproduced in case of this special luminance value. If uniform quantizers with equally spaced quantizing intervals are used for the chrominance signals then 8 bit quantizers are required in order not to produce visible quantizing errors. From Fig. 7 it can be recognized that depending on the luminance level or actual layer many of these 2^{16} quadratic quantizing elements are useless since they are situated outside the reproducible colour space.

Fig. 7 Colour Plane U_{R-Y}, U_{B-Y}, U_Y = 2o4 out of 256

Fig. 9 Colour Plane Quantized with Subjectively Matched Quantizing Elements

II. 1 IMAGE REPRESENTATION IN ORTHOGONAL BASES

For special picture processing algorithm it can be useful to represent the digital picture not in the two-dimensional array of numbers as described before but in another kind of matrix which is more suitable for the subsequent picture processing procedures.

Let $[G]$ be an image matrix of N x N numbers, each number representing the quantized luminance amplitude of one picture element. In a digital computer this matrix can be transformed and represented in the transform domain. Here we will restrict our considerations to unitary transformations $[U^{-1}]$.

A transformation of the image matrix $[G]$ results in a new matrix $[A]$ in the transform domain

$$[A] = [U^{-1}]\,[G]\,[U^{-1}]^{T} \tag{3}$$

where the superscript T denotes the matrix transpose. The inverse transform reproduces the original image $[G]$ as

$$[G] = [U]\,[A]\,[U]^{T} \tag{4}$$

Let $\vec{u}_1, \vec{u}_2, \ldots \vec{u}_N$ be the column vectors of $[U]$

$$[U] = [\vec{u}_1, \vec{u}_2 \ldots \vec{u}_N] \tag{5}$$

then we can write equation (4) in the following form as presented by Andrews in /2/

$$[G] = [\vec{u}_1, \vec{u}_2 \ldots \vec{u}_N] \, [A] \begin{bmatrix} \vec{u}_1^{\,T} \\ \vec{u}_2^{\,T} \\ \vdots \\ \vec{u}_N^{\,T} \end{bmatrix} . \qquad (6)$$

The matrix $[A]$ may be partitioned into the sum

$$[A] = \begin{bmatrix} a_{11} & o & \ldots & o \\ o & o & & \\ & & & \\ o & \ldots & \ldots & o \end{bmatrix} + \begin{bmatrix} o & a_{12} & \ldots & o \\ o & o & & \\ & & & \\ o & \ldots & \ldots & o \end{bmatrix} + \ldots, \qquad (7)$$

and this sum be inserted in equation (6). It follows that

$$[G] = \sum_{i=1}^{N} \sum_{j=1}^{N} a_{ij} \, \vec{u}_i \, \vec{u}_j^{\,T} \qquad (8)$$

Each outer product $\vec{u}_i \, \vec{u}_j^{\,T}$ in equation (8) is a matrix and can be interpreted as an image. We see from equation (8) that the original image $[G]$ can be reconstructed in form of a sum of N x N of these images $u_i \, u_j^T$ each being weighted by a coefficient a_{ij}.

The outer product images of a special transform, the so called Hadamard/Walsh transform, are shown in Fig. 1o.
There are mainly two transformations of particular interest for picture processing, the Hadamard/Walsh

transform and the Fourier transform. In Fig. 11 we see the matrices [U] of these transforms.

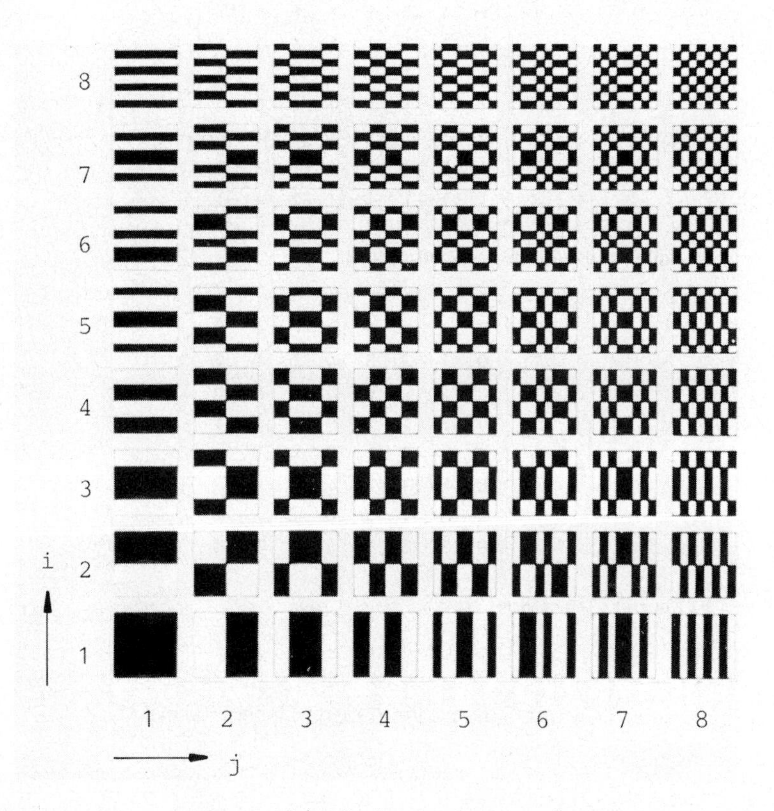

Fig. 1o Hadamard/Walsh Outer Product Expansion (black = +1, white = -1, N = 8)

The Hadamard/Walsh transform is used in picture coding. This matrix only contains elements 1 and -1. The advantage of this transformation is its relatively simple hardware implementation since there are only multiplications with these two elements. For one-and-two-dimensional spatial filtering of images a Fourier transformation of [G] and a digital representation of the pictures in the frequency domain is more appropriate.

$$U = \begin{pmatrix} 1 & -1 & -1 & 1 & 1 & -1 & -1 & 1 \\ 1 & -1 & -1 & 1 & -1 & 1 & 1 & -1 \\ 1 & -1 & 1 & -1 & -1 & 1 & -1 & 1 \\ 1 & -1 & 1 & -1 & 1 & -1 & 1 & -1 \\ 1 & 1 & 1 & 1 & 1 & 1 & 1 & 1 \\ 1 & 1 & 1 & 1 & -1 & -1 & -1 & -1 \\ 1 & 1 & -1 & -1 & -1 & -1 & 1 & 1 \\ 1 & 1 & -1 & -1 & 1 & 1 & -1 & -1 \end{pmatrix}$$

a) Hadamard / Walsh Transform, N = 8

$$U = \left[e^{j\frac{2\pi xy}{N}} \right] = e^{j\frac{2\pi}{8}} \begin{pmatrix} 0 & 0 & 0 & 0 & 0 & 0 & 0 & 0 \\ 0 & 1 & 2 & 3 & 4 & 5 & 6 & 7 \\ 0 & 2 & 4 & 6 & 0 & 2 & 4 & 6 \\ 0 & 3 & 6 & 1 & 4 & 7 & 2 & 5 \\ 0 & 4 & 0 & 4 & 0 & 4 & 0 & 4 \\ 0 & 5 & 2 & 7 & 4 & 1 & 6 & 3 \\ 0 & 6 & 4 & 2 & 0 & 6 & 4 & 2 \\ 0 & 7 & 6 & 5 & 4 & 3 & 2 & 1 \end{pmatrix}$$

b) Fourier Transform, N = 8

Fig. 11 Examples of Unitary Transformations

In this case the elements of the matrix [A] correspond to special frequency components of the image [G]. A low pass filtering e. g. can be performed by only setting special elements of the matrix [A] to zero. The following example may illustrate these processing procedures. In Fig. 12a we have the original digital picture represented in form of matrix [G]. If we apply a Fourier transform to the picture in Fig. 12a we get a matrix [A] which is illustrated in Fig. 12b. The small terms of the matrix [A] in Fig. 12b can be eliminated without effecting the inverse transformation and the reconstructed picture [G]. By these techniques the binary data representing the picture can be reduced. Therefore these transforms are also of interest for picture coding.

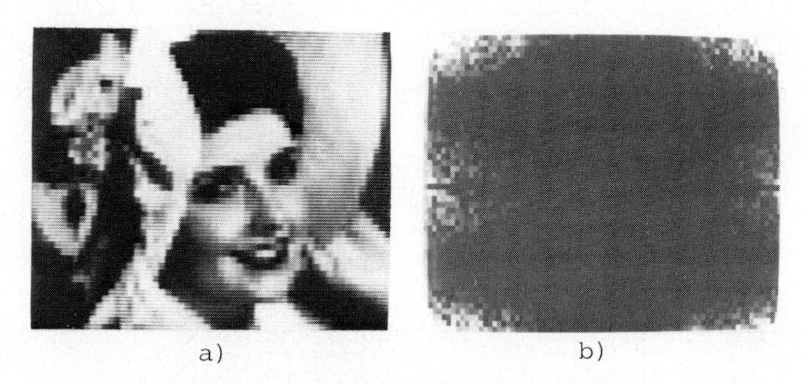

a) b)

Fig. 12a Original Picture in Form of Matrix [G]

Fig. 12b Fourier Transform of Fig. 12a,
The brightness increases with the value of the matrix element, N x M = 3o x 5o

III PICTURE ENHANCEMENT AND RESTORATION

The motivation of picture restoration is, to restore pictures which have been degraded in an imaging process. By these techniques e. g. a focus can be corrected or the blurring introduced by a motion of the camera can be compensated.

For restoring a picture we need some a priori knowledge about the degradation phenomena in the imaging system. Then the imaging system can be described or approximated by some kind of mathematical model. This may be a statistical model or the model of a linear or non-linear system. Also, an a priori information about the objects in the picture and the analysis of the noticeable distortion of these objects help to find this mathematical model.

Let [O] be the matrix of the real picture without distortions and D be an operator designating the mathematical model of the imaging system then the visible picture [G] is given by

$$[G] = D\left\{[O]\right\} + [N] \qquad (9)$$

where [N] denotes additional noise. Now, the idea of picture restoration is to reduce the noise [N] and to perform an inversion of the operation D in order to get [O] or an approximation of [O].

In case of linear distortions the operator D can be written in form of a matrix [D] and the restoration algorithm is

$$[O] = [D^{-1}]\,[G] - [D^{-1}]\,[N] \qquad (1o)$$

If the noise is zero equation (1o) gives an exact solution.

There are many different proposals depending on the a priori knowledge of the imaging system for calculating or estimating D which is the main problem in picture restoration. Readers interested in these processing techniques may find Frieden's contribution to the book of Huang /2/ worthful.

In <u>picture enhancement</u> digital pictures are manipulated to make certain picture information more visible for a human observer or more suitable for a machine processing. A typical example is the edge sharpening of X-ray pictures. Sometimes by enhancement techniques the picture can be changed in such a way that the original natural image gets lost.

The power of picture enhancement shall be demonstrated by an example. Fig. 13a shows a photo of a garage. The objects inside the garage are almost imperceptible for the eye.

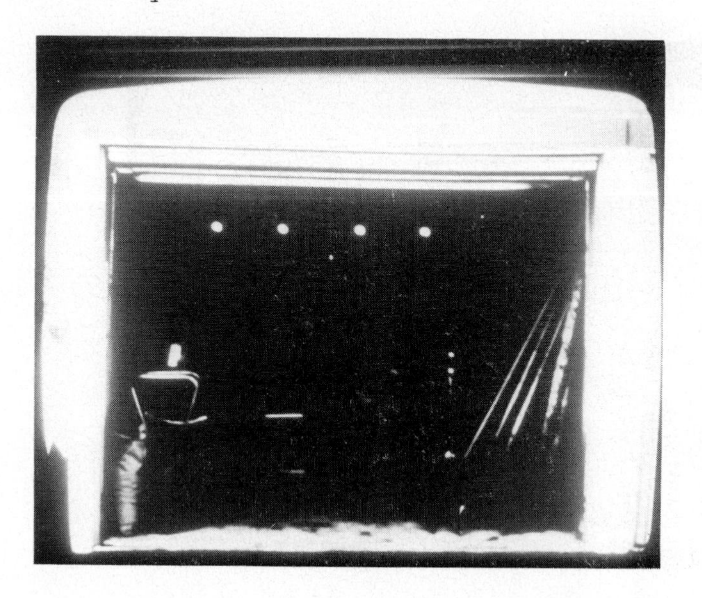

Fig. 13a Original Picture to be Enhanced

To enhance this picture we can use a technique proposed by Wedlich /6/ which is similar to homomorphic filtering /7/. The concept is to reduce the influence of multiplicative distortions in the picture which here is originating from a locally changing illumination of the scene. Wedlich measures the mean of the luminance in the surrounding of an actual picture element and divides its luminance value by the measured mean. It is assumed that the two multiplicative terms have separate spectral power densities. Fig.13b demonstrates the significant enhancement that can be accomplished with this technique.

The aim of the recent research in this field is more to incorporate simple models of the human usual system in these algorithms for improving picture enhancement. For further literature see also /8/.

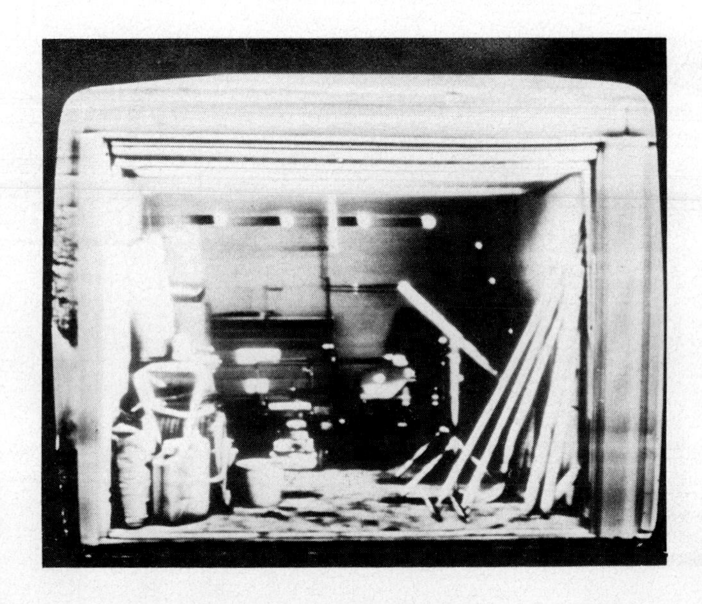

Fig. 13b Enhanced Picture of Fig. 13a (from G. Wedlich /6/)

IV PICTURE ANALYSIS AND FEATURE EXTRACTION

Generally picture analysis and feature extraction are preprocessing stages of further machine processing.

The aim of picture analysis usually is to find a simplified digital representation of a picture containing only that information which is relevant for the subsequent picture processing procedure. One example for picture analysis is finding the contours in a picture and describing the picture by the contours of its objects. Another example is movement detection and describing a picture in form of a two-level picture of moving and unmoving objects.

Feature extraction is a preprocessing step especially for automatic pattern recognition. In pattern recognition the distinction between different patterns is based on features. These features are taken from the picture. According to the feature information the present pattern in question is classified into one of the considered possible pattern classes. The reliability of the recognition process strongly depends on the elaboration and selection of the features which are the main problems of feature extraction.

There are many mathematical concepts and techniques for both picture analysis and feature extraction. The effectiveness of these methods depends on the picture material which varies with the different fields of application. The book of Fu /9/ provides a concise survey about this area including theory and applications.

Here only one example of picture analysis will be presented. For reducing the transmission rate of television pictures only those parts of the present picture have to be transmitted which have changed from frame to frame due to moving objects. The unchanged picture elements and those which have changed owing to noise need not be transmitted. Klie /1o/ uses a movement detector which is based on frame-to-frame differences. If a frame-to-frame difference exceeds a certain "coarse" threshold a moving object is indicated. Additional "fine" thresholds combined with spatial filtering are used to distinguish between noise and moving objects with low contrast by the different correlation of the frame-to-frame differences. In Fig. 14a a single frame out of a moving scene is shown. The black picture elements in Fig. 14b denote moving parts of the scene in Fig. 14a as indicated by the described movement detector.

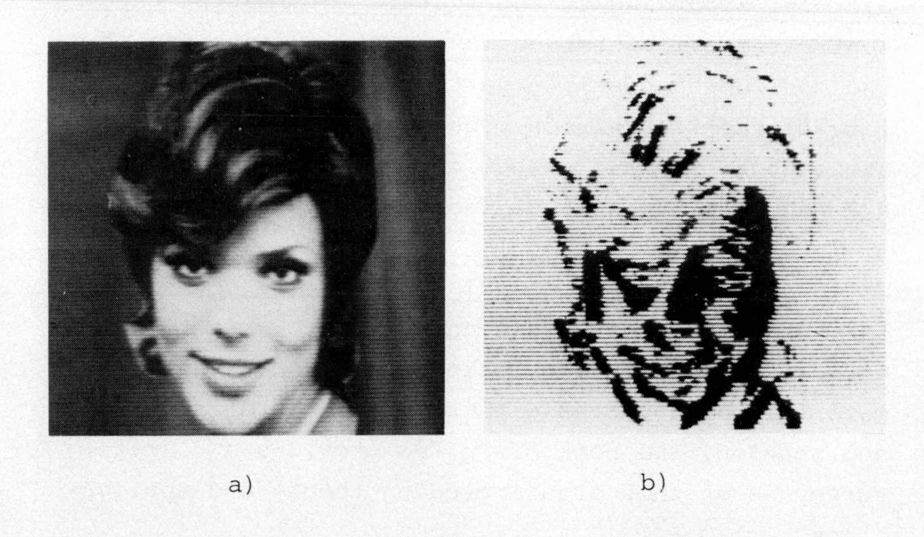

a) b)

Fig. 14a Single Frame out of a Moving Scene

Fig. 14b Output of the Movement Detector (black picture elements indicate changed picture elements of moving objects)

V PICTURE CODING

Picture coding is a special field of source coding. The aim of source coding is to represent the signal of a source with as few binary symbols as possible to achieve a more economical utilization of a given communication or processing system by saving channel capacity, storage capacity or transmission time. The application of source coding is especially interesting in those cases where there is a requirement for large channel capacity or storage capacity as in the case of picture signals. For an easier understanding of the picture coding concepts some general information theory background will be presented before.

V. 1 THEORETICAL BACKGROUND

In 1948 Shannon /4/ developed a mathematical theory for dealing with communication systems. The following theoretical fundamentals of source encoding are based on this so called information theory /11/.

A message H_o of a source can be written as the sum of the entropy H and redundancy R. In picture coding this source is a camera or a scanner and the signal amplitude of a picture element may be considered as a message H_o. If this amplitude is quantized with a K level quantizer we have

$$H_o = \log_2 K \qquad (11)$$

in bit and

$$H_o = H + R \qquad (12)$$

The redundancy R can be removed by a redundancy-reducing code and later reinserted. This coding is reversible. Since the reduction of the entropy H is an irreversible process which causes an information loss, the allowable entropy reduction and the resulting message distortion must be specified by the user through the definition of a distortion criterion d. By this criterion the message fidelity required by the user is fixed. The entropy-reducing code to be used is considered to be optimum if it is matched to source statistics and to the criterion in such a way that it gives the highes possible reduction of the bit rate without exceeding the allowable distortion. The mathematical basis for the analysis of this coding is the rate distortion theory. An interesting example of entropy-reducing coding with a fidelity criterion is amplitude quantization. Since the digitization of an analog picture always includes quantization we normally have a mixture of entropy- and redundancy-reducing coding. Only in the case of digital pictures we can apply pure redundancy-reducing, reversible coding.

Redundancy Reducing Coding

According to the source coding theorem of the information theory a message H_o can be coded by H bit in average. A coding with less than H bit will always cause an entropy or information loss. So the entropy H sets a lower bound for the bit rate of reversible coding.

The entropy H of a sample or a picture element to be

coded is measured by the uncertainity of its appearance. The entropy H per sample from a source with memory, wherein each sample of a sequence is statistically dependent upon N-1 preceding samples is given by

$$H(X_N|X_1,\ldots X_{N-1}) = -\sum_{x_1..x_N} P(x_1..x_N)\log_2 P(x_N|x_1\ldots x_{N-1}) \quad (13)$$

in bit where each x_N represents a quantized K-level sample $a_1,\ldots a_K$ and $P(x_N|x_1,\ldots x_{N-1})$ represents the conditional probability of x_N given $x_1,\ldots x_{N-1}$. Only in the special case of statistically independent samples, where

$$P(x_1,\ldots x_N) = P(x_1)\cdot P(x_2)\ \ldots P(x_N), \quad (14)$$

the entropy per sample can be calculated according to

$$H(X_N) = -\sum_{x_N} P(x_N)\ \log_2 P(x_N) = -\sum_{k=1}^{K} P(a_k)\log_2 P(a_k) \quad (15)$$

and the source may be described by the model of the memoryless source. It can be shown that

$$H(X_N|X_1\ldots\ldots X_{N-1}) \leq H(X_N) \leq H_O \quad (16)$$

The optimum coding with H bit per sample or per picture element is achieved, if each sample is encoded with $-\log P(x_N|x_1,\ldots x_{N-1})$ bit or in the special case of a memoryless source with $-\log P(x_N)$ bit. this means that samples with high probabilities will be coded by codewords of relatively short length and vice versa.

Huffman /12/ developed a procedure for constructing optimal codes with minimum redundancy. For an example see Fig. 15.

SOURCE SYMBOL a_i	PROBABILITY $P(a_i)$	SOURCE CODE	HUFFMAN CODE
a_1	0,500	00	0
a_2	0,250	0L	L0
a_3	0,125	L0	LL0
a_4	0,125	LL	LLL

$$R_u = H_0 - H_u = 2-(0{,}5\cdot 1 + 0{,}25\cdot 2 + 0{,}125\cdot 3 + 0{,}125\cdot 3) = 0{,}25 \text{ bit/Symbol}$$

Fig. 15 Example of a Huffman Code

In case of a source with memory an individual Huffman code for each state $s_j(x_1,\ldots x_{N-1})$ defined by the possible combinations $x_1,\ldots x_{N-1}$ of the source is required. This coding procedure is illustrated in Fig. 16.

To show the lower bound of redundancy-reducing coding the entropy H was calculated using a typical set of test pictures. The samples $x_1,\ldots x_{N-1}$ were taken from the signal past in the surrounding of the actual picture element x_N to be coded, see Fig. 17.

The following entropies H have been measured:
Two level facsimile, resolution 4 x 4 pels per mm^2,
H_o = 1 bit :

$H(X_N|X_1,...X_{12})$ = o,17 bit (typewritten text)

$H(X_N|X_1,...X_{12})$ = o,13 bit (weather map)

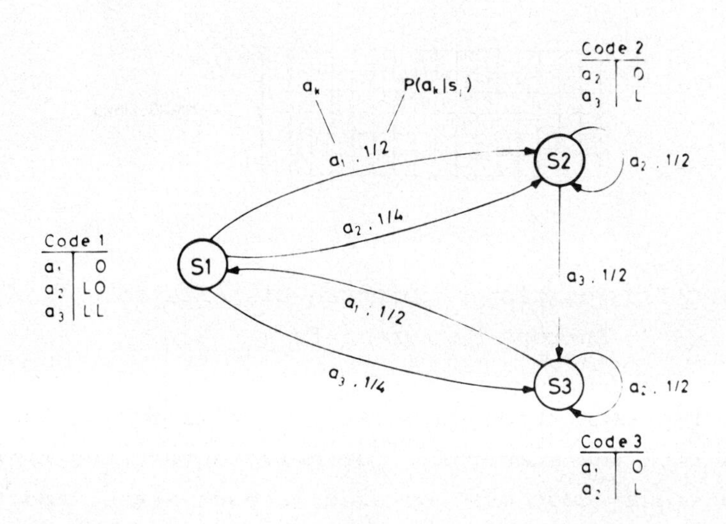

Fig. 16 Grafical Representation for Coding of a Markov Source with 3 States

5 MHz television signals, sampling frequency 1o MHz, H_O = 8 bit:

$$H(X_N|X_{11}\ X_{12}) = 3{,}7 \text{ bit}$$

$$H(X_N|X_3\ X_{12}) = 3{,}2 \text{ bit}$$

These results indicate that a compression factor of about 6 can be achieved for facsimile coding and of about 2,5 for television coding if no information loss is allowed.

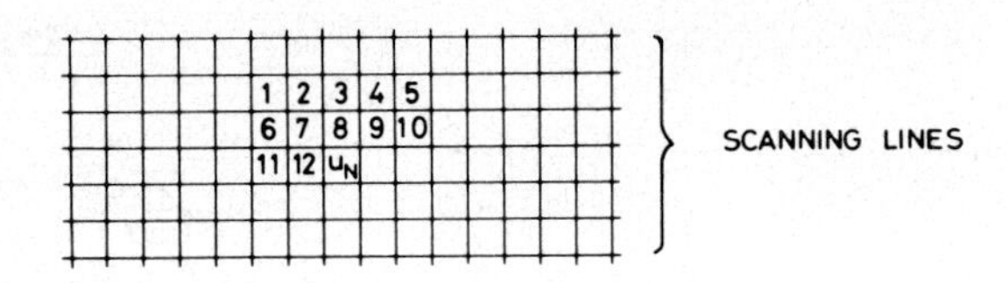

Fig. 17 Position of Picture Elements for Entropy Measurements

The implementation complexity of this coding increases with the number K^{N-1} of different states since an individual code must be used in each state. Additionally this coding produces codewords with variable lengths. The resulting nonuniform bitrate complicates the transmission and storage control. By combining entropy- and redundancy-reducing coding this disadvantage can be avoided and a constant output bitrate per sample can be achieved.

Combined Entropy- and Redundancy-Reducing Coding

In this coding techniques distortion is introduced into the pictures due to the entropy reduction. Therefore the fidelity criterion of the receiver or the maximum allowable distortion d has to be considered. The lower bound for the bitrate R of these coding methods is given by the rate distortion function R(d). A typical characteristic of a rate distortion function is shown in Fig. 18.

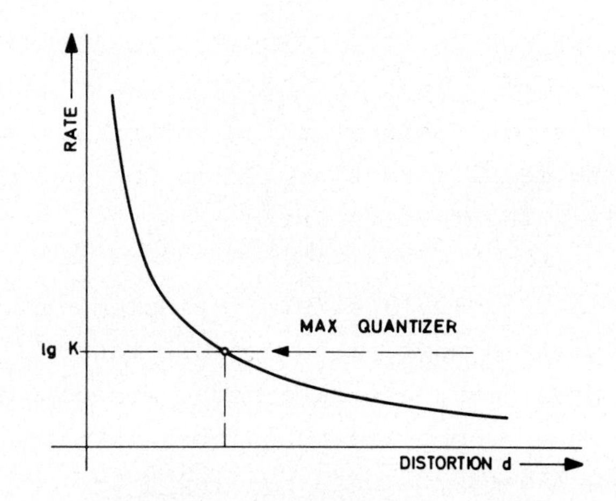

Fig. 18 Typical Characteristic of a Rate Distortion Function

An interesting rate distortion problem is the optimization of a quantizer for a given mean square error criterion. Max /13/ derived an algorithm for designing a K level quantizer which produces the minimum mean square quantizing error. This optimization of the quantizer yields small quantizing intervals for amplitudes in regions of high probability density and relatively large quantizing intervals in regions of low probability density. For memoryless sources this quantizer is the nearest to the rate distortion function among all K level quantizers with fixed codeword lengths of $\log_2 K$. However, it does not reach R(d).

V. 2 PICTURE CODING WITH FIXED CODEWORD LENGTHS

One of the most attractive coding techniques for picture encoding is Differential PCM (DPCM) and its extensions /14/. The reasons are:

(a) DPCM is relatively simple to implement
(b) DPCM gives a uniform output bit rate
(c) DPCM provides a bit rate reduction comparable to that of competing techniques like transform coding and others.

These differential coding schemes are special solutions of the switched quantizer model /15/.

The Switched Quantizer

Let x_N' be the next sample representing one picture element which is to be quantized and encoded. The preceding samples $x_1, \ldots x_{N-1}$ are already quantized, coded, transmitted and reconstructed. So these preceding samples are available in a digital form at the receiver and at the transmitter. If we assume that each digital x_n may accept one out of K possible values, then we have K^{N-1} possible combinations of N-1 preceding samples. Each of these combinations represents one so called state $s_j = s_j(x_1, \ldots x_{N-1})$ of the source.

The switched quantizer is a quantizer that uses an individual quantizing characteristic in each state. To achieve a minimum quantizing noise in the state s_j the quantizing characteristic has to be matched to the probability density $p(x_N|s_j)$ of this state according to rules of Max /13/. After quantizing x_N', coding, transmitting and reconstructing x_N the source moves into a new state. If the optimum quantizing characteristic is used in each state and if the number of quantizing levels is fixed for all states, then we will have the minimum total quantizing noise, a fixed codeword length and a uniform bit rate.

Differential Pulsecodemodulation

In Fig. 19 the block diagram of a digital DPCM system is shown.

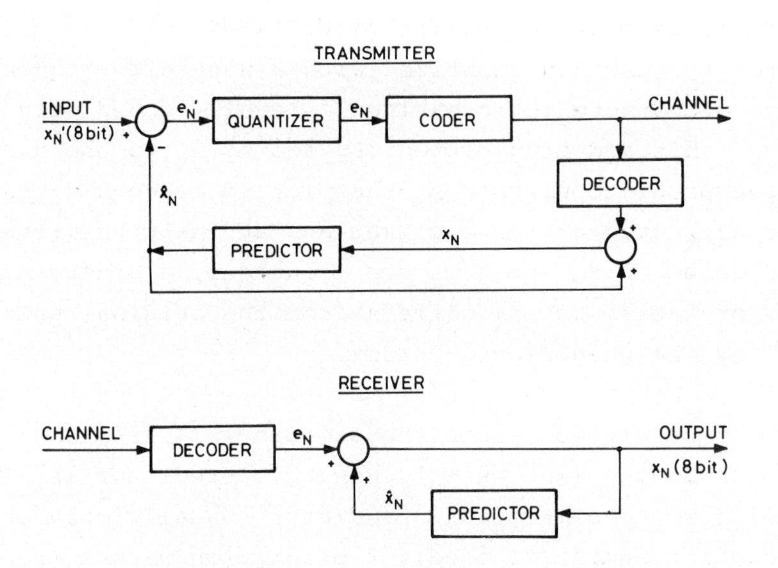

Fig. 19 Block Diagram of a DPCM System

The input signal x_N' of the DPCM system is a PCM video signal. The samples of this PCM signal are quantized with a uniform quantizing characteristic and are coded by binary numbers of 8 bit. For every input sample x_N' the linear predictor generates a prediction value $\hat{x}_N$ which is calculated from N-1 preceding samples

$$\hat{x}_N = \sum_{i=1}^{N-1} a_i \, x_{N-i}. \tag{17}$$

The coefficients a_i are optimized to yield a prediction error e_N'

$$e_N' = x_N' - \hat{x}_N \tag{18}$$

with minimum variance. The prediction error e_N' is then quantized with a non-uniform quantizing characteristic having K' levels, coded with $\log_2 K'$ bit and transmitted. By adding $\hat{x}_N$ to the quantized prediction error e_N the picture element x_N is reconstructed at the receiver and at the transmitter. The value of x_N again is represented by a binary number of 8 bit, but x_N differs from the original sample x_N' by the quantization error.

If the prediction takes the samples x_{N-i} from the same scanning line as x_N' then this predictor is called a one-dimensional predictor. Usually only one preceding sample is used for prediction with a_1 near 1. Fig. 2o shows a processed 1 MHz videophone picture of such a DPCM system, its prediction error and its quantization error. The reconstructed picture shows distorted vertical edges due to a relatively great prediction error with quantizer overload at vertical edges. The resulting quantizing noise becomes more

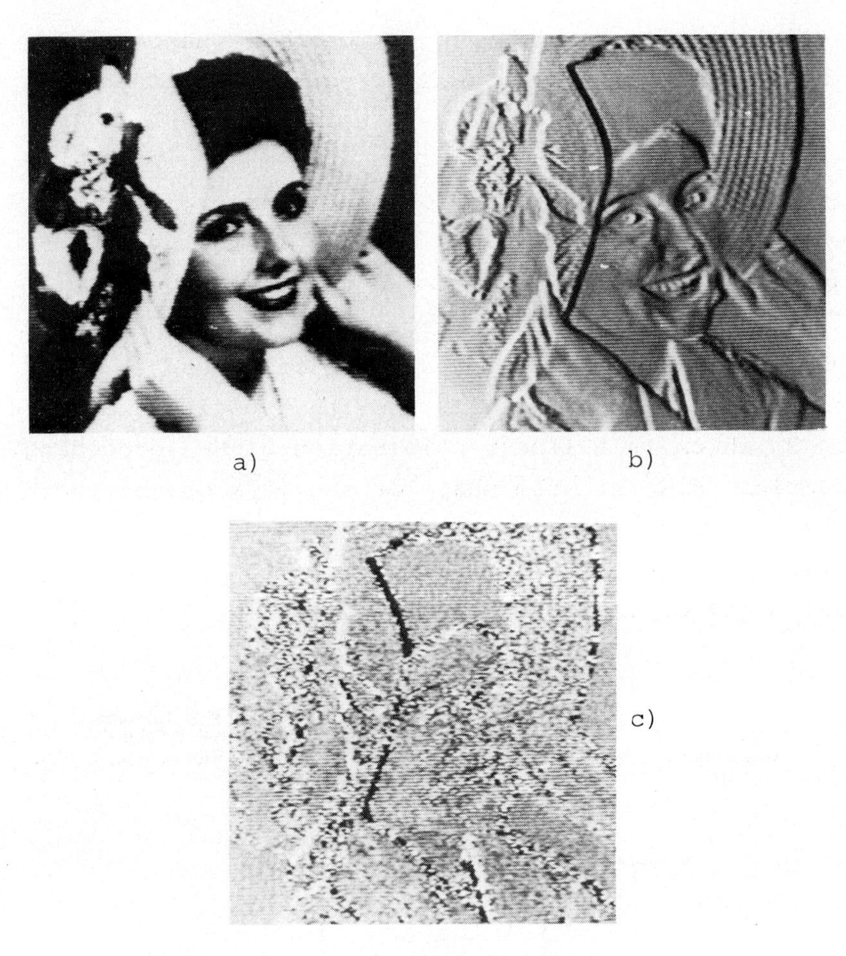

Fig. 2o Results Obtained with a DPCM System Using Previous Sample Prediction $\hat{x}_{N-1}$ and 4 Bit per Sample, resolution 24o x 16o pels

a) Reconstructed picture x_N

b) Prediction error e_N'

c) Quantization error $(e_N' - e_N)$ multiplied by a factor of 8 (positive and negative errors are represented by black and white picture elements while a zero error is grey).

visible in form of busy edges when a sequence of frames is observed on the monitor screen. The overload effect of the quantizer is illustrated in Fig. 21.

In a DPCM system the quantizing characteristic is always shifted in such a way that the zero level of the quantizer is positioned at the prediction value $\hat{x}_N$. If we interpret $\hat{x}_N$ as the source state s_j then the DPCM transmitter with linear prediction uses a shifted quantizing characteristic with a position defined by $\hat{x}_N$ which is a linear combination of N-1 preceding samples. However, the shape of the DPCM quantizing characteristic remains always the same. So the DPCM system is a special suboptimum solution of the general model of the switched quantizer.

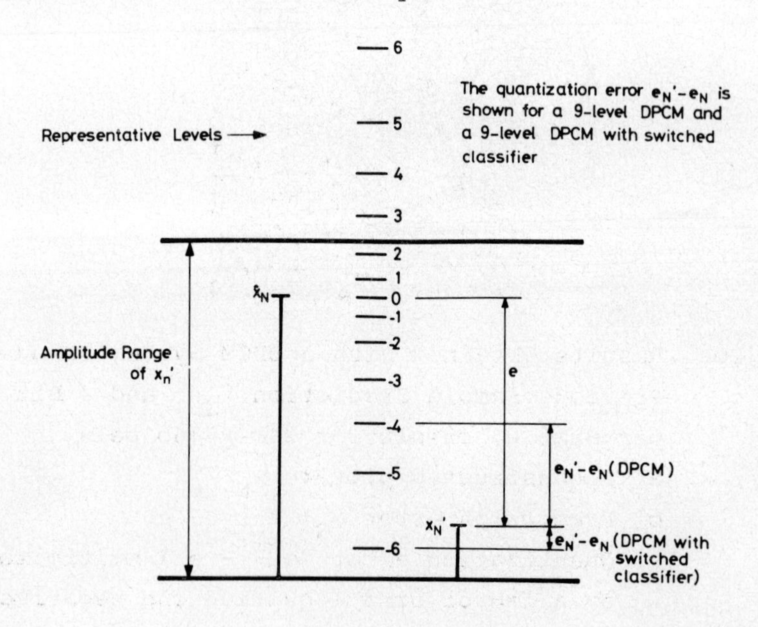

Fig. 21 Graphical Representation of a Shifted Quantization Characteristic

DPCM with a Switched Quantizer and Two-Dimensional Prediction

A DPCM system is not optimum in so far as the quantizing characteristic is not matched to the shape of the conditional probability density function $p(x_N'|s_j)$ of the state. This can be achieved by combining a DPCM system with a state controlled logic called switched classifier as shown in Fig. 22.

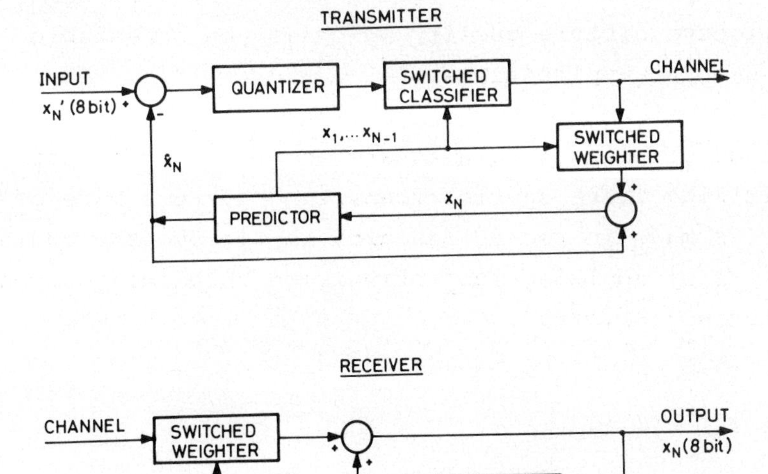

Fig. 22 DPCM System with Switched Quantizer

Furthermore the efficiency of a DPCM system for video signals can be increased by using a two-dimensional prediction. Such a predictior uses additional picture elements of the preceding line for prediction of x_N'. Fig. 23 shows the position of the interesting picture elements which are available for prediction of x_N' with respect to the applied interlaced scanning technique of television signals.

Kummerow /16/ proposes the two-dimensional prediction algorithm

$$\hat{x}_N = 0{,}75\ A + 0{,}75\ E - 0{,}50\ C \qquad (19)$$

and a switched quantizer with 3 quantizing characteristics. Using a similar coding technique Lippmann /17/ achieved the results shown in Fig. 24 which demonstrate that even a coding with 1 bit per sample provides a picture quality which may be acceptable for special applications.

Instead of using a mean square error criterion for optimizing these special quantizing systems more effort is made in recent research to improve the coding systems by applying subjectively weighted distortion criteria /18/, /19/.

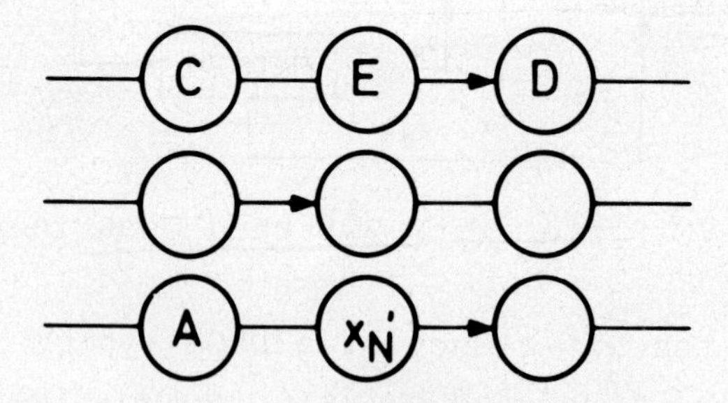

Fig. 23 Position of Picture Elements Used for Prediction (Arrows show direction of scan, element x_N' is the next sample to be quantized and coded)

a)

b)

Fig. 24 Results Obtained with a DPCM-System Using Two-Dimensional Prediction, a Switched Quantizer and 1 Bit per Sample, (resolution 512 x 789 pels)

a) Reconstructed picture

b) Reconstructed picture, Transmission error rate $P_e = 10^{-2}$

References

/1/ A. Rosenfeld, Picture Processing by Computer, New York: Academic Press, 1969

/2/ T. S. Huang, Picture Processing and Digital Filtering, Berlin, Heidelberg, New York: Springer Verlag, 1975

/3/ N. S. Jayant, Waveform Quantization and Coding, New York: IEEE Press Book, 1976

/4/ C. E. Shannon and W. Weaver, The Mathematical Theory of Communication, Urbana: University of Illinois Press, 1949

/5/ L. Stenger, PCM Quantization of Colour Difference Signals, Picture Coding Symposium, Asilomar Cal., USA, Januar 1976

/6/ G. Wedlich, Ein lokaladaptives Bildverarbeitungsverfahren zum Helligkeitsausgleich von ungleich ausgeleuchteten Fernsehbildern, Forschungsbericht des Instituts für Informationsverarbeitung in Technik und Biologie der Fraunhofer-Gesellschaft, Karlsruhe, 1976

/7/ A. V. Oppenheim et al., Nonlinear Filtering of Multiplied and Convolved Signals, Proc. IEEE, vol. 56, No. 8, pp 1264 - 1291, 1968

/8/ B. R. Hunt, Digital Image Processing, Proc. IEEE, vol. 63, No. 4, April 1975

/9/ K. S. Fu, Digital Pattern Recognition, Berlin, Heidelberg, New York: Springer Verlag, 1976

/1o/ J. Klie, A Movement Detector for Digital Television Coding, to be published

/11/ R. G. Gallager, Information Theory and Reliable Communication, New York: John Weley and Sons, 1968

/12/ D. A. Huffman, A Method for Construction of Minimum Redundancy Codes, Proc. IRE, vol. 4o, pp. 1o98 - 11o1, September 1952

/13/ J. Max, Quantizing for Minimum Distortion, IRE Trans.Inform. Theor., IT -6, pp. 7 - 12, 196o

/14/ C. C. Cutler, Differential Quantization for Communication Signals, U.S.Patent 26o5361, July 1952

/15/ H. G. Musmann, A Comparison of Extended Differential Coding Schemes for Video Signals, Proc. of the 1974 IEEE Intern. Zürich Seminar on Digital Communications, March 1974

/16/ Th. Kummerow, Ein DPCM-System mit zweidimensionalem Prädiktor u. gesteuertem Quantisierer, NTG-Fachtagung "Signalverarbeitung",pp 425-439, 1973

/17/ R. Lippmann, DPCM-Bildübertragung mit geringer Störempfindlichkeit, Deutsche Forschungs- und Versuchsanstalt für Luft- und Raumfahrt, Forschungsbericht IB 153 - 74/22, 1974

/18/ J. O. Limb, Visual Perception Applied to the Encoding of Pictures, Proc. 2oth Anniversary Technical Symposium of the Soc. of Photo-Optical Engineers, San Diego, USA, August 1976

/19/ A. Netravali and B. Prasada, Adaptive Quantization of Picture Signals, to be published

Coherant Optical Engineering, F.T. Arecchi and V. Degiorgio (eds.)

OPTICAL FILTERING : ENGINEERING APPROACH AND APPLICATIONS IN MONOCHROMATIC AND WHITE LIGHT

Jean-Pierre GOEDGEBUER and Jean-Charles VIÉNOT

Laboratoire de Physique Générale & Optique, Associé au C.N.R.S. (Holographie et Traitement Optique des Signaux), Université de Franche-Comté, 25030 Besançon Cedex, France

Contents

OPTICAL FILTERING : ENGINEERING APPROACH AND APPLICATIONS IN MONOCHROMATIC AND WHITE LIGHT

Jean-Pierre GOEDGEBUER and Jean-Charles VIÉNOT

I. INTRODUCTION AND FUNDAMENTALS :

Many operations in optics can be described in terms of filtering. For instance Young's fringes may be considered as a sinusoidal filtering of the spectrum of a one slit by the spectrum of the other. From that point of view interferometry, holography, speckle phenomena, etc, may be regarded as filtering.

However, for a Physicist working in optics, "optical filtering" is linked to "optical data processing", i.e. the modification of any information by *control* of its passage through a suitable channel ; any signal containing a certain amount of information, it can be useful to isolate some interesting piece of information from the rest considered as background noise.

In most cases, the interesting information is carried by *characteristic frequency bands*, and the filtering techniques aim at extracting these useful frequencies from the spectrum of the signal. Filtering therefore consists in :

1. carrying out the spectrum of the signal by a Fourier transform
2. filtering the spectral components in some prescribed fashion
3. reconstructing the filtered information by a second Fourier transform.

Coherent filtering mainly allows (1) and (3) to be performed in real-time by means of lenses acting as 2-D Fourier transformers , the information carrier being a coherent beam of light. Several double Fourier configurations, i.e. several double diffraction devices are possible. In Fig.1 only one dimension is considered for simplicity's sake. x and x' are space variables in the object and image planes respectively ; u is a spatial frequency variable, reciprocal of x. The dashed lines represent imaging rays, the full lines formation of the image, and the dots the localization of the

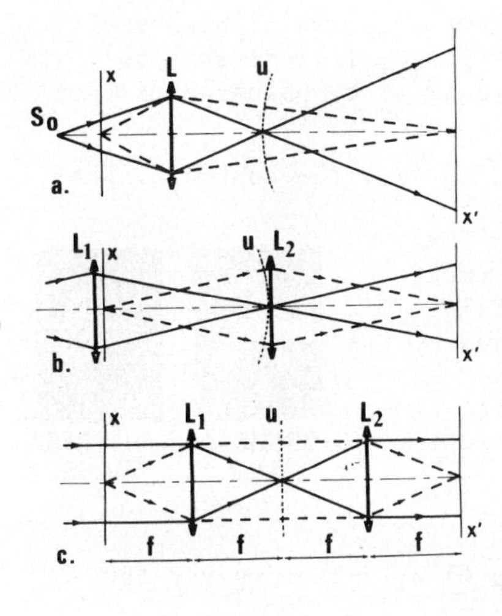

Fig. 1

Fourier spectrum. Filtering is operated by means of transparencies acting as filters, set in the spectrum plane u - the "spherical shape" of the Fourier spectra in Fig. 1 a and 1 b does not intervene as one usually works in narrow spectral regions about the optical axis. Arrangements 1 a and 1 b permit either dilatations or contractions of the spectrum and are used whenever a diffraction pattern has to match the size of a given filter. In Fig. 1 c, the main advantage is to cancel aberrations to using corrected doublets, the spectrum being displayed at the rear and front focal planes of L_1 and L_2.

A filtering operation is shown in Fig 2 : a transparency set in the Fourier plane acts as a filter F (u) on the spectrum S (u), that is mathematically expressed by S (u). F (u). This results in a relative enhancement of a frequency band in the filtered image.

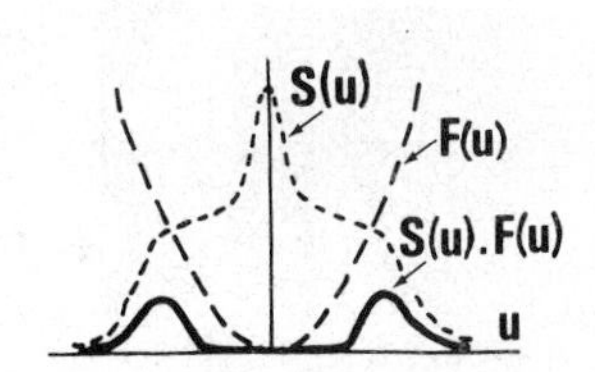

Fig. 2

In Fig. 3 two frequency bands are enhanced and the details stand out of the filtered image. This application to *image deblurring* shows that coherent filtering fits information processing.

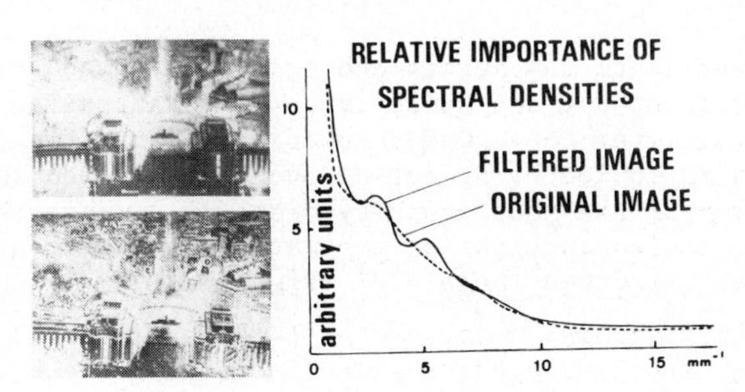

Fig. 3 →

II. APPLICATIONS TO IMAGE PROCESSING : (2,3,4,5)

The applications are closely linked to the structure of the filter which can perform an amplitude and/or phase filtering.

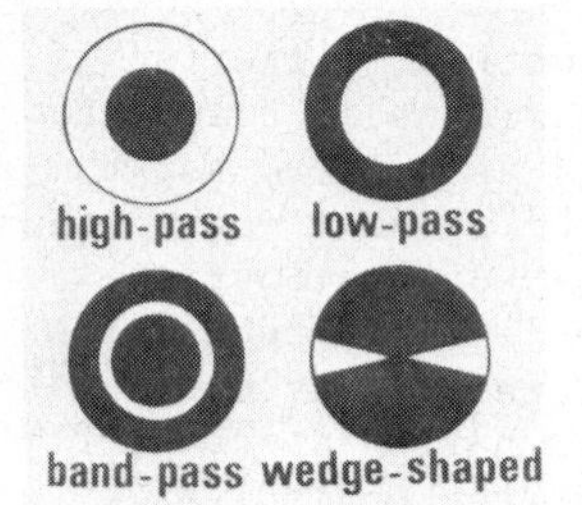

Fig. 4

Fig 4 gives some examples of binary filters. Other types of filters such as computer-generated[6] or hologram filters can be used.

-*contouring* (Fig.5) : the contour lines are extracted by blocking out lower frequencies. Such a high-pass filtering is advantageous when making measurements of the profile of an object.

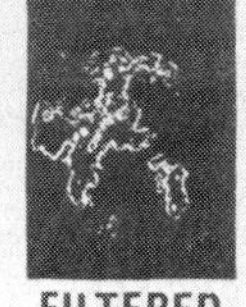

Fig. 5 →

- *enhancement of details* (Fig 3) - band-pass filtering -

- *sampled photograph processing* (Fig 6) by filtering out the periodic structure in the spectrum, this periodic structure being due to the sampling grid.

Fig. 6

- *aerial photograph processing* : Fig.7a

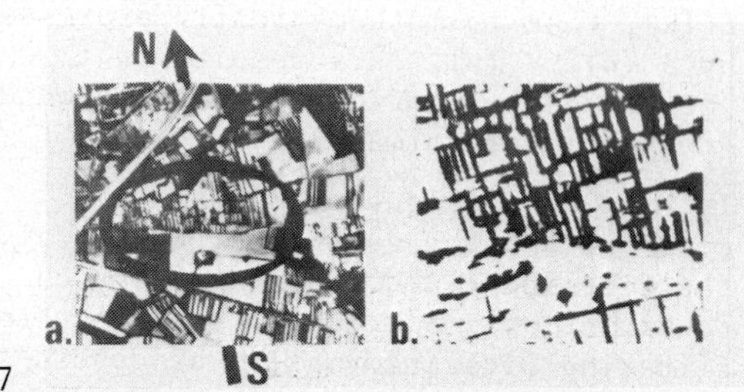

Fig. 7

represents the aerial photograph of a landscape. The problem is to point out any Roman trace of cadastration and separate it from later ones. One assumes that Roman fields are set from North to South. Selecting such a direction in the Fourier spectrum of the photograph by means of a wedge-shaped filter leads to the enhancement of the pattern of Roman cadastration in the reconstructed image [43] (Fig.7b).

- *strioscopy, phase contrast* : the filter is respectively a D-C stop,and a phase plate increasing the optical path of the zero frequency term by $\lambda/4$. These techniques apply to the detection of small phase objects (down to some Å for phase contrast).

- *radar data processing* [7]
- *correction of aberrant images* [8]

- another class of applications deals with *pattern recognition*. Fourier Holograms are used as complex filters. A given information is stored in a Fourier Hologram and is compared to various signals. The response is expressed by correlation terms [9].

III. FILTERING BEHAVIOUR OF FOURIER HOLOGRAMS IN CORRELATORS :

III.1 - Recording of a Fourier Hologram :

An experimental set-up is sketched in Fig 8 a.

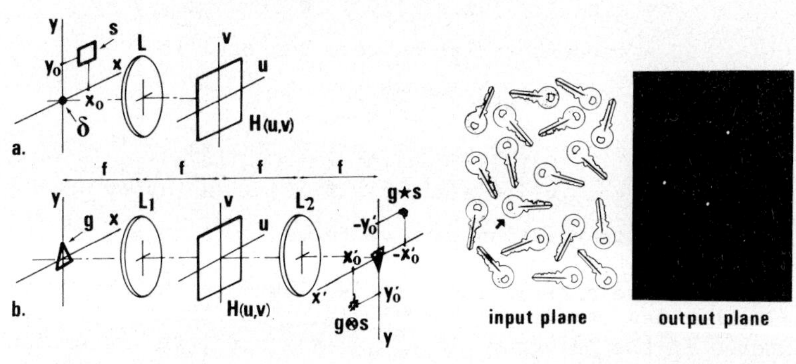

Fig. 8 Fig. 9

A pinhole acting as a reference and the object $s(x,y)$ are placed in the input plane ; the amplitude is :

$a\,\delta(x,y) + s(x,y) \otimes \delta(x - x_o, y - y_o)$, $\otimes$ standing for convolution. A Fourier Hologram $H(u,v)$ is recorded at the output :

$$H(u,v) \propto \left| a + S(u,v) \exp\{-j \frac{2\pi}{\lambda} (ux_o + vy_o)\} \right|^2$$

$$\propto a^2 + |S(u,v)|^2 + a \underset{\nearrow}{S(u,v)^*} \exp\{+j \frac{2\pi}{\lambda} (ux_o + vy_o)\}$$

$$+ a\,S(u,v) \exp\{-j \frac{2\pi}{\lambda} (ux_o + vy_o)\}$$

with $S(u,v) = FT\{s(x,y)\}$ and * denoting a complex conjuguate. (The term $S(u,v)^*$ will be indicated by an arrow to point out its behaviour throughout the filtering operations).

III.2 - correlation technique ; response of a double diffraction device :

Fig. 8b shows how to use a Fourier Hologram as a filter in a double diffraction device. A signal $g(x,y)$ is in the input plane, and the hologram $H(u,v)$ filters the spectrum $G(u,v)$ of $g(x,y)$. At the output, one gets three groups of terms, i.e.

zero term : $g(x',y') \otimes \{a^2\ \delta(x',y') + s(x',y') \star s(x',y')\}$

$= g(x',y') + g(x',y') \otimes \{s(x',y') \star s(x',y')\}$

= geometrical image of g somewhat altered

- 1 term : $g(x',y') \otimes \{a\ s^*(-x',-y') \otimes \delta(-x'-x_o,-y'-y_o)\}$

$= a\{g(x',y') \star s(x',y')\} \otimes \delta(-x'-x_o,-y'-y_o)$

= cross-correlation term centered at

$(x' = -x_o,\ y' = -y_o)$

+ 1 term : $g(x',y') \otimes \{a\ s(x',y') \otimes \delta(-x'+x_o,-y'+y_o)\}$

$=\{g(x',y') \otimes s(x',y')\} \otimes a\delta(-x'+x_o,-y'+y_o)$

= convolution term centered at $(x' = x_o,\ y' = y_o)$

Note that the correlation term is linked to the *resemblance* (10) between the input signal $g(x,y)$ and the $s(x,y)$ signal stored in the hologram. If $s(x,y) = g(x,y)$, the resemblance is maximum and the -1 term is an autocorrelation product that comes out as a narrow bright dot at the output of the correlator. Thus the correlation terms are physically significant.

If the input signal $g(x,y)$ contains $s(x,y)$ and other signals $r_i(x,y)$, the -1 term provides cross- and autocorrelation products, namely $r_i \star s$ and $s \star s$, the latter generating the brightest spot.

III.3 - Applications :

Much work has been devoted to pattern recognition : the problem is to detect a particular signal from among a set of patterns. One first records a Fourier Hologram matched to the desired signal (see next paragraph) and then centers the matched filter in the Fourier plane of the correlator. Filter alignment is critical as re-positionning with respect to the optical axis has to be controlled down to a few microns. Longitudinal re-positionning depends on the focal length f of the lenses (about 0,5 mm for f = 40 cm). In the response plane of the correlator, one or several bright spots indicate which region(s) of the input transparency contain(s) the expected signal.

Examples :

- recognition of a word in a text
- fingerprint identification

- discrimination of signals whose angular orientations are different : in Fig. 9 bright spots appear one after the other when rotating the filter.

- localization of a signal in a radar map ; automatic guidance of aircraft [11,12]. In Fig. 10 the problem is to determine the airplane coordinates, x and y, and direction, θ. A radar map of the landscape below observed from the plane is taken as the input of the correlator. The matched filter is the Fourier Hologram of a reference area in the radar map. A TV camera analyzes the correlation signal whose position in the response plane indicates x and y. The correlation peak is maximized by rotating the filter and the corresponding angular orientation gives θ. The measured x,y and θ coordinates indicate the position of the airplane (accuracy ≃ 2 %) and computer control may operate any correction required.

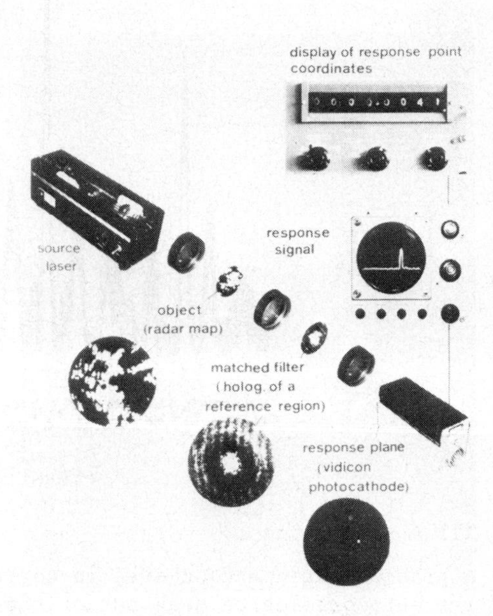

Fig. 10

- other applications deal with the concept of resemblance [10] between patterns of similar shape. In Fig. 11 Hebraïc letters are compared with the first letter "alef" of the Hebraïc alphabet. A photometric analysis of the cross- and autocorrelation terms is performed,and the four different peaks indicate that the four alephs in the text alter slight disparities [13].

- practically many difficulties connected with noise are to be considered. These difficulties may be overcome by *matching* the hologram filter to energies and to characteristic frequency bands especially.

The drawing of any particular character in a text varies with the scribe, the period and the place. Optical information processing not only allows to detect and identify a given letter (pattern recognition) but also to determine the resemblance degree between that letter and the one taken as a reference. Thus clues might be gathered about the origin of the text.

Fig. 11

III.4 - Matching :

A problem often encountered in correlation methods is to bring the autocorrelation peak out of the other responses, i.e. out of background noise.
Suppose a signal s(x,y) is to be extracted from a white noise n(x,y). It can be shown the optimum filter, H(u,v), for maximazing the ratio of peak signal intensity - to - r.m.s. noise is given by [4,9]

$$H(u,v) = \frac{S^{*}(u,v)}{|N(u,v)|^2} \quad \text{where} \quad S(u,v) = FT\{s(x,y)\}$$
$$N(u,v) = FT\{n(x,y)\}$$

In fact, in many applications, the background component is not a white noise and other techniques have to be used. In pattern recognition the signals to be processed may be very similar, and errors may occur. This may be the case in character recognition where the discrimination between auto- and cross-correlation terms may be difficult to perform since the shapes of several alphanumeric symbols sometimes look very much the same, e.g. c and o, C and I ,6 and 9, etc... For instance, let us consider the two letters C and I (Fig. 12). The signal to be detected is I. At the output of the correlator and along the x-direction, the cross-correlation term C $\star$ I is more important than the auto-correlation term I $\star$ I and a very likely error may occur ;

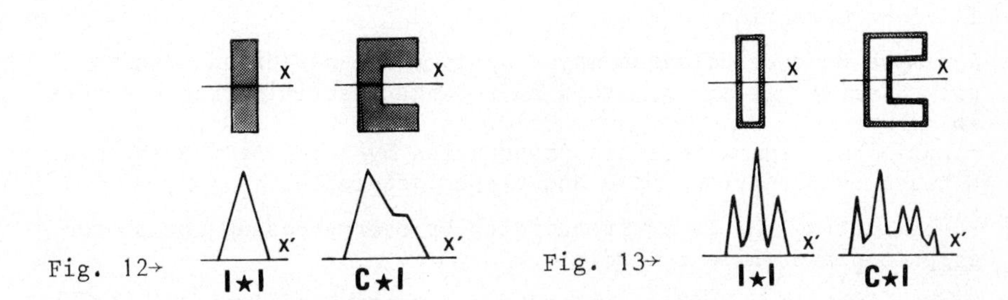

Fig. 12→ Fig. 13→

thus *discrimination must be enhanced.*

Generally the outline of an object plays an outstanding role in its recognition (cf cartoons). On the other hand, high frequencies correspond to fine details and contain most of the differences between two similar signals. By eliminating the lower frequencies one then improves discrimination. As an instance of this, Fig. 13 shows that the cross-correlation term C ★ I is weaker than the autocorrelation product I ★ I as only the contour of the letters I and C is considered. Therefore the *filter has to be saturated with low frequencies* [14,15], *or better, to match the characteristic frequency bands which describe the essential features of the object to be stored* [16,17].

Consequences [15,16] :

(i) a thourough spectral analysis of the signal to be stored in the filter must be performed.

(ii) high-pass, or better, band-pass filters should be used. The problem is to record such holograms matched to the useful frequencies. A *high diffraction efficiency* is needed for these frequencies. Fig. 14 illustrates the steps. The power spectrum S(u) of the object is represented in (a). The characteristic frequency bands of bandwidth ΔN are supposed to be centered at u = ±Nm. Im is the mean intensity in region Nm. Fig. 14 b shows the interference pattern P(u) between the spectrum of the object and the reference wave when recording the hologram. The intensity Ir of the reference wave is adjusted so that Ir = Im, thus the *depth modulation of the fringes is maximum in the region Nm* which insures a high diffraction efficiency for the expected frequencies. Another parameter is *time-exposure* of the plate which has to be long enough to saturate the lower frequency region in the hologram H(u) (Fig. 14 c)

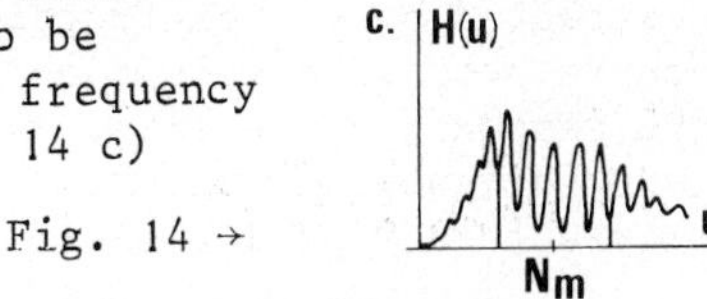

Fig. 14 →

III.5 - Conclusion :

A simple Fourier Hologram may lead to false detections due to poor discrimination. A method to enhance discrimination consists in :

- optimizing characteristic frequencies by matching the energies between the reference wave and these frequencies,

- eliminating the lower frequencies by over-exposing the photographic plate while recording.

Generally, such filters are used for detection of small differences. However they do not suit for signals with significant distorsions ; size, scale change, angular orientation of the objects to be detected are parameters of great importance leading to weaker correlation peaks and to false alarms. A way to overcome these limitations is to select the optimum filter from a set of filters. This is not much convenient (except when using automatic devices) and, therefore, *multichannel systems* are recommended.

IV - OTHER FILTERING TECHNIQUES USED IN OPTICAL PROCESSORS :

IV.1 - Multiplex filtering with multiple holograms :

Fig. 15 sketches two possible configurations to record multiple holograms on a single plate. In (a) each hologram H is recorded with a different carrier frequency. In (b) a single carrier frequency is used ; the angular orientation of the fringes in the hologram varies from an interference pattern to the next ("Theta modulation"[18]) The combination of (a) and (b) permits to increase the number of Fourier Holograms recorded on the plate (up to 100).

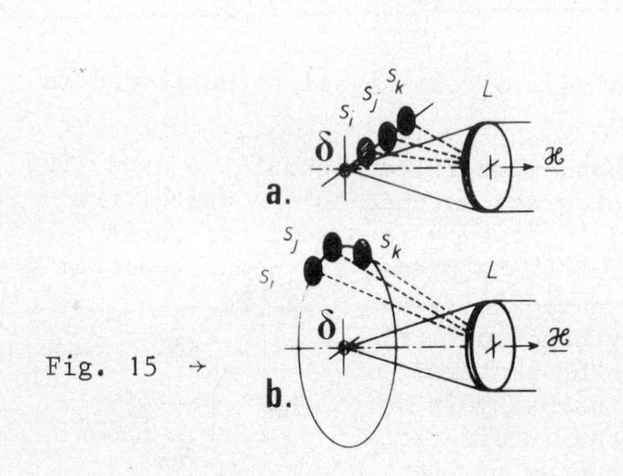

Fig. 15 →

The behaviour of multiple holograms in correlators can be illustrated with 3-D *pattern recognition*, especially in detecting a toy car whatever its orientation [19]. At first one has to record a hologram H in which several well-defined orientations of the object are stored. (Fig. 16 shows the + 1 images reconstructed by diffraction). Then the filter H is set in the correlator ; convolution and correlation terms are observed in the response plane (Fig. 17). The brightest spot indicates the very orientation of the car under test in the correlator. Angular accuracy of 1° is currently obtained.

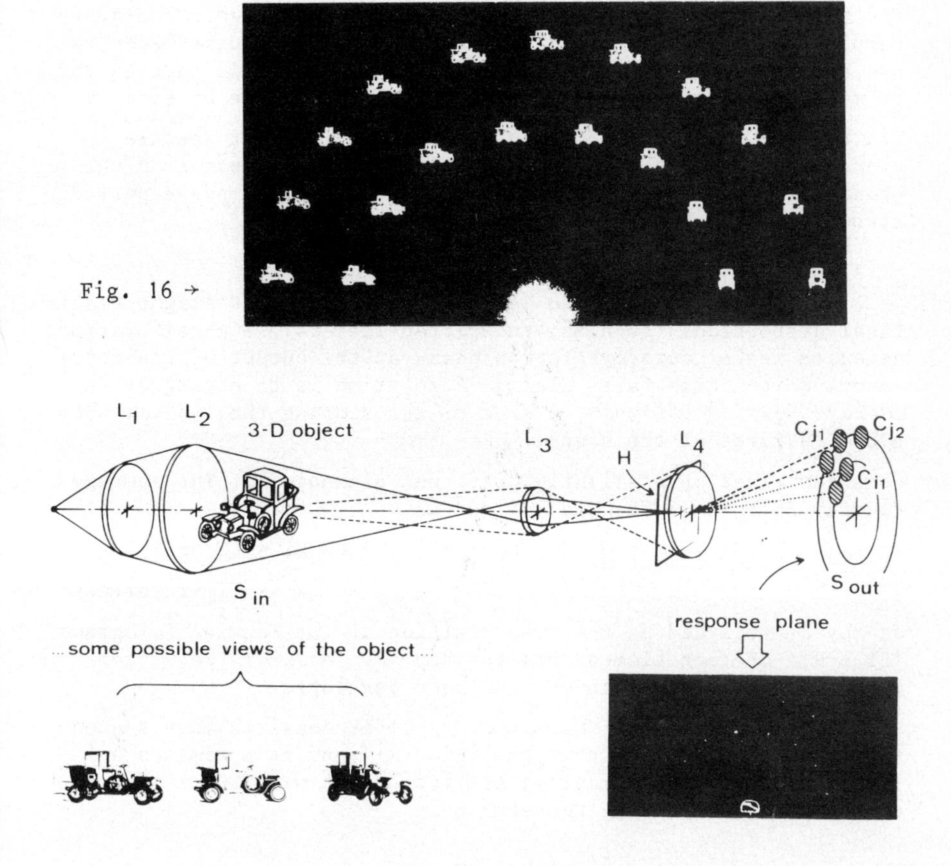

Fig. 16 →

Fig. 17

This technique is a multiplex technique since the input message is simultaneously processed along several channels. In the above experiments no *spatial separation* of the channels has been performed. The technique will be dealt with now.

IV.2 - Sampling methods and multiplex filtering

It was underlined in Sect. III.4 that a high discrimination depends on a high diffraction efficiency of the hologram for some characteristic frequency bands only. When recording a multiple hologram of slightly different objects, the characteristic frequencies of each object may overlap. As a result, the regions in the hologram corresponding to these frequencies are saturated, and a subsequent low diffraction occurs. It can be overcome by recording independent holograms sequentially.

The filtering operations in a correlator are straightforward. By sampling the input signal with a periodic grid, Fourier spectra are displayed in the spectrum plane and filtering is carried out by means of the independent holograms arranged side by side.

Practically, it is difficult to get a set of Fourier spectra containing equal energy. Balancing the energies depends on the choice of the sampling grid, and both amplitude and phase periodic structures have been proposed [20,21, 22,23,24].

IV.3 - Averaging processes :

The objects to be detected may present slight statistical distorsions (such as handwritten letters). Such alterations generate weaker autocorrelation peaks at the output of the correlator, which cause false alarms. A solution is to construct an *average Fourier Hologram*, i.e. a filter storing the typical possible features of the signal.

Consider a set of n slightly different elements fi. The average Fourier Hologram < FH > is defined as

$$< FH > = \frac{1}{n} \Sigma \mid Fi + R \mid^2 \quad \text{with } Fi = FT\{fi\}, \quad R = FT\{reference\}$$

It may be regarded as the superposition of the Fourier Holograms $|Fi + R|^2$ of each element fi, the *carrier frequency being the same*. Two recording methods have been developed.

(i) - *incoherent averaging* [25] : it consists in a sequential recording of each hologram $|Fi + R|^2$, one at a time, on a single plate. These holograms are incoherent with respect to each other, and the resulting pattern is

$$\Sigma|Fi + R|^2 = n\cdot < FH >$$

(ii) - *coherent averaging* [23,26] : each element fi is coupled with a reference ri (Fig. 18 a), the distance between fi and ri being constant. Averaging is carried out via the autocorrelation of the whole set Σ (fi + ri) :

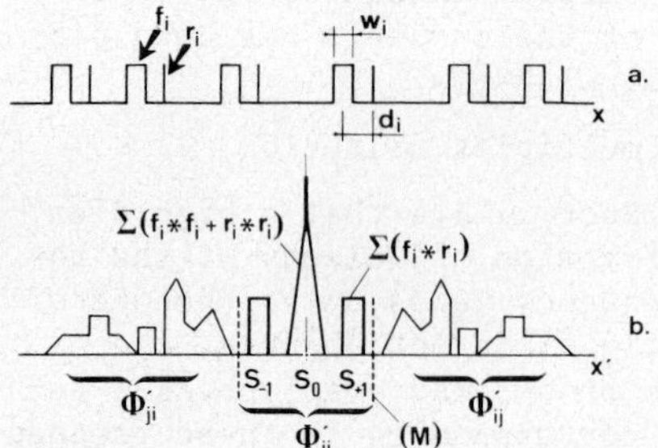

← Fig. 18

$$\Sigma (fi + ri) \star \Sigma (f_k + r_k) = \sum_i \{(fi + ri) \star (fi + ri)\} \text{ central terms}$$

$$+ \sum\sum_{ij \neq i} \{(fi + ri) \star (f_j + r_j)\} \text{ lateral terms}$$

Fig. 18 b is a graphic representation of the autocorrelation of the whole set ; $\Phi'ii$ represents the central terms, $\Phi'ij$ and $\Phi'ji$ the lateral terms.

The Fourier transform of the central term (isolated from the aisle terms by means of a mask) gives < FH >:

$$FT \left[\sum_i \{(fi + ri) \star (fi + ri)\} \right] = \sum_i |Fi + Ri|^2 = n.\langle FH \rangle$$

Therefore the construction of this filter involves building up the autocorrelation optically, and filtering out the unwanted lateral terms (23,26). Though this technique sounds more elaborate than incoherent averaging, it provides a higher diffraction efficiency of the filter (27).

V - REAL-TIME COHERENT FILTERING :

Most coherent optical filtering systems use a photographic film as input material. In spite of their high resolution (3000 mm^{-1} for holographic plates) photographic films are not convenient since they cannot be used for real-time situations. Much work has been devoted to materials suitable for real-time recordings, and to erasable memories, such as *photochromics*, *thermoplastics*, *nematics* and the like (28). This section presents the state-of-the-art in two fields including non-linear optics and spatial modulators.

(i) - *non-linear optics :* As a non-linear medium, e.g. a saturable absorber, is illuminated with a high energy beam, an information can be stored in real-time, the variations of energy causing variations of transparency in the illuminated areas. This technique has been applied to real-time correlators (29,30). In a correlator in use at our laboratory, two high power beams induce a transient Fourier Hologram in cryptocyanine (resolution ≃ 1700 lines/mm). This Hologram acts as a filter on a message carried by a third beam, and, at the output of the processor, a correlation term expresses the resemblance between the message and the signal stored in the filter. Correlation operations are performed in 30.10^{-9} s. This set-up has been used in *sound pattern recognition* (31).

(ii) - *optical transducers :* In these devices, an image is stored either by a video signal (electron beam raster EBR), or by means of a lens imaging a given object onto the light valve (laser beam raster LBR). The stored image is read out by a coherent light beam. As a result, a noncoherent image is transformed into a coherent one in real-time, and it can be seen that such systems replace photographic films advantageously in the input plane of a correlator. Many light valves have been developed : the General Electric "Coherent Light Valve", the Itek "Prom" (32), the LEP "Phototitus" (33). However their low resolution (see Table 1) is the main drawback

	C.L.V.	PROM	PHOTOTITUS
Optical aperture (mm x mm)	26 x 26	40 x 40	35 x 35
resolution (mm^{-1})	125	100	70
dynamic range	40-50 dB	60 dB	60 dB
Modulation raster	EBR	LBR	LBR
Aberrations	–	$\lambda/20$	$\lambda/20$

Table 1

VI - TEMPORAL FILTERING AND OPTICAL PROCESSING IN SPATIALLY COHERENT WHITE LIGHT :

The introduction of monochromatic lasers in optics has made it possible to separate the spatial frequencies from the temporal ones,and been the starting-point of reliable coherent filtering techniques listed previously. It may be of interest to extend such techniques to the *temporal frequencies*,which involves using polychromatic carriers such as white light. Before dealing with *temporal filtering*, a rough representation of white light is given.

VI.1 - Representation of white light :

The time signal f(t) describing white light is not easy to determine since the only physical datum attainable through experiments is the *average power spectrum* $B(\nu)$ (by means of spectroscopic devices ; ν is a temporal frequency). Helpful assumptions are needed :

(i) - at first the statistical aspect of light is disregarded and an amplitude spectrum $b(\nu)$ is introduced :

$$b(\nu) = B(\nu)^{1/2}$$

In Fig. 19 a, for simplicity's sake, $b(\nu)$ is a rect-function centered at $\nu = \nu_o$

(ii) - an average wave-group of white light, k(t) , can be computed :

$$k(t) = \int b(\nu)\exp\{+2\pi j\nu t\}d\nu = TFT^{-1}\{b(\nu)\}$$
$$= \text{sinc}(\pi\Delta\nu t)\exp(-j2\pi_o t)$$

TFT stands for a temporal Fourier transform. The graphical representation of a wave-group is possible via its real part Re{k(t)} (see Fig. 19 b).

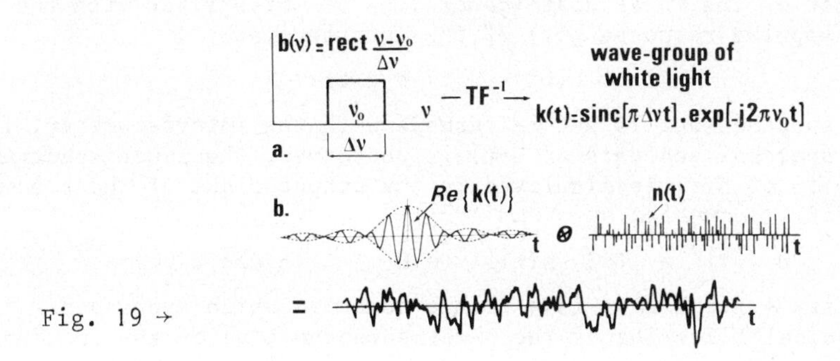

Fig. 19 →

(iii) - taking into account the statistical structure of light, white light may be considered as a random sequence of wave-groups. Mathematically, it may be approximated by the average wave-group k(t) in convolution with a white noise process n(t) :

$$f(t) \simeq k(t) \otimes n(t)$$

VI.2 - Introduction to temporal filtering ; effect of a Michelson interferometer on the chromatic components of white light: Consider a Michelson interferometer whose mirrors are separated by D. If, in a theoretical case, (Fig. 20 a) the input signal is a Dirac pulse of light $\delta(t)$, the response of the interferometer comprises two δ-functions delayed by $\tau = \frac{2D}{c}$ (c : velocity of light). This Dirac pair is the *time impulse response* g(t) of the interferometer and its TFT is the *time transfer function* G (ν) :

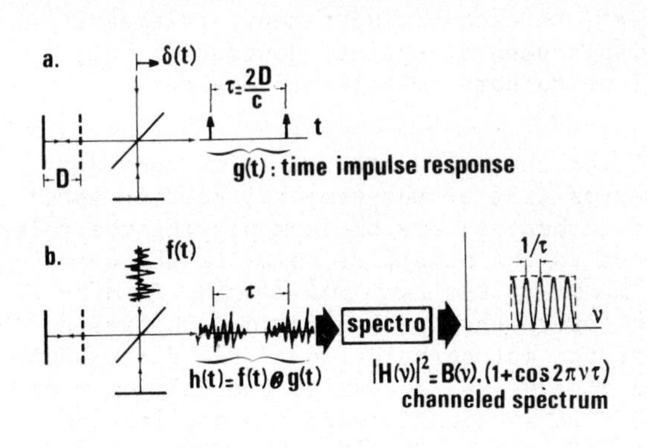

Fig. 20

$$G(\nu) = \text{TFT}\{g(t)\} = \cos(2\pi\nu\frac{\tau}{2})$$

which expresses a *sinusoidal temporal filtering.*

Let now the Michelson interferometer be illuminated with white light (Fig. 20 b). The output signal $h(t)$ is the convolution product of the input disturbance $f(t)$ of white light with the time impulse response $g(t)$ of the interferometer :

$$h(t) = f(t) \otimes g(t)$$

As a spectroscope is set in cascade with the interferometer, i.e. if a spectral analysis of $h(t)$ is performed, the power spectrum $|H(\nu)|^2$ of $h(t)$ is displayed in the output plane of the spectroscope (*channeled spectrum* [34]) :

$$|H(\nu)|^2 = |\text{TFT}\{h(t)\}|^2 = B(\nu)\cdot\{1 + \cos 2\pi\nu\tau\}$$

The term $1 + \cos 2\pi\nu\tau$ is a fringe function which expresses a sinusoidal filtering of the power-spectrum $B(\nu)$ of the incident light : a Michelson interferometer acts as a temporal filter. In fact, it can be extended to any optical device (lens, grating, Fabry-Perot, etc...) which, therefore, can be described in terms of time impulse response and time transfer function [35].

VI.3 - Some applications of temporal filtering ; introduction to temporal holography :

Coming back to the situation depicted in Fig. 20b, the fringe function represents a *coding* of the delay τ between the two signals incident onto the spectroscope slit. In the channeled spectrum, the spacing $\Delta\nu$ of the fringes along the ν-axis is linked to the path-difference 2 D by $\Delta\nu = 1/\tau = c/2D$, and one can perceive possible metrological applications including surface roughness test, thickness measurement, telemeasurement of longitudinal displacements, optical contouring, etc... Measurements range from 1 μm to some cm path-differences [36] (accuracy $\simeq$ 5%o).

Another interpretation of the channeled spectra is to consider them as the coherent superposition of the temporal Fourier spectra of the two delayed disturbances, one of them playing the role of a reference with respect to the other. It comes to the description of *a Fourier Hologram in the temporal domain.* By diffraction in monochromatic light, the spectrogram does behave as a hologram ; an "image" of the autocorrelation of the disturbances is actually reconstructed [37,38], and access to the delay τ - or to the path-difference 2D - is straightforward. An application to surface roughness test is given in Fig. 21. Fig. 21 a is a channeled spectrum-hologram of a rough surface ; its profile is reconstructed in Fig. 21 b (average roughness 1 μm).
This technique also fits information processing devices. (see Ref. 39).

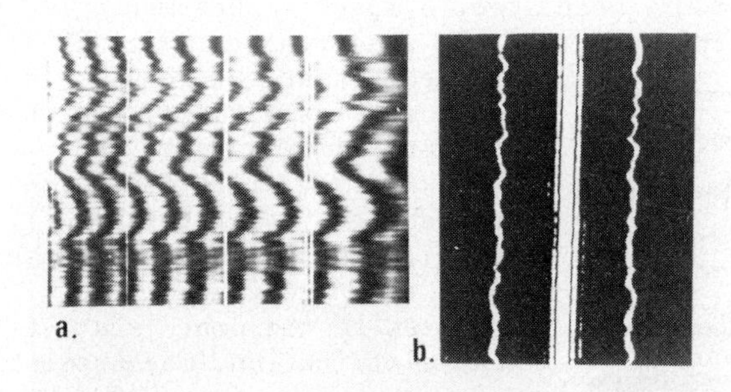

← Fig. 21

VI.4 Introduction to random temporal filtering ; temporal speckle.

The transposition of conventional methods of spatial optics to temporal optics also applies to speckle phenomena. As a spectroscope is in cascade with a random pupil illuminated by a point white light source, the spectral energy distribution displayed at the output of the spectroscope is a speckle-like pattern (40).

Let us consider the theoretical case in which a Dirac pulse of light is reflected back from a rough surface (Fig. 22).

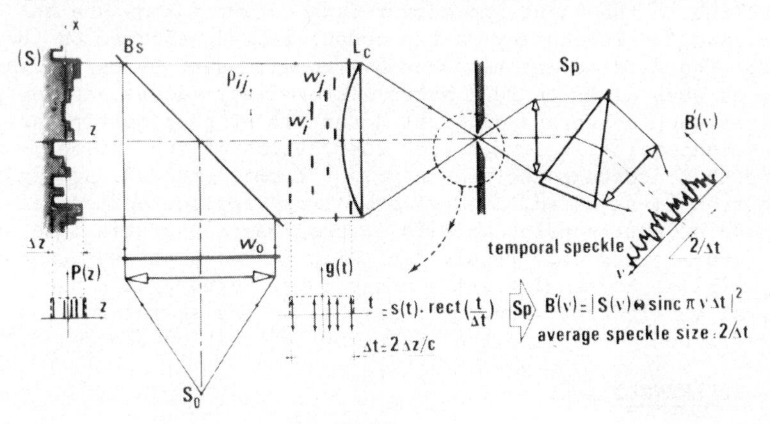

Fig. 22

The impulse response g(t) of the surface S in the z-direction consists of delayed pulses Wi, Wj..., each arising from a different microscopic element of the surface. It can be shown that this random sequence s(t) of pulses is time-modulated by a deterministic function copying the depth statistical distribution P(z) of the accidents of the surface. This random "message" being defined in a limited time interval Δt, its power-spectrum $B'(\nu)$ is expected to be randomly modulated. A speckle phenomenon is actually observed at the output of the spectroscope (Fig. 23).

Fig. 23

It means that the rough surface acts as a random temporal filter. It is to note that speckle statistics depend on the surface parameters [40]. Applications to surface roughness test are in progress (access to the depth statistical distribution, measurement of the r.m.s. of gaussian surfaces, etc...).

CONCLUSION

Some years ago, holography and matched filtering seemed to have settled the optical processing techniques which permit a great amount of unexpensive computations with a high data-rate capability (*). In fact these analogical techniques have drawbacks due to the input and output data conversion on the one hand, and the inaccuracy of the computations performed on the other. The first point has been overcome in part by using input devices such as write/read materials or light valves, and output devices such as arrays of light detectors or photomultipliers. The second point is linked with aberrations of the optical components that decrease the accuracy [41] from 1% to 20%. As high precision is required, *digital processing* replace analogical methods advantageously. Work is in progress along this line and the future trend will likely deal with *hybrid* processes connecting digital and analogical arrangements. Moreover, the use of several wavelengths, e.g. white light, acting as information carriers would increase the amount of information processed [42].

REFERENCES :

1 - P.M. Duffieux, "L'Intégrale de Fourier et ses applications à l'Optique" (Masson Ed.) Paris - 1946 and 1970
2 - A. Maréchal, P. Croce, C.R. Acad. Sci. Paris 237, 607 (1953)
3 - J.E. Rhodes, Am. J. Phys. 21, 337 (1953)
4 - E.L. O'Neil, I.R.E. Trans. Inf. Theory 2, 56 (1956)

(*) These techniques are used in other fields ; an application to pattern recognition in acoustics is given in ref.44 for instance.

5 - L.J. Cutrona, E.N. Leith, C.J. Palermo, L.J. Porcello, I.R.E Trans. Inf. Theory, 6, 386 (1960)
6 - see chapter by A. Lohman
7 - L.J. Cutrona, E.N. Leith, L.J. Porcello, W.E. Vivian, Proc. IEEE, 54, 1026 (1966)
8 - A review of methods for correcting various aberrations is given by J. Tsujiuchi, Progress in Optics (E. Wolf Ed) 1963
9 - A.B. Vanderlugt, IEEE Trans. Inf. Theory, 10, 2, 139 (1964)
10 - J.Ch. Viénot, J. Bulabois, G. Perrin, C.R. Acad. Sci. Paris 263 B, 1300 (1966)
11 - J.Ch. Viénot, J. Bulabois, Rev. Opt. 44, 588 (1965)
12 - J.Ch. Viénot, J. Bulabois, Opt. Acta 14, 57 (1967)
13 - J.M. Fournier, J.Ch. Viénot, Israel J. of Tech. 9, 3, 281 (1971)
14 - S. Lowenthal, Y. Belvaux, Opt. Acta 14, 245 (1967)
15 - R.A. Binns, A. Dickinson, B.M. Watrasiewicz, Appl. Opt. 7, 1047 (1968)
16 - J. Bulabois, A. Caron, J.Ch. Viénot, Opt. Techn. 1, 191 (1969)
17 - J.Ch. Viénot, P. Smigielski, H. Royer, "Holographie optique-Développements - Applications" Dunod Ed. Paris (1971)
18 - D. Armitage, A. Lohman, Appl. Opt. 4, 399 (1965)
19 - J.Ch. Viénot, J. Bulabois, L.R. Guy, Opt. Comm. 2, 431 (1971)
20 - e.g. G. Groh, Opt. Comm. 1, 9 (1970)
21 - e.g. G. Winzer, N. Douklias, Opt. and Laser Tech. 5 (1972)
22 - e.g. H. Damman, K. Görtler, Opt. Comm. 3, 312 (1971)
23 - J.Ch. Viénot, J. Duvernoy, G. Tribillon, J.L. Tribillon, Appl. Opt. 12, 950 (1973)
24 - N. Aebischer, B. Agbani, Nouv. Rev. Opt. 6, 1, 37 (1975)
25 - S.P. Almeida, K.T. Eu, Appl. Opt. 15, 510 (1976)
26 - J. Duvernoy, Opt. Comm. 7, 3 (1973)
27 - La Macchia, J.C. Vincelette, Appl. Opt. 7 (1968)
28 - see chapter by Biedermann
29 - S.H. Lee, K.T. Stalker, Annual Meeting of the O.S.A. San Francisco (1972)
30 - B. Carquille in D.R.M.E. contracts (72.34.112 and 73.34.138) Lab. Optique. Besançon. F.
31 - F. Viénot, C. Bainier, B. Carquille, M. Guignard, to be published in Optica Acta
32 - Imaging characteristics of the PROM are given by S.G. Lipson, P. Nisenson, Appl. Opt. 13, 2052 (1974)
33 - A review of optical relays and their applications is given in Acta Electronica 18, 2 (1975)
34 - Channeled spectra have been known for a long time (Dutour, 1773) see for example H. Bouasse, Z. Carrière, "Interferences" (Delagrave Ed) Paris (1923)
35 - C. Froehly, A. Lacourt, J.Ch. Viénot, Nouv. Rev. Opt. 4, 4 (1973)
36 - A. Lacourt, J.Ch. Viénot, J.P. Goedgebuer, Jap. Jl. Appl.

14 (1975) Suppl. 14.1
37 - J.Ch. Viénot, J.P. Goedgebuer, A. Lacourt, Proc. ICO X Prague (1975) and Appl. Opt. 16,2 (1977)
38 - J.P. Goedgebuer, A. Lacourt, J.Ch. Viénot, Opt. Comm. 16,1, 99 (1976)
39 - A. Lacourt, J.Ch. Viénot, J.P. Goedgebuer, Opt. Comm. 19,2, 68 (1976)
40 - J.P. Goedgebuer, J.Ch. Viénot, Opt. Comm. 19,2,229 (1976)

41 - K. Preston "Coherent Optical Computers" New-York, Mc Graw Hill (1972) ; a comparison of analog and digital techniques for pattern recognition is given in Proc. IEEE, 60,10 (1972)
42 - J.Ch. Viénot, J.P. Goedgebuer, Proc. ICO XI, Jerusalem, Pergamon Press Ed. (1976)
43 - J. Duvernoy, D. Charraut, P.Y. Baurès, to be published in Optica Acta
44 - J. Pasteur, Y. Seyzeriat, to be published in Optica Acta